As widespread social transformations have been paralleled by gains in health and life expectancy through public health and other improvements, a variety of other challenges to health have emerged, particularly in lifestyle related, behaviorally mediated changes in rates of chronic disease. *Hormones, Health, and Behavior* looks at the relationship of human biology and human society, at the intersection of behavior, hormones and health. There is both scientific interest and practical urgency behind the ideas and findings presented here, as the need for a socio-ecological view of function and well-being has become more apparent. This book documents an emerging understanding of how hormones create the linkage between behavior or social life and health. It will inform graduate students and researchers interested in human sciences, human development, anthropology, epidemiology, public environmental and reproductive health.

CATHERINE PANTER-BRICK is Reader in the Department of Anthropology at the University of Durham, UK. Her research interests focus on maternal–child health, working behavior, reproductive ecology, growth, and well-being. She has also edited *Biosocial Perspectives on Children* (1998) with Cambridge University Press.

CAROL M. WORTHMAN is Director of the Laboratory for Comparative Human Biology, and Samuel Candler Dobbs Professor of Anthropology at Emory University, USA. Her research focuses on biocultural processes in human development, reproduction and behavioral biology.

HORMONES, HEALTH, AND BEHAVIOR

A socio-ecological and lifespan perspective

Edited by

C. PANTER-BRICK AND C. M. WORTHMAN

CAMBRIDGE UNIVERSITY PRESS
Cambridge, New York, Melbourne, Madrid, Cape Town, Singapore, São Paulo, Delhi

Cambridge University Press
The Edinburgh Building, Cambridge CB2 8RU, UK

Published in the United States of America by Cambridge University Press, New York

www.cambridge.org
Information on this title: www.cambridge.org/9780521103756

First published 1999
This digitally printed version 2009

A catalogue record for this publication is available from the British Library

Library of Congress Cataloguing in Publication data

Hormones, health and behavior: a socio-ecological and lifespan
perspective/ [edited by] C. Panter-Brick and C. M. Worthman.
p. cm.
Includes bibliographical references and index.
ISBN 0 521 57332 7
1. Psychoneuroendocrinology. 2. Physical anthropology.
3. Clinical health psychology. 4. Human ecology. I. Panter-Brick,
Catherine, 1959– II. Worthman, C. M. (Carol M.), 1948– .
QP356.45.H67 1999
306.4′61–dc21 98-15373 CIP

ISBN 978-0-521-57332-0 hardback
ISBN 978-0-521-10375-6 paperback

Contents

List of Contributors

P. T. Ellison
Reproductive Ecology Laboratory, Peabody Museum, University of Harvard, Cambridge, MA 02138, USA

M. V. Flinn
Department of Anthropology, University of Missouri, Columbia, MO 65211, USA

S. T. McGarvey
Department of Medicine, The Miriam Hospital, University of Brown, 164 Summit Avenue, Providence, RI 02906, USA

C. Panter-Brick
Department of Anthropology, University of Durham, 43 Old Elvet, Durham DH1 3HN, UK

T. M. Pollard
Department of Anthropology, University of Durham, 43 Old Elvet, Durham DH1 3HN, UK

R. M. Sapolsky
Department of Biological Sciences, University of Stanford, Stanford, CA 94305, USA

P. L. Whitten
Department of Anthropology, Emory University, Atlanta, GA 30322, USA

C. M. Worthman
Department of Anthropology, Emory University, Atlanta, GA 30322, USA

1

Contributions of biological anthropology to the study of hormones, health, and behavior

CATHERINE PANTER-BRICK
AND CAROL M. WORTHMAN

1.1 Current issues linking hormones, health, and behavior

This book concerns the relationship of human biology and human society, as viewed by examining the interconnections among behavior, hormones, and health. There is both scientific excitement and practical urgency behind the ideas and findings presented here, as the need for a socio-ecological view of function and well-being has become ever more apparent. During the past two decades, a paradox has emerged in worldwide patterns of health and health risk. On the one hand, international initiatives have succeeded in dramatically increasing life expectancy for many populations by means of interventions that reduce early mortality through technology-based measures such as vaccination, provision of clean water, provision of health care, and improvements in nutrition. On the other hand, a set of health issues has emerged that is less tractable to such measures. These health issues include the emergence of chronic degenerative diseases as major sources of mortality for aging populations, increases in age-specific rates of such diseases concomitant with lifestyle change, and the appearance or resurgence of major infectious diseases with behaviorally-dependent transmission. A common thread running through all these ascendant challenges to health is the importance of patterns of everyday behavior and social relationships in their etiology or transmission. Recognition of the importance of socio-behavioral factors in increasingly significant risks to health has stimulated interest in specifying the pathways by which these factors affect health risk, in order to identify novel health-promoting interventions or realize improved implementation of existing health measures. Such an enterprise is predicated on a fine-grained and thorough understanding of the local contexts that influence health outcomes.

The task of unravelling the relationships between social ecology and health has been impeded by the difficulty of monitoring the physiological processes that characterize either the immediate biological impact or the long-term functional consequences of contextual factors which shape health outcomes. Endocrine measures have proven useful for tracking both such short- and long-term pathways, for hormones serve as indices or markers of the processes that mediate relationships between behavior and health. The endocrine–neuroendocrine system is in large part designed to coordinate various biological processes at disparate body sites over time (e.g., metabolic regulation, growth, reproduction), and to orchestrate physiological responses to shifting inputs and demands. In addition to being responsive to physical demands, this system is exquisitely responsive to cognitive–emotional experiences that thereby influence physiology.

Table 1.1 outlines the action and regulation of the principal endocrine axes discussed in the present volume, and its layout reflects the impressive diversity of endocrine action. Extensive exploration of both the endocrine and neuroendocrine systems has led to a recognition that these occupy a functional continuum and are hence scarcely separable. The distinction was originally based on the blood-borne effects of secretions on remote tissues (classic endocrine activity) versus the cell-mediated and local effects of neurological function (neuroendocrine activity). The functional spectrum over which endocrine and neurological actions interact is reflected in Table 1.1. It includes (a) central nervous system regulation of peripheral gland activity via a set of hormonal messengers with elaborate feedback mechanisms (the hypothalamo-pituitary axes), (b) direct neural regulation of neurotransmitter release by a peripheral gland (epinephrine and norepinephrine release by adrenal medulla), (c) gland-based regulation of circulating levels of nutrients, metabolites, or other biochemicals (pancreatic production of insulin stimulated by blood glucose), and (d) a mobile cell-based distributive system of endocrine action in immune function, in which local stimuli trigger release of cytokines and other substances with autocrine, paracrine, and endocrine effects.

The well-recognized linkages between external and internal conditions forged by the endocrine–neuroendocrine system make it a prime focus for the investigation of links between behavior and health. Expanding knowledge of endocrine physiology and function has converged with continual improvements in endocrine measurement techniques to provide fertile grounds for such investigation.

1.2 Value added by an anthropological approach

Current trends in public health research have created a demand for relevant information from biological anthropology, in particular, and social sciences, in general. Advances in endocrinology have provided the bases for investigating behavior–health linkages through endocrine mediators and moderators (Barr *et al.*, 1994). This volume brings together much of a rapidly expanding anthropological literature that addresses the interrelations among hormones, behavior, and health. The contributions reflect the exciting realization that biological anthropology can provide significant new theories and data that dramatically expand our understanding of the biobehavioral processes known to affect health. Furthermore, such work has demonstrated the capacity to expand our understanding of "normal" human biology across the life course. Analyses of specific local contexts and patterns of temporal variation, using a socio-ecological and lifespan perspective, pose major challenges to notions of panhuman biological uniformity and the search for a single descriptive framework for human biology. The lacuna left by the erosion of a universalist paradigm in human biology is addressed by current efforts aimed to establish conceptual and methodological frameworks for research that take a comparative, lifespan, and biosocial approach. Biological anthropology, with its twin interests in human biology and behavior can, and does, contribute to this set of endeavors in several important ways.

First, anthropology integrates multiple levels of analysis, from ultimate (evolutionary) to proximate (molecular) causes. By contrast, biomedical research has concentrated on identification of how the body works, on the molecular, cellular, and systemic level, while epidemiology has focused on comparative analysis of morbidity, mortality, and risk structures at the population level. Anthropology routinely draws upon evolutionary theory, population-level processes, and individual-level observation for analysis of behavior, physiology, and morphology. Accordingly, the discipline has developed conceptual structures and methodologies to effect an integrated approach that emphasizes the whole organism in specific social contexts over the entire life course. The burgeoning field of evolutionary medicine (Ewald, 1994; Nesse & Williams, 1994; Trevathan *et al.*, 1998), for instance, demonstrates the value of considering human evolutionary history as a basis for hypothesizing how humans are "designed" to work, and thus how changes in human behavior or ecology may result in health risk.

Second, biological anthropology is centrally concerned with human variability, its extent, causes, and consequences. Specifically, it focuses

Table 1.1. *Regulation of principal hormones discussed in this volume*

(*a*) *Hypothalamo-pituitary axes*

Hypothalamus	Anterior pituitary	Target organ	Peripheral hormone	Binding protein	Major function
GnRH	FSH and LH	ovary ovary testis	estradiol (E2) progesterone (P) testosterone (T)	SHBG	reproductive development, regulation, and behavior; sex characters,
GHRH (+) somatostatin (−)	GH	liver and many other tissues	IGF-1, IGF-2	IGFBP-3	growth, organic metabolism,
TRH	TSH	thyroid	T3, T4	TBG	metabolic rate, alertness, neural development, potentiate GH
CRF	ACTH	adrenal cortex	cortisol	CBG	energy availability, immune inhibition, vascular reactivity
CRF (+ ?) dopamine (−) [TRH(+)]	ACTH (+ ?) PRL	adrenal cortex, breast, reproductive organs	Δ4, DHEA, DHEAS none	poorly understood	 breast development, milk production

(*b*) *Direct neural regulation*

Neural input	Target organ	Peripheral hormone	Major function
sympathetic nervous system	adrenal medulla	epinephrine, norepinephrine	cardiovascular function, organic metabolism, stress response

(*c*) *Semi-distributive regulation and production*

Provocative agent	Production site	Hormone produced	Major function
glucose	pancreas	insulin	regulate blood sugar levels

(*d*) *Distributive regulation and production*

Provocative agent	Production units	Hormone produced	Major function
foreign antigen	leukocytes, macrophages	cytokines	immune defence, sickness behavior

Note: Abbreviations used:
ACTH: adrenocorticotropic hormone, CBG: cortisol binding globulin, CRF: corticotropin releasing hormone, Δ4: androstenedione, DHEA: dehydroepiandrosterone, DHEAS: dehydroepiandrosterone sulfate, FSH: follicle stimulating hormone, GH: growth hormone, GHRH: growth hormone releasing hormone, GnRH: gonadotropin releasing hormone, IGF-1: insulin-like growth factor-1, IGF-2: insulin-like growth factor-2, IGFBP-3: insulin-like growth factor binding protein-3, LH: luteinizing hormone, SHBG: sex hormone binding globulin, T3: triiodothyronine, T4: thyroxine, TRH: thyrotropin releasing hormone, TSH: thyroid stimulating hormone
(+): stimulatory, (−): inhibitory, ?: factor uncertain.

attention on two levels of variation, that between and within populations. As much of the work reviewed in this volume underlines, the comparative cross-population approach can assist in identifying factors that accompany variability in function or outcomes (in growth, work, health, reproduction). Biological anthropology has repeatedly demonstrated that the ecologies characteristic of a population can shape its biology in a manner distinctive from others. Therefore, only comparison among populations or subgroups within populations can reveal the differential effects of ecological and other, such as genetic, variables that significantly influence human biology and the population specific "normal range". The need for a population comparative view of human biology becomes especially significant when one considers that most of what we know about human physiology, development, reproduction, endocrinology, and aging is based on studies of members of relatively privileged, healthy, well-nourished post-industrial societies of Europe and North America in the mid–late twentieth century. As such, these societies may represent human biology under rather privileged circumstances, thereby also presenting a very particular and perhaps unusual (in human historical–evolutionary terms) picture of human biology that reflects the particularities of largely favorable and novel ecological circumstances. In this view, gaining insight into health issues that plague the west may depend on a comparative approach that reveals the ways in which patterns of development, function, and aging may be distinctive in these groups and lead to specific health risks. Concurrently, comparative research can expand our view of what constitutes the normal and expectable ranges for functional and morphological parameters across the life span. Such ranges may often be condition specific or ecologically contingent.

Improved models of health determinants increasingly rely on probing individual variation within populations. Basic and clinical research have focused on establishing how the body works in health and illness, and identifying ranges of normal function for humans; the prevailing assumption guiding this work is that such norms will be generalizable or universal across populations. Epidemiological analyses reveal the variation in internal health risk structures within populations and provide the statistical basis for comparison between them. Furthermore, not all individuals who fall within "risk categories" manifest the phenomenon for which they are at risk, for risk merely denotes a greater probability of a negative health outcome (i.e., morbidity or mortality). Identification of the determinants of positive outcomes in individuals at risk (sometimes termed "positive deviance") (Ben-Yehuda, 1990; Johns, 1993; but see Sagarin, 1985), has been increasingly recognized as a means to uncover health-protective factors, not only constitutional or intrinsic (biologically based) but also conditional or extrinsic (social–contextual) variables. Moreover, the complexities of human social ecology (namely, the living conditions created or influenced by social factors such as social behavior, subsistence activity, or social status) mean that individuals do not live in "a" society; rather, they live under specific circumstances determining the microecology that effectively defines the actual experiences, exposures, challenges, inputs, and supports they encounter. Anthropologists therefore concentrate on the socio-cultural factors that generate differential demands, resources, and contexts encountered by members of a society. Individuals within populations encounter specific circumstances that are organized by status (age, gender, social class), specific practices (e.g., reproductive behavior, child care, workloads), attitudes (perceived needs, valued goals), and social constructions of the life course. Hence, members of a society may experience quite different lived microecologies that can vary within (synchronically) and across (diachronically) time.

Third, biological anthropology aims not only to identify variability *per se*, given that people and populations may differ from each other in innumerable ways, but also to ascertain the relative significance of observed variation. Variation is of specific interest when it is demonstrably associated with changes in important outcomes, such as differential growth, fertility, activity, and survival. Of course, the determination of what is "important" is also a matter of value and open to socio-political critique and debate. Nevertheless, identification of relative risk and benefit can assist in the process of evaluating outcomes and weighing alternative interventions. Here, integration of multiple levels of analysis must ensure that an evaluation of significant variation will draw on multiple domains. Such an approach, moreover, allows identification of conflicting outcomes, or trade-offs, in which a positive effect on a desired outcome may be offset by a negative effect in another. Instances from chapters in this book include McGarvey's data linking lifestyle change and enhanced affluence among Samoans to increased risk for obesity, diabetes, and cardiovascular disease, and Ellison's suggestion that heightened gonadal activity resulting from sustained good nutrition and health contributes to increased risk for reproductive cancers among affluent contemporary populations.

Fourth, anthropology places strong emphasis on people's actual, everyday lives. The centrality of action "on the ground" leads to a keener appreciation of the cultural logics and practical constraints that significantly inform human behavior and thus, human ecology and experience.

As with any empirical endeavor, biological anthropology recognizes the need to operationalize functional concepts or hypotheses in real-world terms that break down phenomena into their testable, measurable components. For instance, in this volume, the important but vague concept of stress is repeatedly explored in both non-human and human primate models to elaborate the nature of stress and its distinguishing pathways linking inputs, biobehavioral characteristics, and consequences. Nevertheless, an insistence on integrating specified and validated models into an overall view of the person in specific social contexts that vary over time, supports a drive to integrate what we learn of specific functional relationships into models of increasing complexity, diversity, and thus enhanced general validity. In particular, a concern with lifespan processes, namely, the shifts in functional capacities, risks, and opportunities across the lifespan, lead anthropologists to take a diachronic view of human variation. This view is codified in life history theory, which concerns species-specific patterns of time and resource allocation to the critical tasks of survival, maintenance, growth, and reproduction across the lifespan. Central to anthropological thinking is the recognition of finite resources in time and energy, and of the consequences of limited resources to long-lived creatures living in complex social contexts. Juxtaposition of competing demands with limited resources leads to trade-offs among functional biological domains (e.g., growth versus reproduction), between current and future goals (immediate survival versus future livelihood; present versus deferred reproduction), and across adaptive demands (biological versus social). The specific set of trade-offs faced by individuals or populations arises from temporal and socio-ecological parameters that define the distribution of resources and demands. As noted above, the landscape of trade-offs faced by individuals is shaped by age, sex, and status-specific biological and socio-cultural variables that shift the urgency and distribution of demands and access to needed resources. Here, the anthropological view of humans as intelligent and intentional or planful, but also as constrained and information-limited, may help us understand processes that lead to differential health outcomes. Not everything can be optimized all of the time, so the prime question becomes what to optimize, and when. Maintenance of a lifespan perspective in anthropology foregrounds the significance of trade-offs across time, and the temporal nature of constraints placed upon individual choices. These issues are discussed in Worthman's chapter, particularly with reference to the role of the endocrine–neuroendocrine system in mediating these trade-offs on the biological level.

Fifth, the anthropological contributions in this volume reflect the stimulating effects of methodological innovations and modifications that have paved the way for new approaches to the study of hormones, behavior, and health in everyday contexts in diverse populations across a wide range of settings. Anthropology has repeatedly self-identified as a parasite (or perhaps symbiont) on the technological advances in other research disciplines that borrows those methods and adapts them to anthropological purposes. A new generation of anthropologists has also begun to develop novel methods that substantively add to the existing battery of research methodologies. Underlying this trend has been recognition that methods are and should be driven by the research questions they serve, rather than the other way around. This is especially true in biological anthropology, for methods designed to work in clinical or laboratory settings with excellent infrastructure and high participant compliance may not be suitable for remote settings and/or non-clinical populations unaccustomed to elaborate biomedical protocols. The techniques for measuring hormones and other biomolecules in saliva (Ellison, 1988; Ellison *et al.*, 1993), blood spots (Worthman & Stallings, 1994, 1997) and urine (Campbell, 1994) have greatly expanded the scope of research involving endocrine measures. These laboratory advances underlie much of the work reviewed in this volume.

Another methodological advantage which anthropology brings to the study of dynamics among hormones, health, and behavior, is its strong tradition in behavioral research. This tradition involves fieldwork along with diverse techniques for studying behavior and its cultural, cognitive, and performance measures. Fieldwork with close observation in the study community provides a basis for fine-grained characterization of behavior, while attention to why people do what they do, and react the way they do, permits understanding of the motivational structures underlying behavior. Behavioral research and field residence are labor intensive, and therefore the number of people and geographical spread that can be covered (per investigator) is limited; such research is costly (in time and effort as well as money). In large-scale studies of health-related issues, the cost factor has led to a marked preference for less observational and fine-grained approaches and reliance instead on interview or survey techniques. Such an approach may work well when good models are available to suggest which variables will be important, but may be less productive where operationalized conceptual schemes are limited or absent, or their relevance to a specific situation is questionable. As our empirical models of how things work have become more and more elaborate, the business of operationalizing them for field research has become increasingly challenging. Contributions in this volume (Sapolsky, Flinn, Panter-Brick and Pollard) explicitly illustrate the process of translating a research problem concerning linkages among hormones, behavior, and health, into a workable field project.

Finally and fortuitously, advances in biostatistics over the past 20 years have tremendously strengthened anthropological research by providing powerful techniques for dealing with repeated measures (of individuals over time), truncated series (individual follow-up where target period is not completely observed), categorical variables, multivariate problems, and small sample sizes (e.g., Breen, 1996; Long, 1997; Mayer & Tuma, 1990). Availability of these techniques on the large statistical packages makes them accessible to many researchers.

1.3 Relevance to health

This volume reflects and extends ongoing investigation of the relationship of biological and social factors in differential health and illness. More than 30 years ago René Dubos noted that: "Any medical problem presents itself under two aspects which are sharply different, but complementary. On the one hand, all phenomena of health and disease reflect the biological unity of mankind; on the other hand, all are conditioned by the diversity of the social institutions and ways of life" (Dubos, 1966, p. 367). A perpetual challenge to identification of linkages between factors exogeneous to the body and biological health outcomes has been the spatial and temporal remoteness of cause and effect. Risk factors, or conditions that influence the probability that some variable will be associated with individual health consequences, can be especially difficult to detect because the association may be quite remote or complex. An intriguing instance of this, is the recent work by Barker and colleagues concerning relationships of gestational factors, reflected in birth and placental weights, to long-term risk for diabetes or cardiovascular disease in British cohorts (Hales *et al.*, 1991; Barker *et al.*, 1993; Law *et al.*, 1995). Although they have established statistical linkages on the population level, the mediating mechanisms remain a matter of speculation (Barker, 1997; Dennison *et al.*, 1997) and the possible confounding variables that may covary with birth weight but act as the actual causative factors (e.g., family environment or dysfunction) remain to be empirically determined. In the case of diabetes, endocrine and metabolic measures can be used to ascertain whether there is an association, at or near birth, between birth or placental weight and these markers of diabetic risk. Even if there is no such early association, it may be difficult to establish whether, when, and how some slight gestational effect on developmental trajectory or sensitivity to exogenous factors such as diet or cognitive–emotional demands may lead to emergence of disease decades later.

Behavioral patterns that alter health risk can arise from two sources, that of a particular individual and that of other people. For instance, social stress or heavy workload can directly affect the physiology of the actor, but if the actor is also pregnant, ill, or a producer of resources important for family welfare, then those behaviors will influence the well-being of others. Increasingly, attention has turned to how the behavior of others, be they individuals or large sets of individuals who enact sociocultural schemas affecting structural conditions of life, defines the field of action and experience of any given person or animal. Thus, the ecology of everyday life is largely defined by the behaviors of others. This point is emphasized by Sapolsky's discussion in this volume of the complex relationships among temperament, social instability, and social rank in male baboons. Integral to this sociocultural dimension of primate ecology are emotional–cognitive dimensions that are also increasingly identified as significant determinants of health. Such factors are the focus of the concept of psychological stress which in turn has been closely implicated in both physical and mental health. Several of the chapters in this volume (Sapolsky, Flinn, Panter-Brick and Pollard, and McGarvey) concern behaviorally mediated dimensions of stress and the endocrine–neuroendocrine mediators of the stress–health connection.

As a consequence of its interests in human adaptation, adaptability, and differential well-being, the field of anthropology has attended more closely to functional impairments than to frank pathology. Variation in developmental outcomes, psychobehavioral patterns and competencies, or functional capacities, is of interest insofar as it results in variation in well-being and ill health (the latter may be crudely defined as distress or functional impairment) which, in turn, can be characterized in quantitative or qualitative terms. Again, social expectations entailed by cultural values placed on particular outcomes may lead to a differential evaluation of, or social responses to, individual conditions or behaviors. Thus, behavior differences can carry different implications for well-being in a setting-dependent manner. To use the Samoan examples considered by McGarvey, a psychobehavioral characteristic such as assertiveness brings different health consequences in traditional Western Samoa versus wage labor American Samoa, or in divergent work settings (corporate management versus shop floor) in Western post-industrial societies, by entailing different consequences in terms of efficacy and social appropriateness of given behaviors. Implications of behavior variation, such as in style or degree of assertiveness, can be strongly gender- or age-differentiated within cultures as well as divergently organized between cultures.

Well-being or frank ill-health is therefore socially situated, in both its individual and trans-individual aspects, and must be investigated as such. This is the perspective that informs the concepts and research presented in this volume. Furthermore, health status does not simply arise naturally or inevitably from the constitution of the individual (e.g., genes) or from the action of external forces (e.g., pathogens, or stressful circumstances). Rather, it is produced, as an outcome of the interaction of constitutional with external domains via complex pathways that operate over time, often quite long periods of time. Structures of health risk can vary with age or through secular trends: factors that are health-promoting or neutral at one point may be deleterious at another. This developmental and temporal perspective is also integral to all contributions in the present volume. The socially situated and temporal character of health and health determinants leads to an important conclusion, namely, that global prescriptions for health interventions are not feasible or desirable. Nevertheless, we can identify the critical variables that are likely to be operative, and ascertain the conditions under which they become important. For instance, Boyce and colleagues (1995) have observed that cardiovascular reactivity (heart rate response to challenging tasks) in 3 to 5-year-old American children is associated with high incidence of respiratory illness when such children experience high stress childcare environments, but high reactivity children show lower rates of illness than low reactivity children under low stress childcare settings. Another example emerges from the enormous literature on infant feeding, infant survival, birth spacing, and maternal reproductive health. Although breastfeeding has been repeatedly demonstrated to have significant effects on birthspacing via the impact on maternal reproductive function, the importance of specific dimensions of breastfeeding behavior varies across populations, and the impact also depends on maternal energetic variables (see discussion by Panter-Brick and Pollard). Likewise, the impact of supplementary feeding practices on infant health and survival depends on pathogen load, and short versus long-term trade-offs between growth and survival. Thus, despite an intensive search for an ideal schedule for breastfeeding and supplementation, the main conclusion appears to be that appropriate behavioral strategies vary between, and even within, populations.

Finally, we note the escalating need for attention to issues of mental, not just physical, health (Desjarlais *et al.*, 1995). Although these two dimensions of health are well understood as interrelated, the high priority given to physical health interventions has led to rapid strides in life expectancy concurrent with massive worldwide social transformations, and increasing dislocations have raised substantial challenges for psychosocial adjustment and health. Dynamics among hormones, behavior, and health often intimately involve emotional–cognitive processes and make this an especially appropriate arena for investigating determinants of mental health as well.

1.4. Goals and overall organization

This volume aims, then, to draw together the existing conceptual and empirical threads linking hormones, behavior, and health with specific anthropological contributions that examine the sources of human variation and its consequences across social contexts and through time. We use this lens to focus on endocrine mediators and moderators of interactions between behavior and health, and deploy a socio-ecological and lifespan framework that views behavior as both a determinant and a consequence of the specific ecologies individuals inhabit. The practical goals are to: (1) provide an articulated set of overviews of current knowledge to act as a resource to readers interested in this area, and (2) point to a set of ideas, analytical frameworks, methodologies, and empirical findings that can ultimately assist in guiding choices in resource allocation for preventive medicine and health intervention. Our goal as scientists is ultimately to improve the quality, not just quantity, of life for people at all ages and in diverse contexts.

The organization of this book reflects this set of goals. Each chapter provides details on relevant endocrine function, on its associations with behavior and social ecology, and on connections to health consequences over the life course. The long-term, lifespan developmental character of interactions among behavior, hormones, and health figure as a major theme throughout the volume. Most contributions include in-depth case studies of the authors' research on the problem in a specific setting as well as extensive comparative data from a range of populations. By working through a specific hormone–health–behavior nexus in terms of the relevant physiological, developmental, epidemiological, methodological, and analytical parameters in association with examples from current studies, the authors also illustrate how research in this area is conceived and implemented. Each chapter includes a flow chart outlining the model of hormone–behavior–health pathways under discussion. The content of each contribution is described briefly below.

Chapter 2 offers a comparative perspective on social relations and life history. Robert Sapolsky discusses the biobehavioral bases of individual differences in social relations, in response to resource and social instability among non-human primates across the life course. He surveys models, methods, and findings concerning stress as investigated in a number of species, and uses his extensive fieldwork with baboons in Masai Mara, Kenya, as a case study to work through the emerging, much more nuanced, views of stress, social dominance, and the vicissitudes of life. He also presents intriguing data concerning effects of temperamental differences on social relations and stress across the life course.

In Chapter 3, Carol Worthman evaluates life history theory and the endocrine organization of human development, then considers the implications for variation in that architecture suggested by pronounced contemporary worldwide and historical secular trends to accelerated growth and maturation. She summarizes her comparative work on endocrine trajectories in later- versus earlier-maturing populations, and considers the causes and consequences of variation in maturational timing. In addition, she reviews the hormonally mediated and developmentally organized variation in physical and mental health outcomes over the short and long term.

Relationships among childhood stress, family environment and health represent the complex set of issues addressed by Mark Flinn in Chapter 4. He considers particular challenges for the study of such relationships, and goes on to provide a detailed case study that exemplifies the kind of study design, methodology, and analysis required. His extensive data from a village on Dominica enable him to reflect on the temperamental, family compositional, and physiological pathways by which children's daily life experiences are connected to their affective and physical well-being. This work represents the most sustained, intensive anthropological effort to date to provide fine-grained psychobehavioral research in the context of close ethnographic observation.

Chapter 5, by Catherine Panter-Brick and Tessa Pollard, tackles perhaps the most closely studied nexus of hormone–behavior–health risk, namely work. The topic has been been a focus of attention in Western populations, principally with regard to psychosocial work stress and cardiovascular disease. Its coverage is here considerably expanded to include: (a) research in Western (modern industrial) and non-Western populations, (b) wage/office work versus subsistence labor, (c) unemployment, (d) workload stresses on women's reproductive health, and (e) children's psycho-social stress in school and street work environments. Their review is illustrated with salient examples from the literature.

The burgeoning field of reproductive ecology is surveyed in Chapter 6 by one of its principal architects, Peter Ellison, and related to risks for reproductive cancers in contemporary populations. The etiology of cancer is characterized and the relationship of gonadal steroids to carcinogenesis in target tissues reviewed, with particular reference to breast cancer. In the light of this relationship, he presents his collaborative work documenting wide population variation in patterns of gonadal steroid output and thus lifetime exposure to these carcinogenic agents. Behavioral and socio-ecological sources of this variation are outlined and illustrated with extensive fieldwork with a variety of populations.

Chapter 7 examines reported links between cancer and diet from an evolutionary perspective. Pat Whitten describes the effects of nutrients and other chemical components of plants like phytoestrogens on the reproductive system. She considers the physiological mechanisms predisposing modern populations toward selection of diets that increase cancer risk and illustrates parallels to the mechanisms that organize foraging behavior and reproductive timing in the great apes. She argues that evolutionary adaptations to frugivorous diets predispose humans to select and construct lifestyles that become "unhealthy" when developed to their logical extreme.

In Chapter 8, Steven McGarvey pursues health implications of world-wide social change, often termed "modernization," that entail changing work and dietary patterns, altered everyday human relationships and cash-driven modes of consumption, as well as modified expectations and values. He outlines methodological approaches to tracking these changes in lifestyle and provides detailed information on how such changes are associated with endocrine-mediated metabolic shifts that enhance risk for adiposity and cardiovascular disease. His extensive work on these issues in western and American Samoa provide the basis for a closely considered analysis of how social transformations translate into new patterns of health risk that emerge over the life course.

1.5 Audience

The book is directed to scholars from a range of academic disciplines, including human sciences, human biology and ecology, biological anthropology, endocrinology, developmental behavioral pediatrics, developmental psychology, public health and epidemiology. We hope these will include readers from around the globe who are engaged by scientific and practical concerns over how transforming conditions of life may influence health and well-being. The work presented herein should demonstrate how a human biology approach that closely observes people in their everyday settings, attends to affective and social dynamics, employs a lifespan developmental perspective, compares patterns of variation within and between populations, and integrates these with a sense of trade-offs across competing demands and practical constraints, will singularly advance understanding of the determinants of health. Although these determinants are complex and variable within and between populations, the empirical advances presented in this volume indicate how biological anthropology provides means to unpack these complexities in a way that contributes substantially to scientific and practical policy concerns.

1.6 References

Barr, R. G., Boyce, W. T., & Zeltzer, L. K. (1994). The stress-illness association in children: a perspective from the biobehavioral interface. In *Stress, Risk, and Resilience in Children and Adolescents*, ed. R. J. Haggerty, L. R. Sherrod, M. Garmezy, & M. Rutter, pp. 182–224. Cambridge: Cambridge University Press.

Barker, D. (1997). Intra-uterine programming of the adult cardiovascular system. *Current Opinion in Nephrology and Hypertension*, **6**, 106–10.

Barker, D., Hales, C., Fall, C., Osmond, C., Phipps, K., & Clark, P. (1993). Type 2 (non-insulin-dependent) diabetes mellitus, hypertension and hyperlipidaemia (syndrome X): relation to reduced fetal growth. *Diabetologia*, **36**, 62–7.

Ben-Yehuda, N. (1990). Positive and negative deviance: more fuel for a controversy. *Deviant Behavior*, **11**, 221–43.

Boyce, W., Chesney, M., Alkon, A., Tschann, J., Adams, S., Chesterman, B., Cohen, F., Kaiser, P., Folkman, S., & Wara, D. (1995). Psychobiologic reactivity to stress and childhood respiratory illnesses: Results of two prospective studies. *Psychosomatic Medicine*, **57**, 411–22.

Breen, R. (1996). *Regression Models: Censored, Sample-selected, or Truncated Data.* Thousand Oaks: Sage Publications.

Campbell, K. L. (1994). Blood, urine, saliva and dip-sticks: Experiences in Africa, New Guinea, and Boston. *New York Academy of Sciences*, **709**, 312–30.

Dennison, E., Fall, C., Cooper, C., & Barker, D. (1997). Prenatal factors influencing long-term outcome. *Hormone Research*, **40** (Suppl. 1), 25–9.

Desjarlais, R., Eisenberg, L., Good, B., & Kleinman, A. (1995). *World Mental Health: Problems and Priorities in Low-Income Countries.* New York: Oxford University Press.

Dubos, R. (1966). Man and his environment – biomedical knowledge and social action. *Perspectives in Biology and Medicine*, **9**, 523–36.

Ellison, P. (1988). Human salivary steroids: Methodological considerations and applications in physical anthropology. *Yearbook of Physical Anthropology*, **31**, 115–42.

Ellison, P. (1993). Measurement of salivary progesterone. *Annals of the New York Academy of Science*, **694**, 161–76.

Ellison, P., Panter-Brick, C., Lipson, S., & O'Rourke, M. (1993). The ecological context of human ovarian function. *Human Reproduction*, **8**, 2248–58.

Ewald, P. (1994). *Evolution of Infectious Disease*. Oxford: Oxford University Press.

Hales, C., Barker, D., Clark, P., Cox, L., Fall, C., Osmond, C., & Winter, P. (1991). Fetal and infant growth and impaired glucose tolerance at age 64. *British Medical Journal*, **303**, 1019–22.

Johns, D. P. (1993). Nutritional need or athletic overconformity: Ethical implications for the sport psychologist. *Sport Psychologist*, **7**, 191–203.

Law, C., Gordon, G., Shiell, A., Barker, D., & Hales, D. (1995). Thinness at birth and glucose tolerance in seven-year-old children. *Diabetic Medicine*, **12**, 24–9.

Long, J. (1997). *Regression Models for Categorical and Limited Dependent Variables*. Thousand Oaks: Sage Publications.

Mayer, K., & Tuma, N. (1990). *Event History Analysis in Life Course Research*. Madison, Wisconsin: University of Wisconsin Press.

Nesse, R. M., & Williams, G. (1994). *Why We Get Sick: The New Science of Darwinian Medicine*. New York: Times Books.

Sagarin, E. (1985). Positive deviance: an oxymoron. *Deviant Behavior*, **6**, 169–81.

Trevathan, W., McKenna, J., & Smith, E. eds. (1998). *Evolutionary Medicine*. New York: Oxford University Press (in press).

Worthman, C. (1998). Evolutionary perspectives on the onset of puberty. In *Evolutionary Medicine*, ed. W. Trevathan, J. McKenna, & E. Smith. New York: Oxford University Press (in press).

Worthman, C., & Stallings, J. (1994). Measurement of gonadotropins in dried blood spots. *Clinical Chemistry*, **40**, 448–53.

Worthman, C., & Stallings, J. (1997). Hormone measures in finger-prick blood spot samples: new field methods for reproductive endocrinology. *American Journal of Physical Anthropology*, **103**, 1–21.

2

Hormonal correlates of personality and social contexts: from non-human to human primates

ROBERT M. SAPOLSKY

2.1 Introduction

Primatologists have long been fascinated by the dominance hierarchies of their study subjects. A number of early researchers attributed what was, in retrospect, an inappropriate power to an animal's rank in predicting its reproductive success, degree of attractivity to the opposite sex, willingness to defend other individuals against predators, and so on. Many of these misconceptions were clarified with the emergence of new theoretical constructs, such as the recognition of female choice, of alternative competitive strategies among individuals, or the importance of kin-based models of natural selection. In the wake of some of this early overenthusiasm for the explanatory power of rank, a number of individuals questioned the validity of the rank concept – whether such dyadic assymetries actually occur under most ecological conditions (e.g., Rowell, 1974), and whether the animals themselves "understood" rank in a meaningful way (as opposed to a hierarchy merely being an artificial construct projected by human observers) (e.g., Bernstein, 1981).

Despite these critiques, dominance hierarchies have remained close to the hearts of many primatologists, and are thought to occur among many species and to provide a meaningful way to understand individual differences in the experiences and quality of life of non-human primates. Out of this has emerged a considerable number of studies built around the premise that rank in such hierarchies helps to explain individual differences in the physiology of primates, with important implications for understanding disease patterns and differential fitness. In this chapter, I analyze the current understanding of one type of rank/endocrine correlate, one relevant to the understanding of patterns of stress-related disease. The primary point is to show how, in recent years, a large number of variables have been shown to modify these correlates, reducing the predictive power of rank; one of the most important ones, which will be covered at some length, concerns the role of personality differences in modulating rank/endocrine relations.

2.2 A brief overview of glucocorticoid action

Glucocorticoids are secreted by the adrenal gland in response to a broad array of physical or psychological stressors. Such secretion represents the final step in a neuroendocrine cascade, beginning with secretion of corticotropin releasing factor (CRF) and related secretagogs by the hypothalamus within seconds of the onset of a stressor. This, in turn, causes pituitary secretion of adrenocorticopin hormone which subsequently causes adrenal synthesis and release of glucocorticoids, commencing within a few minutes of the onset of the stressor. Among primates and humans, the dominant glucocorticoid is cortisol (also known as hydrocortisone).

Along with glucocorticoids is the stress-induced activation of the sympathetic nervous system and the release of the catecholamines epinephrine and norepinephrine (adrenaline and noradrenaline). Glucocorticoids and catecholamines are vital for surviving an acute physical stressor (such as evading a predator) (Munck *et al.*, 1984; Sapolsky, 1991). They mobilize energy from storage tissues such as fat and liver in order to supply exercising muscle, and increase cardiovascular tone to accelerate the delivery of energy substrates. In addition, they inhibit long-term anabolic processes such as growth, tissue repair, bone recalcification, digestion, and reproduction; the rationale for this is that such costly and optimistic processes are best delayed until a more auspicious time. In addition, glucocorticoids suppress immunity; this appears to prevent the immune system – often activated in response to a variety of stressors – from inadvertently being overactivated to the point of autoimmunity (Munck *et al.*, 1984). Finally, during the early phases of the stress-response, glucocorticoids and catecholamines increase glucose and oxygen delivery to the brain, enhancing certain features of cognition (McEwen & Sapolsky, 1995).

In contrast, with sustained stress, overexposure to glucocorticoids and catecholamines can be highly pathogenic (for review see Sapolsky, 1991). The short-term mobilization of energy, if extended over time, can produce myopathy, fatigue, and exacerbation of adult-onset diabetes. Sustained stress-induced hypertension is well recognized to be deleterious, primarily through damage to the surface of blood vessels and resulting atherosclerosis. The chronic inhibition of various anabolic processes can produce colitis, osteoporosis, amenorrhea, and impotency. In addition, chronic stress-induced immunosuppression appears to increase vulnerability to a number of infectious maladies. Finally, sustained stress and glucocorticoid exposure can have deleterious and even neurotoxic consequences for parts of the central nervous system, and may constitute a significant pacemaker of brain aging.

Collectively, these observations suggest what would be an ideal profile of these potent stress hormones: under basal, non-stressed circumstances, it appears to be most adaptive to minimize the secretion of glucocorticoids and catecholamines. In the face of a purely psychological stressor (i.e., one requiring no physical output), secretion should ideally remain at the same low level. In the face of a physical challenge to homeostasis, however, the most optimal of responses would involve a rapid and enormous outpouring of secretion. Finally, with the abatement of such a stressor, the ideal profile would involve a rapid return to baseline.

Numerous investigators have tested the hypothesis that particular social ranks among primates are associated with more optimal profiles of these hormones. Relatively few of these studies have examined sympathetic catecholamines. This is because their circulating concentrations can change in response to perturbation within a few seconds, making measurement of true basal levels impossible in animals that are not catheterized. Furthermore, catecholamines degrade relatively quickly, traditionally requiring rapid refrigerated centrifugation, making their study under field conditions even more difficult.

Thus, most of these studies have instead concentrated on glucocorticoids. The majority of them have involved measurement of cortisol concentrations in the bloodstream; this gives, of course, the truest measure of the amount of this hormone reaching target tissues throughout the body, but requires either the anesthetization of subjects, the training of them to present for venepuncture in a non-stressful manner, the rapid immobilization of a caged and conscious animal for blood sampling, or the use of indwelling catheters. A number of studies have made use of urinary measures, which has the advantage of not requiring the subject's disturbance. Rapid changes in circulating cortisol concentrations are, obviously, immediately detectable in blood samples, but are smoothed out and integrated over time in urinary samples; thus, the former approach is far more useful for answering questions requiring minute-by-minute time resolution. Perhaps the most time-integrated measure of glucocorticoid

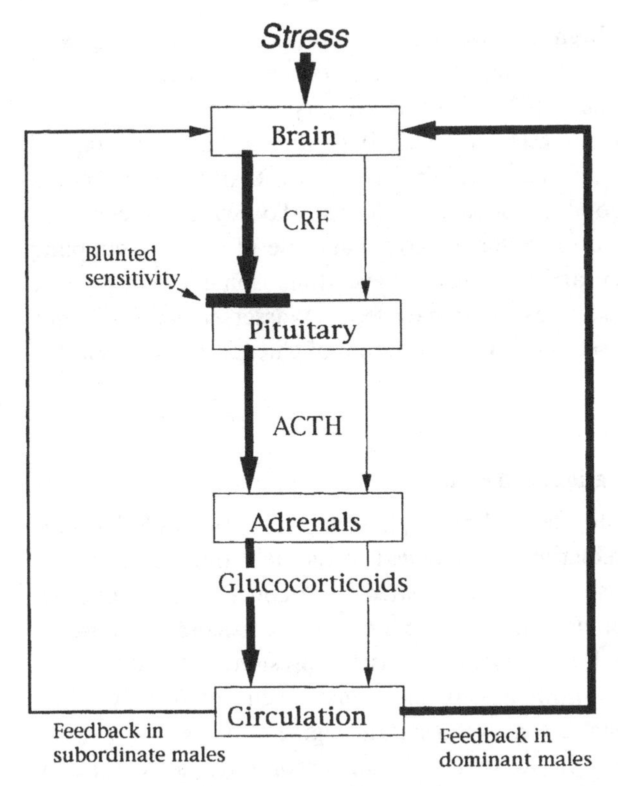

Figure 2.1. Mechanisms underlying the elevated glucocorticoid concentrations in subordinate male baboons. Among both subordinate and dominant animals, the release of glucocorticoids (specifically cortisol, the glucocorticoid of primates and humans) increases in response to stress: the hypothalamus secretes CRF (corticotropin releasing hormone) and related hormones, and these stimulate the pituitary gland to release ACTH (adrenocorticotropic hormone). This causes the adrenal glands to release glucocorticoids into the bloodstream. In dominant males, the brain can accurately detect feedback from the blood, thus readily detecting when a threshold level of glucocorticoids is reached. The hypothalamus then inhibits the secretion of CRF and its relatives, leading to a decline in glucocorticoid release. In subordinate baboons, the brain is less able to accurately detect a feedback signal, and so the brain "believes" that glucocorticoid levels are low even when they are actually high. Consequently, secretion of CRF is markedly increased. The pituitary of subordinates is somewhat insensitive to CRF, but the large amounts reaching the pituitary nonetheless trigger an increase in the secretion of ACTH, which then leads to the chronic hypersecretion of glucocorticoids. (Modified from Sapolsky, 1992.)

hypersecretion has been the weighing of adrenals postmortem, with adrenal hypertrophy indicating sustained glucocorticoid hypersecretion. It has recently become popular to study salivary cortisol concentrations in humans (Kirschbaum & Hellhammer, 1994); this has the advantage of the ease of collection, plus a better time resolution than urinary samples (although not as good as circulating levels). To my knowledge, no rank/endocrine studies of primates have made use of salivary sampling yet. Finally, Wasser and colleagues have done some exciting work measuring gonadal steroids in primate feces (Wasser *et al.*, 1994); this approach has been used only rarely in studies of glucocorticoids (Whitten *et al.*, 1997).

2.3 Cortisol concentrations and rank

Based on the physiological and pathophysiological actions of glucocorticoids, the basic prediction has emerged in the field that social subordinance would be associated with elevated basal concentrations of these hormones and, perhaps, with an impaired glucocorticoid response to stress. The rationale for this prediction is the presumed stressfulness of subordinance – for a low-ranking individual, even supposedly basal, non-stressed circumstances involve a certain degree of stress.

Such stress is thought to reflect a number of physical features of subordinance. This can include nutritional deprivation (most typically during times of severe resource limitation) or increased energy demands (i.e., in situations where subordinant individuals are not food-deprived but might have to work harder for their calories). In addition, subordinant animals may be disproportionately subject to the physical stressor of displaced aggression, or at a higher risk of predation because of their more peripheral status.

Perhaps even more importantly, social subordinance can often be associated with a number of psychological stressors. An extensive literature has demonstrated the components of psychological stress: for the same physical stressor, an individual is far more likely to mobilize a stress-response (and to eventually develop a stress-related disease) if they lack outlets for frustration, if there is no predictive information as to the intensity and duration of the stressor, if there is a minimal sense of control, a perception that the stressor signals worsening conditions, or if the individual lacks social support (Levine *et al.*, 1989). In many primate social systems, subordinance carries with it these precise psychological disadvantages, in that individuals can often not predict or control their access to resources, may be limited in the forms of social support that they can access (for example, being groomed), may be subject to unpredictable displaced aggression, and may lack many outlets that would reduce frustration (such as displacing aggression onto a more subordinate individual).

Given this general picture, it is not surprising that many studies have reported elevated basal cortisol concentrations and/or enlarged adrenals in subordinate individuals. This includes rhesus macaques (Gust *et al.*, 1993), fascicularis macaques (Adams *et al.*, 1985; Kaplan, 1986; Shively & Kaplan, 1984), talapoins (Keverne *et al.*, 1982), olive baboons (Sapolsky, 1990), squirrel monkeys (Manogue *et al.*, 1975), tree shrews (Fuchs & Flugge, 1992, 1995), and mouse lemurs (Schilling & Perret, 1987).

Similar patterns have been reported among non-primate species, including rats (Barnett, 1955; Popova & Naumenko, 1972; Schuurman, 1981; Raab *et al.*, 1986; Korte, 1990; Sakai *et al.*, 1991; Blanchard *et al.*, 1993a, b, 1995; Dijkstra, 1992; McKittrick, 1995), mice (Davis & Christian 1957; Southwick & Bland, 1959; Bronson & Eleftheriou, 1964; Louch & Higginbotham, 1967; Archer, 1970; Leshner & Polish, 1979; Schuhr 1987), hamsters (Huhman *et al.*, 1990, 1992), guinea pigs (Sachser, 1987), wolves (Fox & Andrews, 1973; McLeod *et al.*, 1996), infant hyenas (Frank *et al.*, unpublished results), rabbits (Farabollini 1987), pigs (McGlone *et al.*, 1993; Fernandez *et al.*, 1994), fish (Fox *et al.*, 1997), and even sugar gliders (Mallick *et al.*, 1994).

Potentially, the glucocorticoid hypersecretion of subordinate individuals could reflect some peripheral regulatory change, such as decreased clearance of the hormone from the circulation. In the sole test of this possibility among primates, clearance rate was found not to differ by rank (Sapolsky, 1990, for olive baboons). Instead, the hypersecretion is likely to be driven at the level of the brain. Subordinate individuals appear to hypersecrete CRF from the hypothalamus (Sapolsky, 1989, for olive baboons; such a conclusion is based on a rather indirect technical approach for inferring CRF levels, in that direct measurement requires sacrifice of subjects). Furthermore, both direct and indirect measures indicate elevated ACTH concentrations in subordinate animals (Farabollini, 1987, for rabbits; Huhman, 1990, for hamsters; Sapolsky, 1993*a*, for olive baboons). This general picture of hypersecretion at all levels of the adrenocortical axis is commensurate with the notion that subordinate individuals are subject to the highest rates of stressors (i.e., that hypersecretion is initiated at the level of the brain, the site at which such stressors are sensed).

In addition, subordinate individuals appear to be resistant to negative feedback regulation of the adrenocortical axis. As with most endocrine systems, the end product (in this case, glucocorticoids) has an inhibitory effect upon more rostral components of the system (i.e., secretagog release from the brain), and such feedback regulation is essential to normal circulating levels of these hormones. Subordinate baboons are resistant to this feedback inhibition (Alberts *et al.*, 1997). Thus, hypersecretion can result from both an overabundance of stressors activating the system, and from a blunted regulatory capacity to inhibit the system in the absence of stress (Figure 2.1). These two causes of hypersecretion intertwine, in that sustained stress will downregulate glucocorticoid receptors in feedback sites in the brain, damping sensitivity to feedback regulation (for review see Sapolsky & Plotsky, 1990).

Thus, a large number of studies from varied primate and non-primate species suggest that subordinance is associated with glucocorticoid hypersecretion, and some evidence suggests that subordinate individuals pay a pathogenic price for this hypersecretion, including suppression of high density lipoprotein: low density lipoprotein (HDL:LDL) cholesterol ratios (which increases the risk of cardiovascular disease) and numbers of circulating lymphocytes (Sapolsky, 1993*a*). Despite this seemingly clear picture, there are numerous studies which have failed to find glucocorticoid hypersecretion in subordinates. The extremely informative variables that explain most of these discrepancies occupy the remainder of this review.

2.4 Glucocorticoids, rank, and the societal context of that rank

As a truism that should be obvious to all primatologists by now, rank can only be understood in the context of the primate society in which it occurs. The basic structure of hierarchies (e.g., whether hereditary or plastic, whether linear or imbedded with circularities), the behaviors needed to attain and maintain dominance (e.g., aggression versus affiliation), the intensity and frequency with which those behaviors are expressed, and the alternatives available to dominance will all vary by species, and within species over time and across ecological clines.

Thus, it is not surprising that the link between subordinance and basal glucocorticoid hypersecretion should come with many exceptions. One example concerns the frequency with which aggression plays a role in reinforcing the position of dominant individuals. Among Old World primates, for example, individuals in stable dominant positions rarely have the highest rates of aggression, and instead typically rely upon bluff and psychological intimidation; dominant individuals who must frequently aggress against subordinates in order to maintain their dominance are typically in the process of dropping in the hierarchy (Altmann *et al.*, 1995). A very different story is seen in a recent report (Creel *et al.*, 1996) concerning wild populations of dwarf mongooses and of wild dogs. In both cases, dominant individuals have significantly higher rates of aggression than do subordinates and, in fact, dominant individuals must continuously reinforce their position with aggressive reassertion of their rank. In such circumstances, it is not surprising that the authors found the dominant individuals to have the highest basal glucocorticoid concentrations. A similar theme is seen among rats in a visible burrow system; normally, it is the subordinate individuals who have the highest glucocorticoid concentrations (Blanchard *et al.*, 1995); however, if a colony is composed of only rats pre-selected for their aggressiveness, it is now the dominant individuals who are most hypersecretory (Blanchard *et al.*, 1995).

The exceptions just discussed concern what it takes to be dominant. Just as important is what it is like to be subordinate in a particular setting, in that subordinance does not always carry the same physical or psychological costs. For example, among female marmoset monkeys, being low-ranking is not particularly onerous, and is basically a position held until an older, more dominant relative dies; in such a system, subordinant animals are not subject to particularly high rates of physical or psychological stressors, and do not have elevated glucocorticoid levels (Abbott *et al.*, 1997). As an example taken from within a single species, while subordinant female rhesus monkeys were reported to be hypercortisolemic in one troop, they were not in another which had significantly higher levels of reconciliative behavior (Gust *et al.*, 1993). As another example, subordinate male baboons were reported to be hypercortisolemic over the course of a number of years of study of a wild population (Sapolsky, 1993*a*). During the tragic East African drought of 1984, dominant animals spent a significantly greater portion of their time foraging than in other years, and thus had less time to target subordinates for approach–avoid interactions or displacement aggression; during that drought season, cortisol concentrations in subordinates were significantly lower than in other years. Ironically, for such individuals, the ecological challenge of a drought was, indirectly, a social blessing (Sapolsky, 1986).

In many ways, the issues of what it involves to be dominant or subordinate in a particular primate society is best encompassed in the dichotomy between stable and unstable hierarchies. During times of stability, in which dominant individuals are safely ensconced in their positions and can maintain them with a minimum of effort or aggression, dominance carries with it the psychological advantages of large degrees of control, predictability, social support, and outlets for frustration. The primate studies cited in the previous section showing lower basal cortisol concentrations among dominant individuals were all derived from animals in such stable hierarchies.

In contrast, during times of instability (such as following group formation in captive animals, or during hierarchical reorganization in wild populations due to the death, injury, or transfer of key individuals), the dominant individuals are typically at the center of the tensest rank competition. Rankings typically shift frequently, rates of aggression and of dominance interactions increase, rates of social outlets decline. Under such circumstances, dominant animals lose the psychological advantages of their rank and, arguably, are even subject to the highest rates of psychological stressors. At such times, dominance is no longer associated with low basal cortisol concentrations (Table 2.1; Chamove & Bowman, 1976, Keverne *et al.*, 1982, for talapoin monkeys; Coe *et al.*, 1979; Mendoza *et al.*, 1979, for squirrel monkeys; Gust *et al.*, 1991, for rhesus monkeys; Sapolsky, 1983*a*, for baboons). In most captive studies, approximately a year is required after group formation for hierarchies to stabilize and the initial pattern of hypercortisolism among dominant individuals to switch to hypercortisolism of subordinates (Sapolsky, 1993*b*).

Thus, depending on the physical and psychological costs of dominance and/or subordinance in a particular primate society, the adrenocortical correlates of rank can be diametrically opposite. This theme continues in the next section.

2.5 The personal experience of rank and its societal context

Complex social primates do not merely live in societies that illustrate certain abstract principles, such as "the stressful consequences of societal-wide rank instability" or "the salutary effects of a highly reconciliative social group". Instead, they experience these factors in highly personal, individualized ways. Not surprisingly, such personal experience can modify the rank/glucocorticoid relationships discussed above.

One variable would be how often or intensely an animal experiences the advantages and disadvantages of its station in life. For example, in a study of female rhesus monkeys, those subordinate females who were subject to the highest rates of dominance or aggressive interactions or the lowest rates of affiliative interactions, had the highest cortisol levels (Gust *et al.*, 1993); subordinate animals less subject to the indignities of their rank or with more affiliative support were not hypercortisolemic. In a similar

Table 2.1. *Physiological correlates of dominance in stable versus unstable hierarchies*

Stable	Unstable
Dominant males have the lowest basal cortisol concentrations	
YES	NO
M. talapoin (Keverne *et al.*, 1982)	*M. talapoin* (Keverne *et al.*, 1982)
P. anubis (Sapolsky, 1983)	*P. anubis* (Sapolsky, 1983)
S. sciureus (Manogue *et al.*, 1975)	*S. sciureus* (Coe *et al*; 1979; Mendoza *et al.*, 1979)
	M. mulatta (Chamove & Bowman, 1976)
Dominant males have largest increases in cortisol secretion during stress	
YES	NO
S. sciureus (Manogue *et al.*, 1975)	*S. sciureus* (Coe *et al.*, 1979)
P. anubis (Sapolsky, 1983)	*P. anubis* (Sapolsky, 1983)
Dominant males are more sensitive to glucocorticoid feedback regulation	
YES	NO
P. anubis (Sapolsky, 1983)	Not examined
P. cynocephalus (Sapolsky *et al.*, 1997)	Not examined

Note: (Adapted from Sapolsky, 1993*b*).

theme indirectly related to cortisol secretion, among wild female baboons, there was a dose–response relationship between the frequency with which females suffered displaced aggression at the hands of a particularly aggressive transfer male and their extent of immunosuppression (Figure 2.2). As another example, captive barbary macaques were found to have elevated cortisol concentrations at the time of their estrus swellings; however, the smallest increments were found in the subset of females who were groomed the most and who were subject to the lowest rate of aggressive displacement by males (Wallner, 1996).

As another index of the importance of personal experience, in studies of wild fascicularis macaques (van Schaik, 1991), and of male wolves in large outdoor enclosures (McLeod *et al.*, 1996), basal cortisol concentrations did not correlate with rank. Instead, the highest concentrations were seen in animals whose ranks were most unstable (the beta male in the wolf study, and recent transfer animals among the macaques); such instability was occurring in the context of hierarchies that were, overall, quite stable. As an example of the potentially long-lasting effects of personal experience, captive macaques raised by mothers who had anxiety-provoking foraging

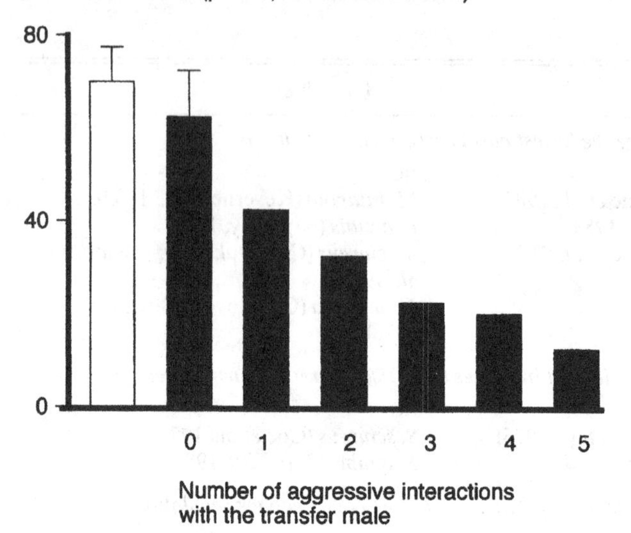

Figure 2.2. Relationship between numbers of circulating lymphocytes (white blood cells) and frequency with which female baboons were attacked by an aggressive transfer male. The open bar indicates average lymphocyte counts prior to the transfer of the aggressive male into the troop. (Modified from Sapolsky *et al.*, 1997.)

demands placed on them showed elevated cerebrospinal fluid concentrations of CRF as adults (Coplan *et al.*, 1996).

A final example is a particularly subtle one. In this study, basal cortisol concentrations among wild male baboons were analyzed as a function of the instability of an individual's rank (i.e., the rate at which he was having reversal interactions with animals immediately above or below him in the hierarchy). The prediction was that, independent of the stability of the overall hierarchy, an animal whose own rank was precarious and unstable should be hypercortisolemic, relative to the typical values found for that rank. This prediction was partially confirmed, in that higher rates of unstable interactions with males immediately behind the subject in the hierarchy were indeed associated with hypercortisolism. In contrast, higher rates of unstable interactions with males immediately ahead were not associated with elevated glucocorticoid levels (Sapolsky, 1992*b*). While initially puzzling, this is, in fact, quite logical – instability with those immediately subordinate indicates that one is in danger of being displaced from one's rank, an obvious stressor. In contrast, instability with those immediately ahead in the hierarchy indicates a chance of moving up in the hierarchy – a promotion, rather than a stressor.

Thus, the studies just cited concerning the intensity or frequency with which an animal is exposed to social stressors show that it is not the abstract state of having a stressful rank, or living in a stressful society, which best predicts cortisol concentrations; rather, it is one's very concrete personal experience of that rank and society. Moreover, the study latterly cited indicates that it is not just personal experience which is important (e.g., the rate at which one is involved in unstable interactions). The *meaning* of such interactions (e.g., an advance versus a decline in the hierarchy) can be critical also.

2.6 Personality as a modulator of adrenocortical status

Few who have spent any time observing social groups of primates fail to be struck by their strong, individualistic personalities, and a number of studies have explored issues of temperament and personality in primates (for review see Clarke & Boinski, 1995). A handful of studies have now examined the endocrine correlates of some of these personality differences.

One body of studies has emerged from the superb work of Suomi and colleagues with captive rhesus monkeys (for review see Suomi, 1997). They have shown that within a few months of age, infant rhesus show stable personality differences along a reactivity continuum – how readily they behaviorally and physiologically react to novelty. "High reactor" individuals become anxious, agitated, and withdrawn when placed in a novel environment (in contrast to the exploratory behavior seen in animals at the other end of the spectrum), and are particularly shy when placed with novel peers. Such novel situations provoke marked glucocorticoid and adrenocorticotrophic hormone (ACTH) secretion in these high reactors. Moreover, when separated from their mothers, high reactors become atypically behaviorally depressed, and show marked physiological responses (glucocorticoid hypersecretion, enhanced sympathetic tone, and immune suppression).

These personality differences appear to be long-lasting among these animals, persisting into adulthood. Moreover, these stylistic differences are observed in free-living rhesus and effect the life history trajectory of these individuals. The shy, high reactive animals leave their natal troops at significantly later ages than do their peers. After such transfers, high reactive males use more conservative and ultimately less successful strategies for joining a new troop, whereas high reactive females are less likely than average to emerge as competent, primiparous mothers.

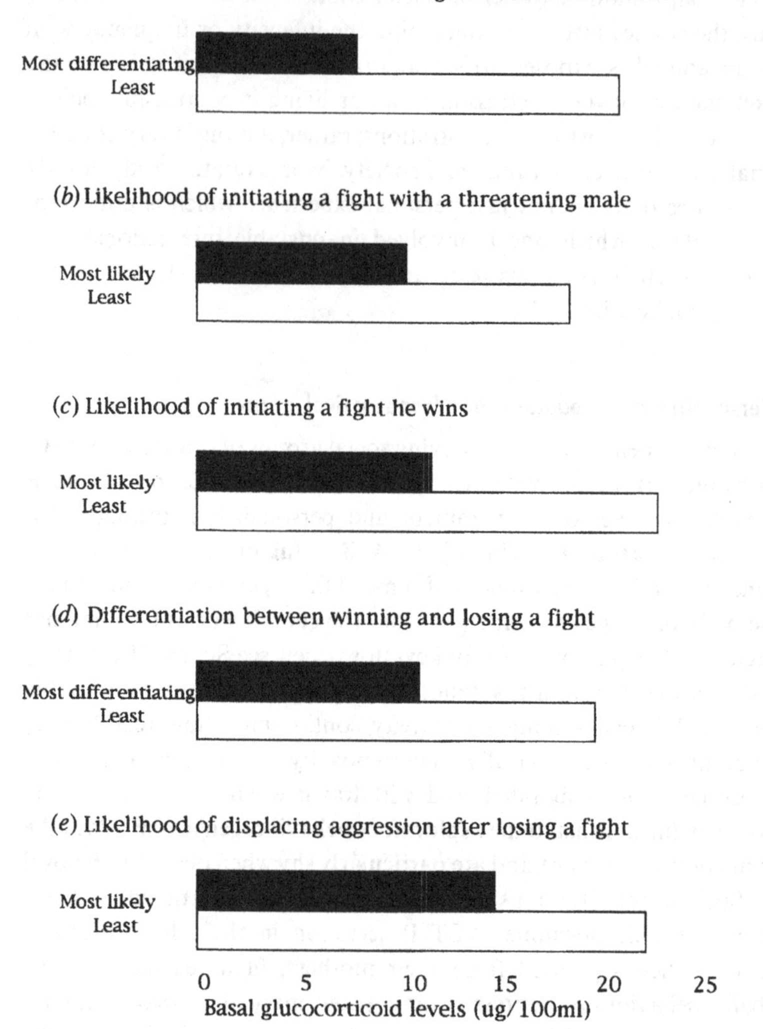

Figure 2.3. Impact of personality traits on circulating glucocorticoid levels. Dominant male baboons with certain personality traits (top bar of each pair) have lower basal glucocorticoid levels in the bloodstream than do other dominant males (lower bar of each pair), which suggests that attitude is a more important mediator of physiology than rank alone. Dominant males who can distinguish between the threatening and neutral actions of a rival have glucocorticoid levels that are about half as high as those of other dominant males (*a*). Similarly, low glucocorticoid levels are found in males that start a fight with a threatening rival instead of waiting to be attacked (*b*); know which fights to pick and so are likely to win fights they initiate (*c*); distinguish between having won and lost a fight (*d*); or take out their frustration on subordinates when having lost a fight (*e*). (Modified from Sapolsky, 1992.)

Some similar themes have emerged from my own work on endocrine correlates of personality differences among wild baboons. Among high-ranking males, a number of personality traits predict low basal cortisol concentrations, independent of rank. One cluster of traits is related to aspects of male–male competition, and lower basal cortisol concentrations are observed among individuals who show the greatest degree of social discrimination – being able to differentiate (in terms of their reactivity) between neutral and threatening interactions with rivals, between low-grade and high-grade threatening interactions, and between positive and negative outcomes of such interactions. Furthermore, such low cortisol males are able to maintain some degree of social control during threatening interactions, and can effectively displace frustration if losing a fight (Figure 2.3; Sapolsky & Ray, 1989; Ray & Sapolsky, 1992; Sapolsky, 1992). Similar themes are seen in the analysis of personality/cortisol correlates among the low-ranking males in this population (Virgin & Sapolsky, 1997). Collectively, these low cortisol males appear to be the least behaviorally reactive (much as in Suomi's 1987 and 1997 studies), have the highest degree of social control, predictability, and outlets for frustration.

A second, independent cluster identified another group of low-cortisol males. These were individuals with the highest rates of grooming and being groomed by female "friends" (see Smuts, 1986, for a discussion of this term in the context of non-human primates) and the highest rates of interactions with infants. Thus, these low cortisol individuals appeared to have the highest rates of social support (Figure 2.4; Sapolsky *et al.*, 1997; Ray & Sapolsky, 1992). A recent study indicates that this subset of individuals have an atypically high likelihood of "successful aging" (defined as having a large number of affiliative relationships in old age, and being able to remain in their troop, rather than being driven into transfering troops due to the harrassment of younger dominant males) (Sapolsky, 1996*a*).

Some of these themes are demonstrated in comparing the histories of two male baboons that I have studied. Saul transferred into the troop in late adolescence and was fairly peripheral to most social interactions, tending to be the first individual emerging from the forest each morning, maintaining an atypically high distance between himself and other animals, taking part in grooming only rarely. Due to large body size and an extremely aggressive fighting style, he became the number one ranking male early in adulthood; his rise to the top was marked by an absence of any cooperative coalitions with other males. His tenure as the alpha male

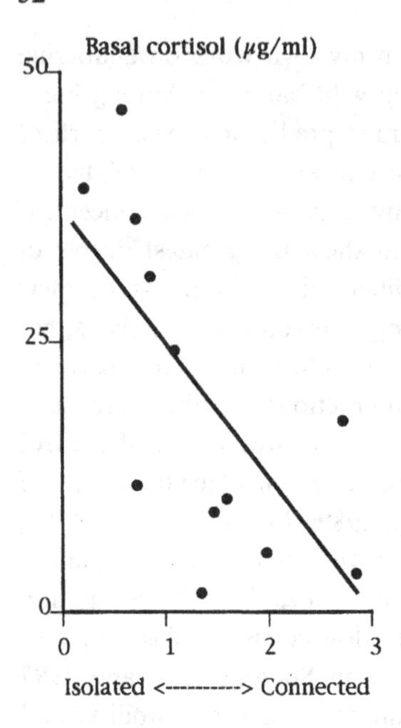

Figure 2.4. Basal circulating concentrations of cortisol are predicted by social connectedness for adult males. Each dot represents the value for one animal. Social connectedness is an index of the extent to which males score high (connected) or low (isolated) on a scale of social measures. These include rates of affiliative interactions with adult males and with adult females; rates of agonistic interactions with adult males and with adult females; proportion of time spent within 3 meters of another animal; the average number of such animals within 3 meters; number of reciprocal grooming relationships; the proportion of available sexual consortships with adult females obtained by the individual. (From Sapolsky *et al.*, 1997.)

lasted 2 years, which is quite long among male baboons, and was characterized by a minimum of social grooming with females, a minimum of interaction with infants, and a continuation of his very aggressive fighting style in the face of often rather minor provocations and challenges from other males. His reign as alpha was terminated by the formation of a cooperative coalition of six males who left him grievously injured in a fight; examination of him a month later revealed a broken arm and a dislocated shoulder; he limped badly for the rest of his life. He dropped precipitously in the hierarchy and was subject to a great deal of harassment by members of that coalition of six; within a year, he transfered troops, where he occupied a spot near the bottom of the hierarchy for his few remaining years.

A sharp contrast is provided by a contemporary of Saul's. Joshua also transferred into the troop as an adolescent, and quickly showed a behavioral style that included developing close affiliative, non-sexual relationships with a number of females. During the period of hierarchical instability that followed the formation and rapid dissolution of the coalition of six males mentioned in the previous paragraph, he briefly emerged as the alpha male as a result of coalitional support from another individual. Within 6 months, he had voluntarily relinquished the position – giving a series of overtly subordinate gestures to the number two-ranking male despite an absence of any challenge by that individual. Joshua instead spent most of his social time interacting with affiliated females and their infants, many of whom were likely to have been his offspring (due to high rates of covert matings throughout his adulthood with female friends). He showed a marked disinterest in male–male competition, giving low-intensity subordinance gestures or walking away from challenges by other individuals. He outlived his male cohort within the troop by a minimum of 3 years, and spent his aged years in that troop, still maintaining high rates of affiliative interactions with females and infants.

These studies suggest that, independent of rank, personality can strongly influence glucocorticoid secretion. Specifically, lower basal cortisol concentrations occur in animals who are less reactive to novelty, who are most socially adept and controlling of tense intermale competition, and who have the most forms of social outlets and support. Moreover, both field and laboratory studies suggest that some of these traits can be stable over time. From where do these personality differences arise? Suomi and colleagues have done careful studies suggesting some small but significant genetic contributions to these traits. More importantly, such studies indicate that a stronger contribution comes from the style of the mothering, with highly reactive and anxious mothers generating offspring with similar traits (Scanlon *et al.*, 1985; Suomi, 1987); moreover, offspring of high reactive mothers can have those traits significantly attenuated if they are foster-mothered by a particularly nurturant female (Suomi, 1997). In a variant on this, recent work has shown that females who are stressed or anxious during their gestation period are more likely to produce hypercortisolemic, reactive offspring (Clarke, 1996), showing that environmental mothering effects can include intrauterine ones.

What are the more proximal mechanisms that make for the widely different personality styles? Any discussion of this must be fairly speculative, invoking only the haziest of possible neurobiological underpinnings. Broadly, it strikes me that the "low cortisol" profiles involve at least four features: an atypical tendency towards forming affiliative relationships, low anxiety, a high degree of "social intelligence", and marked impulsivity control. The tendency towards affiliation could be reframed as marking individuals who gain particular reinforcement or "pleasure" from such interactions. This could involve something as mechanistic as a particularly high density of tactile receptors in the skin that transduce grooming into an unusually pleasant sensation; the generality of the behaviors of these individuals beyond grooming makes such a peripheral explanation seem unlikely, however. More centrally, the affiliative pattern could reflect differential responses of limbic pleasure pathways; while the heterogeneity of responsiveness of these pathways has been documented for decades and would plausibly involve individual differences in responsiveness to sociality as a reinforcer, this has never been explicitly examined. As another possible biological marker, recent work has revealed some neuroendocrine features of parental and affiliative behavior in animals, centered around the actions of oxytocin (Carter *et al.*, 1992; Insel & Shapiro, 1992), and this may be relevant to understanding some of the differences in affiliative behavior shown in these studies.

In a somewhat less speculative vein, individual differences in anxiety-prone behaviors are relatively well understood in the context of numbers, distribution and affinity of receptors for benzodiazepines (the minor tranquilizers such as valium or librium), and levels of the still poorly characterized endogenous benzodiazepines. Dramatic strain differences in anxiety-prone behaviors and susceptibility to stress-related diseases have been shown to involve differences on this neurochemical level (as well as sensitivity of these endpoints to numerous environmental factors) (Sanger *et al.*, 1994). Some of my own current work with these wild baboons examines individual differences in benzodiazepine neurochemistry.

There is considerable emphasis in some quarters at present on the notion of "emotional intelligence" in both humans and non-humans (Cheney & Seyfarth, 1990; Goleman, 1995), and this seems highly intermixed with impulsivity control as a factor in these individual differences. Readers familiar with the behavior of male baboons in groups will recognize some of the low cortisol personality males as being those rare individuals who can consistently walk away from potential fights, being relatively unprovokable. An extensive neuropsychological literature suggests this to be the province of the pre-frontal cortex and its projections to the limbic system.

These are relevant to understanding highly affective measures such as provocability, more cognitive issues such as ability to carry out reversal or extinction tasks without perseveration, and the intermixing of affect and cognition in issues of long-term planning, foresight and gratification postponement. The most dramatic demonstrations of the importance of these pathways arise from "frontal disinhibition" patients who have sustained damage to these regions (such as the famed Phineas Gage who, in the 1840s, had massive frontal damage following an accident, and whose dramatic transformation into an aggressive, disinhibited individual marked him as the first such patient) (McAllister, 1992; Cummings, 1993; Damasio *et al.*, 1994). The frontal cortex occupies a disproportionately greater percentage of both the cortex and the brain as a whole in apes as compared to other primates, and in primates as compared to non-primates. Most neurobiologists would consider that individual differences within species in "frontality" of behavior will have some material correlates in this part of the brain, but this has not been studied in primates, to my knowledge. An extensive neurochemical literature demonstrates the importance of serotonin in the functioning of these pathways, in that low levels of forebrain serotonin or of peripheral markers of serotonin metabolism are associated with aggression, impulsivity and disinhibition (Stein *et al.*, 1993; Bell & Hobson, 1994); some preliminary studies of these wild baboons suggest differences in whole blood serotonin levels that may be relevant to individual differences in provocability (R. Sapolsky, M. Raleigh and G. Brammer, unpublished results).

It becomes fascinating to consider the evolutionary implications of some of these personality differences. There is much reason to think that a number of the personality features associated with low cortisol concentrations should enhance fitness. First, dominant baboons with the low cortisol traits of being the most discriminating about the intensity of male–male competition (as discussed above) remain in the dominant cohort significantly longer than rank-matched males who were least adept at those competitive skills (Sapolsky & Ray, 1989; Ray & Sapolsky, 1992); this is likely to translate into enhanced fitness. Moreover, low cortisol males with the highly affiliative personality styles are the ones most likely to be the beneficiaries of female choice (Smuts, 1986), and numerous studies have now shown the fitness advantages of this alternative reproductive strategy. Thus, what are the long-term implications of these personality differences that, at least within the framework of these studies, appear to carry some natural selective advantages? One can imagine directional selection occurring, in which social primates are being selected for increasing degrees of sophistication in discriminating among differing social scenarios and increasing degrees of impulse control and capacity for delayed planning; this certainly appears to be a trend in moving from monkeys to apes to humans (and to be accompanied by increasing neuroanatomical complexity to the frontal cortex). Perhaps some of the low cortisol personality types discussed, now constituting a minority of primate populations, will ultimately constitute the norm in future generations.

Conversely, one can imagine scenarios of situational dependency and balanced selection in which, under certain circumstances, "high cortisol" personality traits would be advantageous. For example, in the human realm, certain high cortisol personality profiles are associated with increased risk of certain stress-related maladies; despite that, one can imagine certain settings (i.e., Germany in the 1930s for any number of minorities) where it would, in fact, have been highly adaptive to have been the anxious, neophobic, high-reactor who fled. Thus, the fitness implications of some of these endocrine and personality differences are far from clear, and require careful, further study.

2.7 Humans, hypercortisolism, and individual differences

The example given in the preceding paragraph offers a transition to the issue of individual differences in stress-responses in humans. Can such variability be understood in the context of any of the issues raised in this review? I briefly touch on this possibility.

A number of investigators have examined some of these endpoints in the context of human "rank". These have involved an array of paradigms, focusing on endpoints such as occupational status in large corporations, outcomes of athletic competitions, rankings in the military, scores on personality inventories or, in a more ethological bent, assessments of degrees of "social dominance" by peers in small groups (see Elias, 1981; D. Hellhammer *et al.*, unpublished results; Meyerhoff *et al.*, 1988). These strike me as generally too artificial to be of much value in understanding human patterns of stress-related disease, for the simple reason that any given individual is likely to be part of a large number of (valid or invalid) hierarchies, and with differing "rank" in each. Furthermore, the human capacity for rationalization, internalization of standards, and so on, greatly confounds such simple rankings. A number of investigators have argued, in a way that I find most convincing, that the most valid "ranking" system for making sense of human health patterns is socioeconomic status, which is an enormously powerful predictor of mortality and morbidity risk (even after controlling for variables such as differential access to medical care) (Pincus & Callahan, 1995).

The more subtle issues of the societal setting and the personal experience of rank seem more informative about humans. One classic example should illustrate this point. During the blitzkrieg of World War II, central London was hit nightly, whereas in the suburbs, the bombing was far more sporadic, occurring perhaps once a week. As a measure of the impact of an unstable and unpredictable social setting, there was a significantly greater increase in the incidence of ulcers in the suburban population than in the urban one (Stewart & Winser, 1942).

Not surprisingly, there is tremendous richness in considering the links between personality and stress-responses in humans (for review see Sapolsky, 1996*b*). These studies have not only considered glucocorticoid profiles, but also those of catecholamines (given the greater ease of obtaining meaningful measures from a cooperative human than a non-human primate), as well as physiological or pathophysiological endpoints sensitive to these stress hormones.

There are a number of ways of conceptualizing these studies. In a way that I find most informative, elevated basal levels of glucocorticoids or catecholamines in humans are found in a number of acute or chronic states in which there are discrepancies between the stressor that a person is confronted with and the sort of coping responses mobilized.

One example occurs with major depression, in which half the sufferers are typically found to have significantly elevated glucocorticoid concentrations and chronic activation of the sympathetic nervous system (see Carroll, 1982; Roy *et al.*, 1985; Sapolsky & Plotsky, 1990). A classic cognitive picture of depression is one of "learned helplessness" where, as a result of some painful experience of loss of control, the individual has overgeneralized this to a consistent response style. In the face of stressful challenges, depressed individuals are less likely to mobilize a coping response of any sort and, if they happen to chance upon an effect response, are less likely to sustain it (Seligman, 1975). Importantly, if such helplessness is experimentally generated in laboratory animals with prolonged uncontrollable aversive stimuli, there results a similar chronic overactivation of the stress-response (Seligman, 1975). The neuroendocrine mechanisms underlying the hypercortisolism of depressives have been studied in tremendous detail, and involve changes at the level of the brain, pituitary, and adrenal glands. Remarkably, a large percentage of those mechanisms underlie the hypercortisolism in subordinate baboons (for review see Sapolsky, 1993*a*), emphasizing the possible cognitive and affective parallels between human depression and social subordinance in a non-human primate.

Two other personality styles come from the other end of the spectrum, namely individuals who overactivate coping responses in the face of stressors. The first case would be individuals with any of a number of anxiety disorders, in which there is a perception of life as filled with frequent, unpredictable, and uncontrollable stressors demanding a constant hypervigilance. Not surprisingly, anxiety disorders are typically associated with basal hypersecretion of stress hormones (see Curtis, 1970; Watson & Clarke, 1984). A similar, if more specialized version of this is seen with Type A individuals. Refinement of the Type A concept has shown that the core of these people's maladaptiveness is to see the world as full of malevolent challenge, and to mobilize pointless coping responses that are built around a corrosive, cynical hostility. The Type A personality was originally defined in the context of a pathologic endpoint (i.e., cardiovascular disease). More recent studies have shown this to involve the intermediary step of basal hypersecretion of stress hormones (for review see Williams, 1989).

Perhaps the most interesting and challenging linkage between a personality type and an overactive stress–response is seen in individuals with "repressive" personalities. These individuals are, by definition, neither depressed nor anxious, and typically describe themselves as unstressed, happy, and productive (a description borne out with more formal testing (Brandtstadter *et al.*, 1991)). Instead, these are individuals who are best described as living regimented, unemotional lives of discipline and conformity. Personality tests show them to have a particular dread of social disapproval, a tendency to suppress negative affect (i.e., the defining feature of their "repressiveness"), and an aversion to ambiguity (e.g., an atypically strong tendency to endorse statements with "never" or "always'"); related to this, they report and perceive only dominant emotions, being unable or uncomfortable with blendings of multiple emotions. These individuals have markedly elevated basal cortisol levels and hyper-reactive sympathetic stress-responses (Weinberger *et al.*, 1979; Brown *et al.*, 1996). In effect, they are highly stressed by the process of producing a life without any stressors.

Repressive personalities present an important final point for this review. These are not individuals exposed to a great deal of stress with which they are coping maladaptively. Instead, their physiological profiles seem to indicate that, sometimes, it can be stressful to laboriously construct a world in which there are no stressors. This is very subtle indeed and reflects, I suspect, some unique attributes of the human psyche. The same point is made in a recent study showing the pathophysiological power of suggestion, in which half of a population of normotensive volunteers were told that they were, in fact, hypertensive and who thus presented with elevated blood pressure when next examined (Rostrup & Ekeberg, 1992); no non-human primate is likely to fall for this. A theme throughout this chapter has been the increasing sophistication with which we understand social interactions and their physiological sequelae among non-human primates. The assumption, of course, in these studies is that they will inform us about the human condition, as well they do. The human studies cited here suggest to me that, nevertheless, certain provinces of behavioral physiology are uniquely human.

2.8 Conclusions

Despite the ever expanding links between stress and disease, most individuals cope reasonably well with stress, and some bodies and psyches do so superbly. This chapter has reviewed some of the key individual behavioral and social differences among male olive baboons that give rise to differences in stress-related physiology (summarized in Figure 2.5), and has tried to place those differences in an evolutionary context. A tremendous emphasis in the field, stretching back decades, has been on the importance of social rank in influencing individual differences in physiology and stress-related disease. A prime revisionist emphasis of this chapter has been how *little*, in fact, rank *per se* predicts any of those endpoints. Instead, it seems virtually meaningless to think about the physiological correlates of rank outside the context of a number of other modifiers – the sort of society in which the rank occurs (with societal style differing by species, by ecosystem within species, and by circumstances relevant only transiently to a particular group), the individual's personal experience of both the given rank and society, and the powerful role of personality as a perceptual filter of outside experience. This dovetails nicely with the de-emphasis of rank in other niches of primatology – examinations instead of alternative strategies to overt rank-related competition, to the role of early mothering style in bringing about personality differences, or to multi-group issues related to primate "culture". It leads to a final, somewhat obvious point – if we are endlessly struck with the complexity of these issues as they apply to non-human primates, the complexity expands exponentially when considering humans. This makes for a daunting challenge in trying to understand

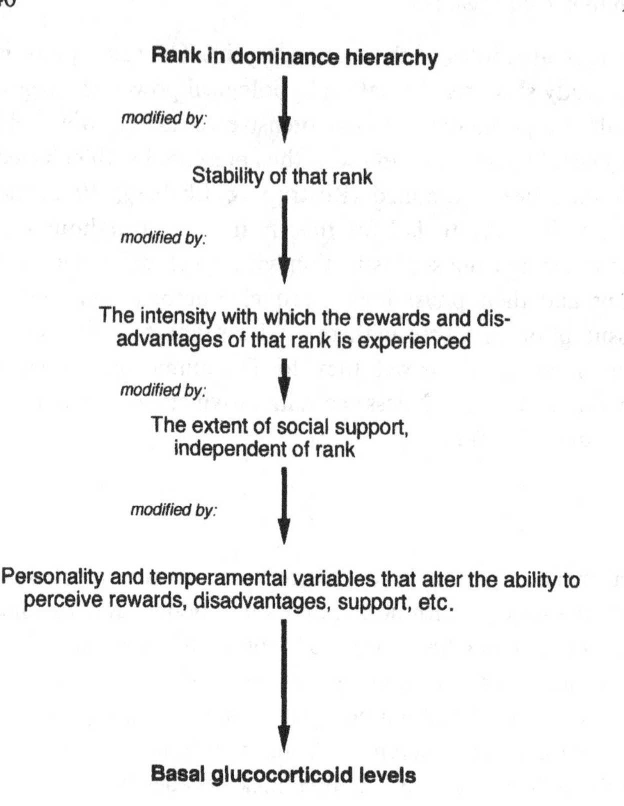

Figure 2.5. Social modifiers of the magnitude of the stress-response in male olive baboons, as measured by basal glucocorticoid levels.

differing vulnerabilities to stress-related diseases among humans. It seems critical to take on this challenge, for a simple reason. Thanks to the advances in Westernized medicine that spare us many of the diseases that plagued our ancestors, most of us will have the luxury of succumbing to a stress-related disease and most, I suspect, would prefer that to occur later rather than sooner.

Acknowledgments

These studies were made possible by the long-standing generosity of the Harry Frank Guggenheim Foundation, and the MacArthur Foundation.

2.9 References

Abbott, D., Saltzman, W., Schultz-Darken, N., & Smith, T. (1997). Specific neuroendocrine mechanisms not involving generalized stress mediate social regulation of female reproduction in cooperatively breeding marmoset monkeys. In Kirpatrick, B., ed., *The Integrative Neurobiology of Affiliation. Annals of the New York Academy of Sciences* (1), p. 211.

Adams, M. (1985). Psychosocial influences on ovarian endocrine and ovulatory function in *Macaca fascicularis. Physiological Behavior*, **35**, 935.

Adams, M., Kaplan, J. Koritnik, D., & Clarkson, T. (1985). Ovariectomy, social status, and atherosclerosis in cynomolgus monkeys. *Arteriosclerosis*, **5**, 192–201.

Altmann, J., Sapolsky, R., & Licht, P. (1995). Baboon fertility and social status. *Nature*, **377**, 688.

Archer, J. (1970). Effects of aggressive behavior on the adrenal cortex in laboratory mice. *Journal of Mammology*, **51**, 327.

Barnett, S. (1955). Competition among wild rats. *Nature*, **175**, 126.

Bell, R., & Hobson, H. (1994). 5-HT1A receptor influences on rodent social and agonistic behavior: A review and empirical study. *Neuroscience and Biobehavior Review*, **19**, 325.

Bernstein, I. S. (1976). Dominance, aggression and reproduction in primate societies. *Journal of Theoretical Biological* **60**, 459.

Bernstein, I. S. (1981). Dominance: The baby and the bathwater. *Behavior and Brain Science*, **4**, 419.

Blanchard, D. C., Sakai, R. R., McEwen, B., Weiss, S. M., & Blanchard, R. J. (1993*a*). Subordination stress: Behavioral, brain, and neuroendocrine correlates. *Behavioural Brain Research* **58**, 113.

Blanchard, D. C., Spencer, R. L., Weiss, S. M., Blanchard, R. J., McEwen, B., & Sakai, R. R. (1995). Visible burrow system as a model of chronic social stress: behavioral and neuroendocrine correlates. *Psychoneuroendocrinology*, **20**, 117.

Brandtstadter, J., Baltex-Gotz, B., Kirschbaum, C., & Hellhammer, D. (1991). Developmental and personality correlates of adrenocortical activity as indexed by salivary cortisol: Observations in the age range of 35 to 65 years. *Journal of Psychosomatic Research*, **35**, 173.

Bronson, F., & Eleftheriou, B. (1964). Chronic physiologic effects of fighting on mice. *General and Comparative Endocrinology*, **4**, 9.

Brown, L., Tomarken, A., Orth, D., Loosen, P., Kalin, N., & Davidson, R. (1996). Individual differences in repressive-defensiveness predict basal salivary cortisol levels. *Journal of Personality and Social Psychology*, **70**, 362.

Chamove, A., & Bowman, R. (1976). Rank, rhesus social behavior, and stress. *Folia Primatology*, **26**, 57.

Carroll, B. (1982). The dexamethasone suppression test for melancholia. *British Journal of Psychiatry*, **140**, 292.

Carter, C., Williams, J., Witt, D, & Insel, T. (1992). Oxytocin and social bonding. *Annals New York Academy of Sciences*, **652**, 204.

Cheney, D., & Seyfarth, R. (1990). *How Monkeys See the World: Inside the Mind of Another Species*. Chicago: University of Chicago Press.

Clarke, A. (1996). Long-lasting effects of prenatal stress on adaptive and affiliative behavior in adolescent rhesus monkeys. In *The Integrative Neurobiology of Affilitaion*, ed. I. Lederhendler, C. Carter, & B. Kirkpatrick. p. 10. New York: Academy of Sciences.

Clarke, A., & Boinski, S. (1995). Temperament in nonhuman primates. *American Journal of Primatology*, **37**, 103.
Coe, C., Mendoza, S., & Levine, S. (1979). Social status constrains the stress response in the squirrel monkey. *Physiological Behavior*, **23**, 633.
Coplan, J., Andrews, M., Rosenblum, L., Owens, M., Friedman, S., Gorman, J., & Nemeroff, C. (1996). Persistent elevations of CSF concentrations of CRF in adult nonhuman primates exposed to early-life stressors: Implications for the pathophysiology of mood and anxiety disorders. *Proceedings of the National Academy of Sciences of the USA*, **93**, 1619.
Creel, S., Creel, N., & Monfort, S. (1996). Social stress and dominance. *Nature*, **379**, 212.
Cummings, J. (1993). Frontal–subcortical circuits and human behavior. *Archives of Neurology*, **50**, 873.
Curtis, G., Fogel, M., McEvoy, D., & Zarate, C. (1970). Urine and plasma corticosteroids, psychological tests, and effectiveness of psychological defenses. *Journal of Psychiatric Research*, **7**, 237.
Damasio, H., Grabowski, T., Frank, R., Galaburda, A., & Damasio, A. (1994). The return of Phineas Gage: Clues about the brain from the skull of a famous patient. *Science*, **264**, 1102.
Davis, D., & Christian, J. (1957). Relation of adrenal weight to social rank of mice. *Proceedings of the Society of Experimental Biology and Medicine*, **94**, 728.
Dijkstra, H. (1992). Hormonal reactions to fighting in rat colonies: prolactin rises during defence, not during offence. *Physiology and Behavior* **51**, 961.
Elias, M. (1981). Cortisol, testosterone and testosterone-binding globulin responses to competitive fighting in human males. *Aggressive Behavior*, **7**, 215.
Farabollini, F. (1987). Behavioral and endocrine aspects of dominance and submission in male rabbits. *Aggressive Behavior*, **13**, 247.
Fernandez, X. (1994). Agonistic behavior, plasma stress hormones, and metabolites in response to dyadic encounters in domestic pigs: interrelationships and effect of dominance status. *Physiology and Behavior*, **56**, 841.
Fox, H., White, S., Kao, M., & Fernald, R. (1997). Stress and dominance in a social fish. *Journal of Neurosciences*, **17**, 6463.
Fox, M., & Andrews, R. (1973). Physiologic and biochemical correlates of individual differences in behavior of wolf cubs. *Behavior* **46**, 129.
Fuchs, E., & Flugge, G. (1992). Psychological stress and its neurophysiological consequences. *ISPNE*, **23**, 67.
Fuchs, E., & Flugge, G. (1995). Modulation of binding sites of CRH by chronic psychosocial stress. *Psychoneuroendorinology*, **20**, 33.
Goleman, D. (1995). *Emotional Intelligence*. New York: Bantam Press.
Gust, D., Gordon, T., Hambright, K., & Wilson, M. (1993). Relationship between social factors and pituitary-adrenocortical activity in female rhesus monkeys (*Macaca mulatta*). *Hormones and Behavior*, **27**, 318.
Gust, D., Gordon, T., Wilson, M., Ahmed-Ansari, A., Brodie, A., & McClure, H. (1991). Formation of a new social group of unfamiliar female rhesus monkeys affects the immune and pituitary adrenocortical system. *Brain Behavior Immunity* **5**, 296.
Huhman, K. (1990). Effects of social conflict on POMC-derived peptides and glucocorticoids in male golden hamsters. *Physiology and Behavior*, **47**, 949.
Huhman, K. (1992). Hormonal responses to fighting in hamsters: Separation of physical and psychological causes. *Physiology and Behavior*, **51**, 1083.
Insel, T., & Shapiro, L. (1992). Oxytocin receptors and maternal behavior. *Annals of*

the NY Academy of Sciences, **652**, 122.
Kaplan, J. (1986). Adrenal responsiveness and social status in intact and ovariectomized *Macaca fascicularis. American Journal of Primatology*, **11**, 181.
Keverne, E., Meller, R., & Eberhart, J. (1982). Dominance and subordination: Concepts or physiological states? In Chiarelli, V., Corruccini, J. (eds). *Advanced Views in Primate Biology*. p. 136, Springer-Verlag.
Kirschbaum, C., & Hellhammer, D. (1994). Salivary cortisol in psychoneuroendocrine research: recent developments and applications. *Psychoneuroendocrinology*, **19**, 313.
Korte, S. (1990). Behavioral and neuroendocrine response to psychosocial stress in male rats: The effects of the 5-HT 1A agonist ipsapirone. *Hormones and Behavior*, **24**, 554.
Leshner, A., & Polish, J. (1979). Hormonal control of submissiveness in mice: Irrelevance of the androgens and relevance of the pituitary–adrenal hormones. *Physiology of Behavior*, **22**, 531.
Levine, S., Wiener, S., & Coe, C. (1989). The psychoneuroendocrinology of stress: A psychobiological perspective. In *Psychoendocrinology* ed. F. Brush, & S. Levine, p. 46, New York: Academic Press.
Louch, C., & Higginbotham, M. (1967). The relation between social rank and plasma corticosterone levels in mice. *General Comparative Endocrinology* **8**, 441.
Mallick, J., Stoddart, D., Jones, I. & Bradley, A. (1994). Behavioral and endocrinal correlates of social status in the male sugar glider (*Petaurus breviceps* Marsupialia: Petauridae). *Physiology and Behavior*, **55**, 1131.
Manogue, K., Leshner, A., & Candland, D. (1975). Dominance status and adrenocortical reactivity to stress in squirrel monkeys (*Saimiri sciureus*). *Primates*, **16**, 457.
McAllister, T. (1992). Neuropsychiatric sequelae of head injuries. *Psychiatric Clinics of North America*, **15**, 395.
McEwen, B., & Sapolsky, R. (1995). Stress and cognitive function. *Current Opinions in Neurobiology*, **5**, 205.
McGlone, J. J., Salak, J. L., Lumpkin, E. A., Nicholson, R. I., Gibson, M., & Norman, R. L. (1993). Shipping stress and social status effects on pig performance, plasma cortisol, natural killer cell activity, and leukocyte numbers. *Journal of Animal Science*, **71**, 888.
McKittrick, C. (1995). Serotonin receptor binding in a colony model of chronic social stress. *Biology of Psychiatry*, **37**, 383.
McLeod, P., Moger, W., Ryon, J., Gadbois, S., & Fentress, J. (1996). The relation between urinary cortisol levels and social behaviour in captive timber wolves. *Canadian Journal of Zoology*, **74**, 209.
Mendoza, S., Coe, C., Lowe, E., & Levine, S. (1979). The physiological response to group formation in adult male squirrel monkeys. *Psychoneuroendocrinology*, **3**, 221.
Meyerhoff, J., Oleshansky, M., & Mougey, E. (1988). Effects of psychological stress on pituitary hormones in man. In Chrousos, G., Loriaux, D., & Gold, P. (eds.). *Mechanisms of Physical and Emotional Stress*. pp. 463, 465, New York: Plenum Press.
Munck, A., Guyre, P., & Holbrook, N. (1984). Physiological actions of glucocorticoids in stress and their relation to pharmacological actions. *Endocrine Review*, **5**, 25.

Pincus, T., & Callahan, L. (1995). What explains the association between socioeconomic status and health: Primarily medical access of mind–body variables? *Advances*, **11**, 4.
Popova, N., & Naumenko, E. (1972). Dominance relation and the pituitary–adrenal system in rats. *Animal Behavior*, **20**, 108.
Raab, A., Dantzer, R., Michaud, B., Mormeded, R., Taghzouti, K., Simon, H., & Le Moal, M. (1986). Behavioural, physiological and immunological consequences of social status and aggression in chronically coexisting resident–intruder dyads of male rats. *Physiology and Behavior*, **36**, 223.
Ray, J., & Sapolsky, R. (1992). Styles of male social behavior and their endocrine correlates among high-ranking baboons. *American Journal of Primatology*, **28**, 231.
Rostrup, M., & Ekeberg, O. (1992). Awareness of high blood pressure influences on psychological and sympathetic responses. *Journal of Psychosomatic Research*, **36**, 117.
Rowell, T. (1974). The concept of social dominance. *Behavioural Biology*, **11**, 131.
Roy, A., Pickar, D., Linnoila, M., & Potter, W. (1985). Plasma norepinephrine in affective disorders: Relationship to melancholia. *Archives in General Psychiatry*, **42**, 1181.
Sachser, N. (1987). Short-term responses of plasma norepinephrine, epinephrine, glucocorticoid and testosterone titers to social and non-social stressors in male guinea pigs of different social status. *Physiology and Behavior*, **39**, 11.
Sakai, R., Weiss, S., Blanchard, C., Blanchard, R., Spencer, R., & McEwen, B. (1991). Effect of social stress and housing conditions on neuroendocrine measures. *Social Neuroscience Abstracts*, **17**, 621.1.
Sanger, D., Benavides, J., Perrault, G., Morel, E., Cohen, C., Joly, D., & Zivkovic, B. (1994). Recent developments in the behavioral pharmacology of benzodiazepine (omega) receptors: evidence for the functional significance of receptor subtypes. *Neuroscience Biobehavior Reviews*, **18**, 355.
Sapolsky, R. (1983). Endocrine aspects of social instability in the olive baboon. *American Journal of Primatology*, **5**, 365.
Sapolsky, R. (1986). Endocrine and behavioral correlates of drought in the wild baboon. *American Journal of Primatology*, **11**, 217.
Sapolsky, R. (1989). Hypercortisolism among socially-subordinate wild baboons originates at the CNS level. *Archives in General Psychiatry*, **46**, 1047.
Sapolsky, R. (1990). Adrenocortical function, social rank and personality among wild baboons. *Biological Psychiatry*, **28**, 862.
Sapolsky, R. (1991). Hormones, the stress response and individual differences. In Becker, J., Breedlove, M., Crews, D. (eds.) *Behavioral Endocrinology*, p. 287, MIT Press.
Sapolsky, R. (1993*a*). Endocrinology alfresco: Psychoendocrine studies of wild baboons. *Recent Progress in Hormone Research*, **48**, 437.
Sapolsky, R. (1993*b*). The physiology of dominance in stable versus unstable social hierarchies. In *Primate Social Conflict*, ed. W. Mason & S. Mendoza, p. 171. New York: SUNY Press.
Sapolsky, R. (1992*a*). Stress in the wild. *Scientific American*, January, 116.
Sapolsky, R. (1992*b*). Cortisol concentrations and the social significance of rank instability among wild baboons. *Psychoneuroendocrinology*, **17**, 701.
Sapolsky, R. (1996*a*). Why should an aged male baboon transfer troops? *American Journal of Primatology*, **39**, 149.
Sapolsky, R. (1996*b*). The price of propriety. *The Sciences*, July/August, 14.

Sapolsky, R., & Ray, J. (1989). Styles of dominance and their physiological correlates among wild baboons. *American Journal of Primatology*, **18**, 1.
Sapolsky, R., & Plotsky, P. (1990). Hypercortisolism and its possible neural bases. *Biological Psychiatry*, **27**, 937.
Sapolsky, R., Alberts, S., & Altmann, J. (1997). Hypercortisolism associated with social subordinance or social isolation among wild baboons. *Archives of General Psychiatry*, **54**, 1137.
Scanlan, J., Sutton, S., Maclin, R., & Suomi, S. (1985). The heritability of social dominance in laboratory reared rhesus monkeys. *American Journal of Primatology*, **8**, 363.
Schilling, A., & Perret, M. (1987). Chemical signals and reproductive capacity in a male prosimian primate (*Microcebus murinus*). *Chemical Senses*, **12**, 143.
Schuhr, B. (1987). Social structure and plasma corticosterone level in female albino mice. *Physiology and Behavior*, **40**, 689.
Schuurman, T. (1981). Hormonal correlates of agonistic behavior in adult male rats. *Progress Brain Research*, **53**, 420.
Seligman, M. (1975). *Helplessness: On Depression, Development and Death.* New York: W. H. Freeman.
Shively, C., & Kaplan, J. (1984). Effects of social factors on adrenal weight and related physiology of *Macaca fascicularis*. *Physiology and Behavior*, **33**, 777.
Smuts, B. (1986). *Sex and Friendship In Baboons.* Aldine Press, Hawthorne.
Southwick, C., & Bland, V. (1959). Effect of population density on adrenal glands and reproductive organs of CFW mice. *American Journal of Physiology*, **197**, 111.
Stein, D., Hollander, E., & Liebowitz, M. (1993). Neurobiology of impulsivity and the impulse control disorders. *Journal of Neuropsychiatry and Clinical Neuroscience*, **5**. 9.
Stewart, D., & Winser, D. (1942). Incidence of perforated peptic ulcer: Effect of heavy air-raids. *The Lancet*, 28 February, 259.
Suomi, S. (1987). Genetic and maternal contributions to individual differences in rhesus monkey biobehavioral development. In *Perinatal Development: A Psychobiological Perspective.* ed. N. Krasnegor, E. Blass, M. Hofer, W. Smotherman, p. 168, New York: Academic Press.
Suomi, S. (1997). Early determinants of behaviour: Evidence from primate studies. *British Medical Bulletin*, **53**, 170.
van Schaik, C. (1991). A pilot study of the social correlates of levels of urinary cortisol, prolactin, and testosterone in wild long-tailed macaques (*Macaca fascicularis*). *Primates*, **32**, 345.
Virgin, C., & Sapolsky, R. (1997). Styles of male social behavior and their endocrine correlates among low-ranking baboons. *American Journal of Primatology*, **42**, 25.
Wallner, B. (1996). Female post estrous anogenital swelling and male-female interactions in barbary macaque. In *The Integrative Neurobiology of Affilitaion.* ed. C. Carter, B. Kirkpatrick, & I. Lederhendler, *New York Academy of Science*, p. 45.
Wasser, S., Monfort, S., Southers, J., Wildt, D. (1994). Excretion rates and metabolites of oestradiol and progesterone in baboon (*Paipo cynocephalus cynocephalus*) faeces. *Journal of Reproductive Fertility*, **101**, 213.
Watson, D., & Clark, L. (1984). Negative affectivity: The disposition to experience aversive emotional states. *Psychologic Bulletin*, **96**, 465.
Weinberger, D., Schwartz, G., & Davidson, R. (1979). Low-anxious, high-anxious,

and repressive coping styles: Psychometric patterns and behavioral and physiological responses to stress. *Journal of Abnormal Psychology*, **88**, 369.

Whitten, P., Stavisky, R., Aureli, F., Russell, E. (1997). Response of fecal cortisol to stress in captive chimpanzees (*Pan troglodytes*). *American Journal of Primatology*, **44**, 57.

Williams, R. (1989). *The Trusting Heart: Great News About Type A Behavior*. New York: Random House.

3

Epidemiology of human development

CAROL M. WORTHMAN

3.1 Introduction

The past 150 years have witnessed a quiet revolution in human development that still sweeps across the globe today: children nearly everywhere are growing faster, reaching reproductive and physical maturity at earlier ages, and achieving larger adult sizes than perhaps ever in human history. Although, in biological terms, the recent secular trends to accelerated growth and maturation may be nearly as dramatic as the widespread reduction in infectious illness and death, their social, psychological, behavioral, and health implications remain poorly understood. Yet we could think of the great improvements in public health and nutrition that apparently underlie the secular trends as a massive intervention in human development with ramifying consequences, likely positive, possibly negative. Rather belatedly, the possibility that developmental factors play a role in the contemporary emergence of chronic morbidities such as obesity, hypertension, diabetes, and cancer (Ellison, Chapter 6; McGarvey, Chapter 8) is now being considered. Martorell and colleagues have shown that early differences in nutrition predict later differences not only in adult size but also parental competence (Rivera *et al.*, 1995; Martorell *et al.*, 1996). Barker and his co-workers have developed the notion of early metabolic programming, based on epidemiological studies in British samples that link birth and placental weights to later risk for metabolic and cardiovascular disorders (Barker, 1997*a*, *b*). Remarkably, we do not have an integrated model of human developmental physiology that can assist us in thinking about how ontogenetic changes may mediate environmental effects on adult health outcomes.

This chapter argues that patterns of variation in human development reflect variation in endocrine function, organized by conditions of ontogeny, and that this underlying physiological variation contributes to long-term outcomes in differential function and health. Thus, it proposes pathways linking the epidemiology of human development to the epidemiology of adults. A central focus of human biology is identification and explanation of patterned phenomena through time and space, in terms of human biological variation within and across populations, as well as its sources and sequellae. Epidemiology has similar concerns, but its focus is disease rather than human variation, and it is not concerned with the proximal processes that mediate the patterns of risk that it identifies. Human biology's overarching paradigm is evolutionary theory, and alongside the adaptationist paradigm has emerged increasing concern with specific proximate and intermediate pathways by which human biological systems and wider socio-ecological circumstances interact. Adaptation and variation do not happen instantaneously, they must arise through some process; hence, human biologists have extensively concerned themselves with growth and development (e.g., surveys in Falkner & Tanner, 1985). Yet anthropological models of human development themselves remain remarkably underdeveloped, and would benefit from a thorough overhaul in the light of current work.

This chapter combines human biology, life history theory, and epidemiology to consider variation in human development, particularly as it pertains to health outcomes. I argue that hormones play central roles in the physiological architecture of the human life course, as well as in mediating the relationship between socio-ecological context and functional adjustments determining life history parameters such as adult body size, age at puberty, and longevity. More simply, hormones are key elements in the biological machinery that makes growth and maturation, adult function, and aging happen when and how they do. Recent work by human biologists, epidemiologists, and others has documented tremendous worldwide shifts in maturation rates, body size, and life expectancy, produced by widespread public health interventions and socioeconomic transformations that have reduced mortality, particularly early mortality, improved nutrition, and tranformed lifestyles (World Bank, 1993; Eveleth & Tanner, 1990; Schroeder & Brown, 1994).

Despite close attention to the growth and health outcomes of social and economic change, the underlying ranges of normal function that mediate such outcomes have been less carefully considered. In general, the assumption has been that unless the physiology is unduly challenged (e.g., by high altitude, or massive increases in habitual sugar intake), the functional ranges of physiological systems will not vary across human populations.

Although rate of juvenile maturation varies, the presumption has been that different amounts of time are taken but the same adult endpoint is reached: adult outcomes have been presumed essentially invariant on the population level. This assumption has faced mounting empirical challenges.

Current life history theory supplies a good starting place for thinking about endocrine architecture of life history and phenotypic plasticity by identifying biological constraints that inform the evolution of development, thereby indicating the adaptive challenges that developmental biology is designed to resolve. But in addition to such theoretical approaches, research based in physiological measures, such as hormones, is essential because it probes the proximate mechanisms by which linkages to ultimate causation are forged (Finch & Rose, 1995).

3.2 Life history theory

Life history theory attempts to unravel the evolutionary reasons why creatures lead such different lives, from the single-celled asexually reproducing amoeba to the multi-celled sexually reproducing whales, or the short-lived tiny mouse to the long-lived enormous elephant. In this literature, life history constitutes the taxon-specific array of developmental (e.g., age at weaning, adult body size), reproductive (e.g., time to first reproduction, interbirth interval, litter size), and survival (e.g., lifespan) characteristics that define the life course. Features characteristics of species reflect underlying strategies for allocation of two scarce resources, time and energy, among the biologically important tasks of productivity or growth, maintenance, and reproduction. Maintenance is the energy cost of repair, metabolism, and renewal of body systems, and represents investments in and intrinsic determinants of survivorship. Productivity is the net energy available for growth and reproduction above and beyond the costs of maintenance: rapid growth requires high energy availability, which may be generated by lowering maintenance or increasing energy intake. Among determinate growers such as humans who at adulthood cease linear growth, though not growth in mass, age at reproductive maturity defines the point at which allocation of productivity switches from use for growth to use for reproduction. Significantly, primates have markedly smaller productivity values, as inferred from slow juvenile growth rates, than do mammals generally, which suggests that they have high maintenance costs. Axiomatic for life history theory is the allocation rule, which holds that limited resources (time, energy) used for one purpose cannot be used for another. The allocation rule sharpens the boundaries and trade-offs among

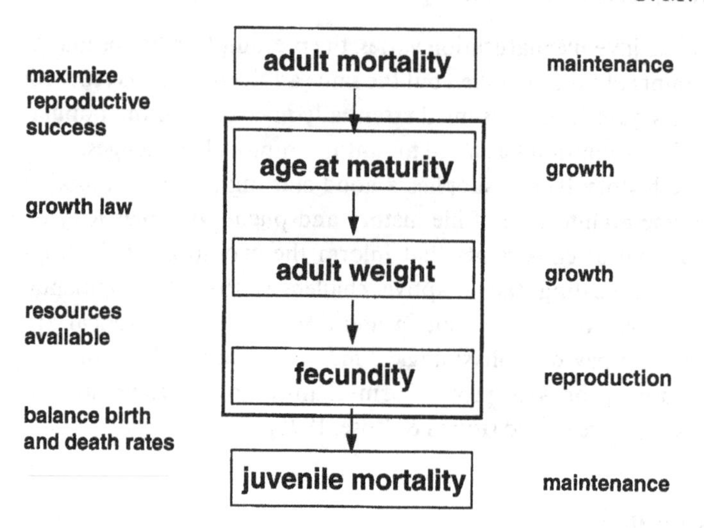

Figure 3.1. Hierarchy of factors determining life history strategy. (After E. Charnov, *Life History Invariants*, 1993.)

competing life history demands for growth or maturation, survival, and reproduction. Trade-offs are the balance of costs and benefits between sets of adaptive demands, such as that of continuing growth, with the reduced mortality that larger body size may bring, versus commencing reproduction, with the fitness advantage that earlier reproduction may accrue.

Schedules of resource allocation thus are viewed as fundamental to life history construction (Charnov, 1993); the hierarchy of resource allocation priorities established by cross-taxon comparative analysis of life history parameters is shown in Figure 3.1. The central set of boxes comprises the hierarchy of life history parameters, graded top to bottom by importance, based on proportion of life history variation explained. The column on the left indicates the principal adaptive constraint that informs the trade-offs between life history parameters in adjacent boxes. The column on the right denotes the allocation domain characteristic of the adjacent life history parameter. Modern adaptationist studies of behavior have focused almost exclusively on fitness, or reproductive outcomes. The consistent observation that mortality rates closely predict life history parameters has stimulated increased attention to maintenance and underscores the importance of mortality factors in shaping human development. Indeed, adult mortality rates have been found to be the most important predictors of other life history parameters. Some (Charnov, 1993) maintain that the relevant mortality rates are those linked to external causes that represent selective pressures independent of any adaptive capacity of the organism. Nearly all mortality risk arises from organism–environment interactions and relates to properties of the organism, its behavior or physiology; thus, maintenance – what the organism does to sustain function and reduce mortality risk – is listed as the allocation domain most pertinent to age-specific mortality rates.

Life history theory has made significant progress in explaining taxonomic variation in how organisms put their lives together, but the theory retains shortcomings that suggest that other levels of biological analysis will be required to relate it to organismic biology. First, the allocation rule may be somewhat flexible, such that assumption of strict trade-offs in resource allocation seriously distorts results from life history models. Indeed, one can imagine that there is high selective pressure for organisms to find means to circumvent the allocation rule by using time or energy so as to fulfil more than one adaptive function at once. Given the well-recognized limitations on the amount of information that can be conveyed genetically, in comparison to the large amount encoded in environments, one of the great insights in modern developmental biology has concerned how and how much ontogeny relies on environmental scaffolding, including the behaviors of conspecifics and the conditions those behaviors create (Changeux, 1985; Edelman, 1987; Hofer, 1996). Thus, the resources devoted by individuals to developmental-, reproductive-, and maintenance-related demands can also constitute important information that shapes the phenotypes of others and therefore represents a significant component of heredity.

Second, the appropriate unit of analysis for life history theory remains vague. As it is grounded in comparative data on species from throughout the animal kingdom, the unit of analysis is minimally at the population level and above, not at that of the individual. Whether life history theory will apply to explaining variation within species is an open empirical question. The assumption has been that natural selection constraints will retain their relative force across all levels of organic organization. This assumption may be true, but in any event our understanding of Darwinian processes remains incomplete enough to make the question of just how selection or adaptation works at different organizational levels a subject of lively debate. The debates are especially vigorous in relation to the role in ontogeny of genetics versus other modes of inheritance and organizational processes (Gilbert *et al.*, 1996; Sterelny *et al.*, 1996). The Modern Synthesis subsumed ontogeny under genetics and established change in gene frequency as the *sine qua non* of evolution. Yet the well-recognized fact remains that other forms of information transmission such as learning can play a large role in formation and replication of phenotype. This is especially true of behavior, and the role of learning in transmission of behavioral phenotypes is particularly pronounced in humans. Nonetheless, attempts to develop dual inheritance models for biologically based (i.e., genetic) and experience-based inter-generational transmission or replication of phenotypes (Durham, 1991; Boyd & Richerson, 1985) have not yet been so successful as to spawn a new generation of empirical research or organize existing bodies of data. Notwithstanding the theoretical gaps, a host of work, empirical and conceptual, suggests that the study of development will provide an important foundation for understanding how biological systems reproduce and convey information with modification through time (Johnston, 1987; Cairns *et al.*, 1990; Gottlieb 1991; Gilbert *et al.*, 1996; Sterelny *et al.*, 1996). Such work further suggests that the coaction of environment and organism irreducibly drives developmental processes (Oyama, 1985), and that developmental process has been a focal target of evolution (Gould, 1977; Raff, 1996). Gilbert and colleagues (1996) recently noted that the discovery of "homologies of process" in ontogeny strongly augment the older "homologies of structure" as evidence of adaptive design and thus, of evolution.

A third limitation of life history theory is that it concerns ultimate evolutionary causes and is biologically static, i.e., it seeks to identify the constraints on macro-evolutionary processes. It does not, nor does it aim to, indicate the adaptations or proximal biological means by which life history patterns and their implied schedules of resource allocation are achieved in the living organism. How do life history parameters become translated into physiology of the living individual? What kind of biological adjustments are required to live fast and die young, or to live long and age slowly? Is variation in adult mortality rates a predictor of variation in the developmental course and reproductive performance *within* species, including humans? These questions raise fascinating issues for the biologist concerned with incorporating evolutionary perspectives into the study of organisms across the lifespan (Finch & Rose, 1995).

As members of a rather slowly developing, long lived species, humans would seem to exemplify one end of the spectrum of life history strategies. Yet we also know that humans show considerable within and between population variation in life history parameters such as birth weight, growth rate, timing of puberty and onset of reproduction, age-specific fertility and mortality rates, and life expectancy. At issue is the extent to which this variation is driven by environmental as opposed to genetic variation. Given the importance of environmental inputs for organizing development, on the molecular through the psychobehavioral level, we may expect to find that variation in life history parameters reflects facultative adjustments to environmental variation, filtered by person–environment interactions also conditioned by genetic capacities and constraints. Capacity for plasticity under diverse circumstances through such person–environment interactions is captured by the concept of a reaction norm and belies genetic determinism. Reaction norms concern the range of phenotypes expressed by a given genotype across a range of environments and thus conceptually attempt to bring plasticity into the range of life history analysis (Stearns & Koella, 1986; Stearns, 1992). Again, reaction norms do not concern the actual biological mechanisms that mediate such phenotypic variation.

3.3 Endocrine architecture of life history

A ready candidate can be offered in response to the question of how life history strategies are effected in the living organism. Hormones likely play central roles in this process (Finch & Rose, 1995). They negotiate the short- and long-term balance of resource allocation among growth, reproduction, and maintenance. Hormones determine productivity by modulating metabolism and setting internal regulatory parameters, they regulate the pace of growth and the timing of developmental transitions such as puberty, and they continually mediate the interface between the individual and its environment by transducing everything from stress to workload. Endocrine regulation may also be centrally involved in facultative adjustment of life history parameters. A vivid example concerning metabolism and net energy storage is provided by evidence for association of fetal undernutrition with reduced adult glucose tolerance and enhanced risk of diabetes (Barker *et al.*, 1993*b*; Phillips *et al.*, 1994). Barker and colleagues further argue for specific early growth patterns as responses to early environments with differential sequelae in risk for specific adult diseases. For instance, neonates short for head circumference, a proxy measure of fetal growth retardation, show long-term blunting of insulin-like growth factor (IGF) response to growth hormone (GH), and hyperresponsiveness to growth hormone releasing hormone (GHRH) that indicate probable receptor-mediated GH resistance (Barker, *et al.*, 1993*a*; but see Law *et al.*, 1995).

Consequently, hormones both mediate resource partitioning and offer a means to modulate long-term life history priorities through organizational or regulatory adjustments to environment during ontogeny. A further major aspect of endocrine action in development is as a pacemaker for growth, maturation, and aging. Puberty is an obvious instance, in which endocrine–neuroendocrine mechanisms both establish the childhood phase and initiate the process leading to reproductive competence (Worthman, 1993). Humans are unusual even among primates for the duration of the juvenile (post-weaning, pre-adult) period. Gonadal quiescence, and possibly adrenal androgen damping, are required to establish and maintain juvenility. Although the precise mechanisms responsible for sustained gonadal quiescence and for the onset of puberty remain uncertain, it is clear that these are located in the brain and operate via regulatory pathways through the hypothalamus. The brain, furthermore, acts as the pacemaker for puberty once it commences, and is required to maintain adult function thereafter.

Hormones, then, play crucial roles in effecting life history processes in the day-to-day prioritization of resource allocation, as well as the long-term scheduling of growth, reproductive effort, and aging. Hence, measurement of hormones provides a window onto key determinants of ongoing function, adaptation, and differential well-being. Furthermore, comparative study of individual and population endocrine variation sheds light on the ecological determinants of such variation and provides a much broader basis on which to characterize "normal" human biology.

3.4 Endocrine trajectories in human development

Consideration of the possible roles hormones play in the physiological architecture of human adaptation and life history requires a rather different view of endocrinology than is prevalent in the western bioscience and clinical literatures. Respectively, these literatures have focused on the identification, characterization, regulation, and molecular biology of hormones and hormone action, and on establishing normative ranges for expectable values. Research and clinical practice have operated on an assumption of human biological unity, i.e., that what obtains for structure, function, maturation or aging in one subset of humans will hold for all others, except under pathological or unusual conditions. The assumption of biological unity has indeed proved valid for establishing a species-general biology that supports worldwide success in exporting the biomedical model. Pertinent to the case in hand, ample evidence indicates that many aspects of endocrinology are universal: insulin is everywhere the principal regulator of blood sugar, testosterone is universally the primary steroid product of the adult testis under regulation by gonadotropins, and iodine deficiency always affects thyroid function.

Alongside a demonstrably universal human biology exists a well-recognized degree of individual variation that at times is very large indeed. For example, the clinically established normal ranges for follicule stimulating hormone (FSH) in adult men extend over fivefold differences of 4–25 IU/L, although this is a brain-driven gonadal regulatory hormone. Anthropologists, in particular, have begun to explore how human ecologies influence endocrine function and, conversely, how causes and consequences of variation in endocrine function may have different consequences under different socio-ecological circumstances. Comparative endocrine research has gradually established that, while the endocrine factors and major functional pathways are nearly universal, endocrine function may vary for specific axes and their interrelationships, along the following lines:

1. age-specific normative ranges,
2. concentrations and distributions of receptors and receptor variants,
3. concentrations of binding proteins,
4. functional setpoints along regulatory pathways or axes, such as among brain, pituitary, and gonad,
5. schedules and trajectories of endocrine output across the life course, and
6. possibly, regulatory dynamics across endocrine axes, such as the ability of hypothalamo-pituitary–adrenal activity to downregulate growth hormone and insulin outputs.

Some of this cross-population variation may be based on genetic diversity, but much of it is likely organized by the specific ecologies in which humans grow up and function. Relatively little is known of the developmental pathways by which endocrine function may be influenced or adjusted. In this chapter, I focus on the data available on this point, and on rethinking endocrinology in developmental terms. Such an analysis requires a thorough understanding of developmental change and regulation of hormone action. Accordingly, the following sections provide an integrated summary of the vast literature on endocrine physiology at various ages, with respect to a set of key endocrine axes.

The lifespan course of endocrine activity in Western populations is outlined below along four axes or domains: hypothalamo-pituitary–gonadal, hypothalamo-pituitary–adrenal, hypothalamo-pituitary–thyroid (thyrotropic), and growth and metabolic regulation (somatotropic) (see overview in Table 1.1.) A summary of lifespan trajectories for these axes will provide the basis both for formulation of ideas about when and how endocrine-regulatory variation may arise, and for comparative analysis in subsequent sections. Inclusion of several axes, rather than the usual focus

Table 3.1 *Overview of developmental and functional parameters of endocrine axes under discussion*

Axis	Pulsatility	Circadian rhythm	Shifts in regulation and circulating hormone levels				Sex difference
			Infancy	Childhood	Puberty	Adulthood	
HPG	∴	∴	✓	✓	✓	F✓/M≅	∴ ∴
HPA/cortisol	∴ ∴	∴	✓	–	–	–	–
HPA DHEAS	–	–	✓	✓	✓	✓	≅
Thyrotropic	∴	∴	✓	✓	✓	–	–
Somatotropic	∴ ∴	∴ ∴	●	✓	✓	≅	≅

Note: – absent; ∴ present; ∴ ∴ pronounced; ✓ regulatory shift; ≅ continuous slope; ● data unavailable; F females; M males.

on a single one, offers the opportunity to compare and constrast different modes of hormonal function as well as to evaluate the relationships between axes. Data have been integrated mainly from work on North Americans and, where that is unavailable, on Western Europeans, to construct a functional picture for populations with roughly similar (though still diverse) cultural and physical ecologies. The endocrine profiles for these intensely studied populations furthermore provide a basis for comparative study of endocrine function across populations or subgroups with contrasting cultural practices, activity patterns, psychological stressors, pathogen loads, nutritional status, or other circumstances. Data sources have been selected on the basis of assay quality and comparability to other sources, of sample size, and of appropriateness or representativeness of the sampling strategy. Where possible, 95% interval has been compiled to indicate degree of individual variation in hormone values.

The following surveys also aim to support a broader view of endocrine action and regulation by expanding coverage beyond the key peripheral hormone/s of a given endocrine axis, such as progesterone, testosterone, or cortisol. Research on normal individuals in everyday settings has tended to focus on these hormones for practical and methodological reasons. The biological implications of a given concentration of such hormones are predicated on several other regulatory dimensions which require consideration in order to probe variation in endocrine function in relation to differential health outcomes. Major parameters modulating endocrine action to be discussed below include the following.

1. Central regulatory hormones or provocative stimuli: For hormones that are centrally regulated, concentrations of the relevant pituitary hormone yield a picture of brain activity and feedback that contextual-

izes the activity of the peripheral target gland. For instance, if levels of gonadal steroid hormones are low and those of gonadotropins likewise low, one may infer a condition of gonadal down-regulation through increased feedback sensitivity at the level of the brain rather than the gonad. Such a situation, if chronic, implies that setpoints in the brain–gonad feedback chain have been altered. Similarly, endocrine activity with decentralized regulation mediated by provocative stimuli, such as blood sugar for insulin or foreign material (antigen) for cytokines, can be better interpreted if both the stimulus and the response are measured. Only through such analysis along the chain of endocrine regulation can variation in function be traced and understood.

2. Circadian variation: Many hormones follow a circadian rhythm. Most studies in naturalistic settings have treated circadian variation as a confounding variable to be controlled, rather than as a dimension of potential meaningful variation. As suggested below, several axes not only exhibit distinctive circadian patterns of activity, but also evince a developmental course in that pattern. Shifts in circadian profiles and/or their developmental course constitute another potential source of population and individual variation. The close entrainment of major GH release to sleep onset, for example, makes this axis vulnerable to practices resulting in sleep disruption.
3. Ultradian variation: A few hormones show very little circadian or ultradian (moment to moment) variation, but for many, release is episodic and pulsatile. The pattern of pulsatility in central regulatory hormones such as gonadotropins or growth hormone is crucial to their effect on target tissues. For most axes, acute changes in activity represent responses to environmental, behavioral, metabolic, or cognitive demands. The most "reactive" of the centrally regulated axes included in the present survey is the hypothalamo-pituitary–adrenal axis, in which activity is driven by exposure and individual reactivity to provocative stimuli and directly determines the amount of cortisol released by the adrenal.
4. Binding proteins: In circulation, many hormones are bound to carrier proteins that protect them from degradation but also render them less accessible for binding to receptors in target tissues. Binding proteins often show distinct developmental, circadian, and even ultradian profiles that, depending on the relative concentration of target hormone or other competing hormones, alters their impact on hormone action. Although individual differences in levels of carrier protein have been linked to endocrinopathies such as hirsutism in women, little research has explored the potential contributions of varying concentrations of

binding protein to individual or population variation in endocrine function, or in outcomes such as relative growth or health.

5. Inter-axis cross-talk: Increasingly, endocrine research has documented interactive effects among axes that likely represent the endocrinology of functional trade-offs among competing physiological and adaptive demands. Endocrine mediation of short- and long-term trade-offs in the allocation of scarce biological resources represents a rich and important vein for future research, but it is beyond the scope of this chapter to do more than outline a few examples of axial cross-talk.

In summary, this exercise should help in formulation of a more unified developmental picture of endocrinology. It provides a platform for future comparative research not only by presenting what is known about function and regulation, but also by indicating sensitive developmental periods or bottlenecks revealed by intervals of change or reorganization. Furthermore, it draws attention to areas of potential investigation by clarifying the temporal and regulatory dynamics of several hormonal axes.

3.5 Hypothalamo-pituitary–gonadal (HPG) axis

The HPG axis comprises one of the classic models of endocrine function: it is centrally controlled through the hypothalamus via its secretion of gonadotropin releasing hormone (GnRH) that, in turn, stimulates release of follicle stimulating hormone (FSH) and luteinizing hormone (LH) by the anterior pituitary. The amount and pattern of release of FSH and LH drive gonadal activity, including the production of gonadal steroids, of which the dominant form in females is estradiol (E2) and in males, testosterone (T). The regulatory loop is closed by the feedback effect exerted by circulating levels of gonadal steroids on hypothalamic GnRH and anterior pituitary release of the regulatory protein hormones. The HPG axis plays two prominent roles. First, it maintains and regulates adult reproductive function and, second, it controls reproductive development and senescence. Insofar as timing of reproductive events is a key element of life history strategy and differential reproductive success constitutes relative biological success of an organism, the axis lies at the heart of adaptive strategy and hence has commanded the keen interest of biological anthropologists.

Age-related change

The lifespan trajectory of HPG activity (Figure 3.2) illuminates neuroendocrine–endocrine underpinnings of human life history, including a

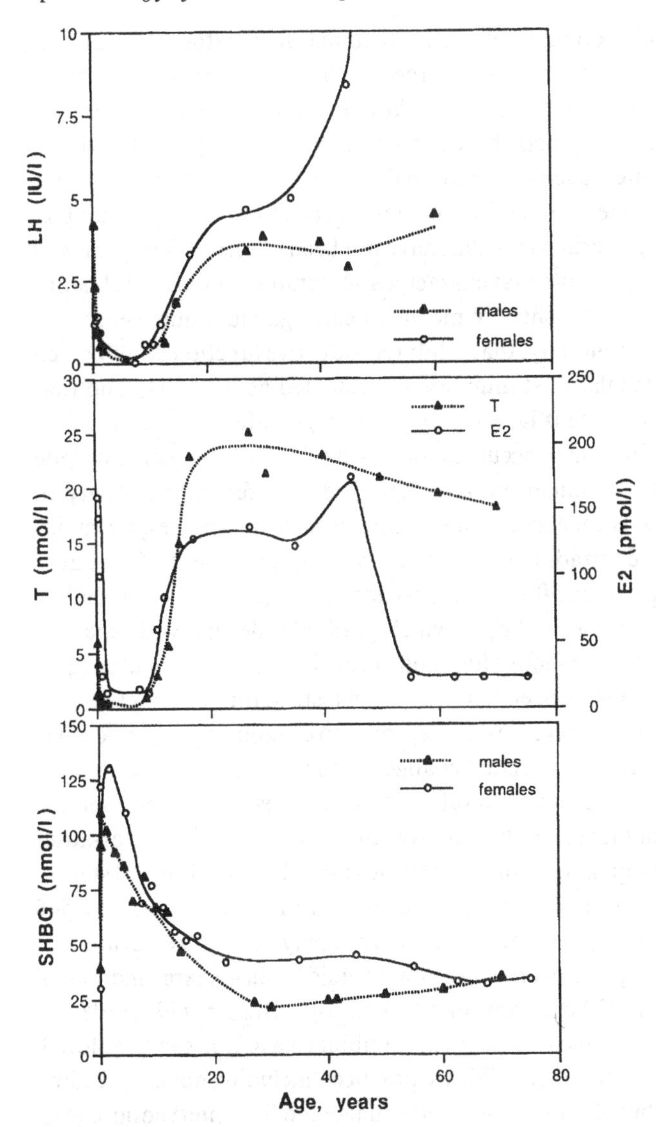

Figure 3.2. Activity of the human reproductive axis across the lifespan. Top panel: circulating concentrations of LH in men and women from birth through senescence. Middle panel: gonadal hormones (T in males, E2 in females) at all ages. Bottom panel: circulating concentrations of SHBG, the gonadal steroid carrier protein. On all panels, values for postmenarcheal women are based on mid-follicular measures. (Data sources: LH: Burger *et al.*, 1991; Kwekkeboom *et al.*, 1990; Veldhuis *et al.*, 1992; Wu *et al.*, 1996. E2: Apter *et al.*, 1989; Reame *et al.*, 1996. Ghai & Rosenfield, 1994; Klein *et al.*, 1996; Vermeulen *et al.*, 1996. SHBG: Belgorosky *et al.*, 1986, 1987, 1988; Bolton *et al.*,1989; Cunningham *et al.*,1985; Field 1994; Goodman-Gruen *et al.*, 1996.)

prolonged juvenile period of reproductive immaturity, rather late onset of puberty and first birth, and early reproductive senescence in women. Humans show inter- and intra-population variation in the timing of puberty that both has genetic bases and represents the impact of environmental circumstances such as gestational factors, nutrition, and infections. Similarly strong genetic or environmental effects on reproductive aging in general, or menopause in particular, have not been observed. The profile of HPG activity by age shows that characteristic features of human life history strategy are created by centrally mediated early gonadal quiescence and then reactivation at puberty, that adult reproductive function is supported by ongoing hypothalamic stimulation (represented here by LH), and that reproductive senescence is largely mediated by gonadal aging. Reproductive "switch-off" in women occurs mainly through follicle depletion despite accelerated central stimulation (although declining central sensitivity to ovarian feedback is reflected in increasing gonadotropins despite maintained or increased estradiol outputs in early peri-menopause). Menopause occurs on average in the fifth or sixth decade and is generally succeeded by massive outputs of gonadotropin which gradually decline with age. No such truncation of reproductive function occurs in men, but testicular aging is possibly reflected in age-related declines in testosterone and increases in gonadotropins. Sex differences in reproductive aging are most clearly captured by divergent age-related changes in LH.

This age profile presents several notable features. First, consider the marked postnatal burst in HPG activity, during which adult or near-adult levels of gonadotropins and gonadal steroids are achieved. Gonadal steroid ouput peaks around 3–4 months and declines to miniscule levels by 5–6 months of age, although gonadotropin output may be slightly elevated for the first 2 years. Reports of cross-sectional studies indicate large individual variability in levels of these hormones at these ages (Burger *et al.*, 1991), but the causes and consequences of such variability have not been explored. One might consider whether childcare practices, including not only feeding and nurturance, but also morbidity and ritual or medical interventions (e.g., circumcision, tatooing) may affect what appears to be an organization period for HPG activity and thereby contribute to variation in timing of puberty, in HPG feedback sensitivities, and thus in robustness of adult gonadal function (Landauer & Whiting, 1981).

Second, sex differences in pubertal onset are centrally mediated, and thus likely reflect inbuilt differences in life history strategy. Premature gonadal senescence in women, on the other hand, seems to be due largely to ovarian follicular depletion rather than to some central factor. Brain refractoriness to ovarian feedback is reflected in pre-menopausal hypersecretion of gonadotropin despite rising estradiol; massive pituitary tropic hormone output follows cessation of ovarian activity. Absence of gonadal depletion in males leads to sustained reproductive function well beyond middle age.

Third, the lifespan profile of the carrier protein for both T and E2, sex hormone binding globulin (SHBG), is distinct from that of its target hormones. Circulating SHBG rises three- to fourfold in the first 3 months after birth, reaching a plateau in boys at around 3 months of age that is sustained until levels begin to fall in year 3. Girls, on the other hand, attain higher levels that are maintained into year 5 and decline in mid-childhood. SHBG concentrations decline in both sexes throughout childhood and puberty, and plateau in early adulthood. Whereas women experience falling SHBG during the postmenopause, resulting in an increased proportion of free androgens or estrogens, men undergo a gradual rise that would tend to exacerbate effects of age-related declines in testosterone by leaving a smaller portion unbound and bioavailable. Nevertheless, a large sex difference (females greater than males) in adult levels of SHBG is evident from puberty until the sixth decade. As absolute molar amounts of gonadal output are greater in males than females, except during pregnancy, this observation suggests that males experience greater amounts and variation in circulating free gonadal steroid. That the gonadal steroid binding protein is so high during infancy and childhood seems paradoxical, given the very low circulating levels produced at these ages. Two possible functional explanations come to mind. For one, high SHBG may strongly buffer the hypothalamus and pituitary from gonadal feedback, so that the very small regular gonadotropin pulses organized by the brain occur in the absence of gonadal intervention. Declining SHBG may contribute to the development of intensifying HPG feedback interactions over the prepubertal and pubertal periods. For another, based on Whitten's analysis of the ubiquity and potency of dietary compounds that mimic actions of gonadal steroids (Chapter 7), one might suggest that high SHBG, particularly in girls, heavily buffers this axis from disturbance by phytoestrogens in the diet. Given that humans are omnivores with high intakes of diverse plant materials, elevated infant and juvenile SHBG may protect the HPG axis from untimely exposure to gonadal steroid action.

Circadian patterns of pulsatility

HPG activity is largely regulated through pulsatile release of GnRH, in a pattern that exhibits characteristic changes in amplitude and frequency across the life course and comprises a kind of "Morse code" for gonadal regulation. That is, the pulsatility conveys information necessary to organize gonadal activity, such that achievement of the same levels through continuous infusion does not organize fully competent gonadal activity. Brain activity, reflected in pulse frequency and amplitude of LH, exhibits distinctive profiles by sex across the lifespan (Figure 3.3), although the picture is further complicated in adulthood by characteristic progression of gonadotropin pulsatility across the mature ovarian cycle (Marshall & Griffen, 1993), a cyclicity not evident in males. Although pulsatile release of GnRH is evident in childhood, amplitude of the pulses is very low; nocturnal amplification of gonadotropin pulses constitutes the first sign of pubertal onset (Cemeroglu *et al.*, 1996; Wu *et al.*, 1991). Although variation in study design limits availability of comparable data by sex and age, it appears that progressive increases in pulse amplitude constitute the major change in brain output at puberty, considerably outweighing the simultaneous increases in number. Among adults, pulse number appears to be similar in females and males, while pulse amplitude is greater in men than women until ovarian senescence provokes a marked increase of amplitude in women at menopause. The complex, timing-dependent, and progressive manner of brain–gonadal interactions across the menstrual cycle likely strongly contributes to the disruptability of ovarian function through energetic (exercise or nutrition-mediated) or psychological insults. Sperm output, by contrast, appears much less susceptible to similar behaviorally, socially, and ecologically mediated perturbations. Via their impact on testosterone, physical and social stress appear to influence male behavior more strongly than sperm count (see Sapolsky, Chapter 2). In men, LH pulse frequency increases while amplitude and duration decrease with age; age-related increases in FSH associate with decreased pulse duration, but bases for this pattern remain uncertain (Veldhuis *et al.*, 1992).

3.6 Hypothalamo-pituitary–adrenal (HPA) axis

The adrenal produces a wide array of hormones that play diverse physiological roles across the lifespan. During gestation, the fetal adrenal acts in concert with the placenta as the principal endocrine unit driving pregnancy and fetal development. After birth, the fetal zone of the adrenal regresses, leaving the cortex and the medulla as two functionally distinct compartments. The former produces mineralocorticoids, glucocorticoids, and adrenal androgens that regulate important functions, from electrolyte

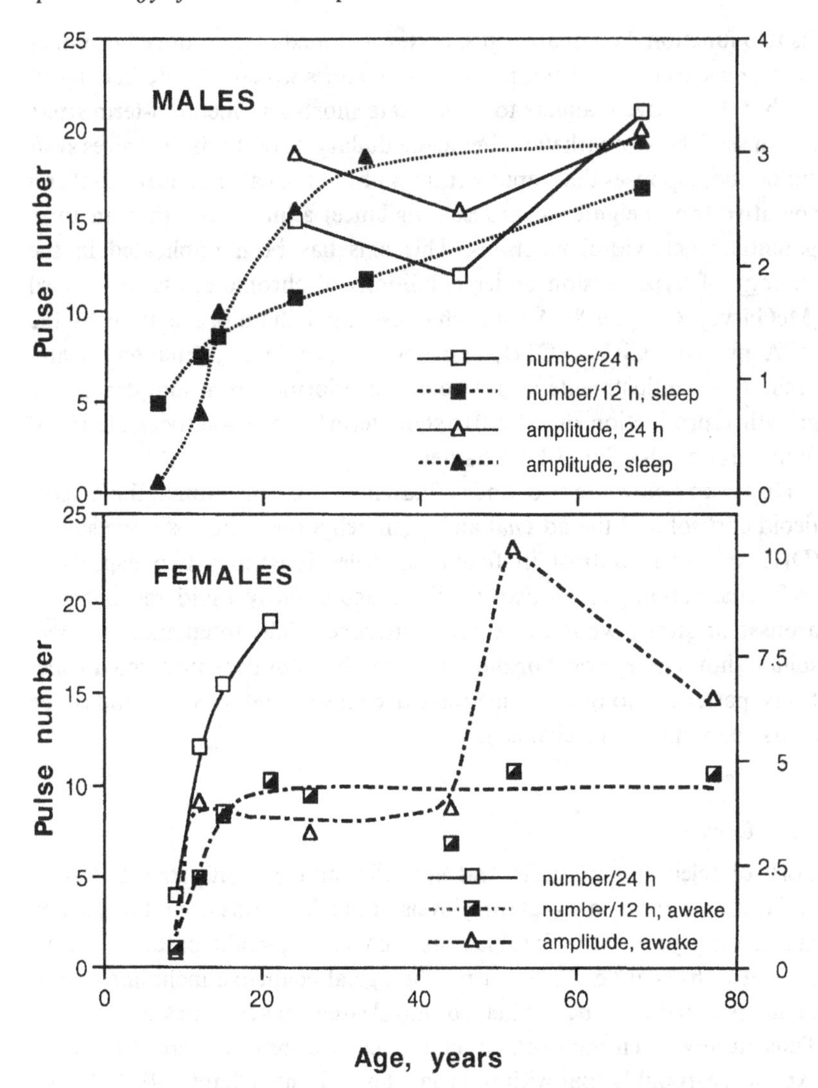

Figure 3.3. LH pulse number (left axis) and amplitude (right axis) in men (upper panel) and women (lower panel) by age. (Data drawn from Apter *et al.*, 1993; Genazzani *et al.*, 1997; Lee *et al.*, 1988; Mulligan *et al.*, 1995; Veldhuis *et al.*, 1992, Wu *et al.*, 1996.)

balance to glucose metabolism and stress response. The latter acts rather like a neuroendocrine structure and releases the adrenergic neurotransmitters epinephrine and norepinephrine into the bloodstream.

The adrenal acts as an important agent in interactions of hormones, health, and behavior, because it mediates responses to adaptive challenges.

Its two functional compartments, cortex and medulla, produce hormones in response to physical and psychological stress which operate in congruent but distinctive manners to orchestrate short- and medium-term stress responses. The sympathetic-adrenal–medullary (SAM) axis organizes swift endocrine responses that rapidly entrain physiological (e.g., heart rate) and cognitive (e.g., heightened arousal, vigilance) adjustments that, in turn, potentiate behavioral reactions. This axis has been implicated in the etiology of hypertension under conditions of chronic stress or arousal (McGarvey, Chapter 8). A somewhat less rapid chain of reactions by the HPA involves CRF, ACTH, and cortisol (Table 1.1, Chapter 1) and orchestrates adjustments in physiological priorities from long-term (e.g., growth, reproduction, digestion) to short-term (vigorous activity, alertness) functions (see also Sapolsky, Chapter 2).

Discussed below are two kinds of adrenocortical hormones, the glucocorticoid cortisol and the adrenal androgen dehydroepiandrosterone sulfate (DHEAS). The contrast in functional roles illustrates two aspects of endocrine action, for cortisol mediates acute, fairly rapid responses to arousal or stress, while the adrenal androgen dehydroepiandrosterone-sulfate shows no episodic or diurnal variability, but does undergo distinctively patterned long-term age-related changes that possibly index life course pacemaking mechanisms.

3.6.1 Cortisol

Cortisol release is pulsatile and episodic, and exhibits strong diurnal variation. Changes in circulating levels of this hormone are influenced by numerous psycho–behavioral factors such as sleep-wake patterns, eating, vigorous physical activity, and psychological challenge including worry, excitement, performance anxiety, or novel situations (van Eck *et al.*, 1996). Thus, at any given moment, circulating cortisol reflects impact of recent experience, roughly that within the last hour (Flinn, Chapter 4). Although it is most generally conceived as a stress hormone, cortisol might be more accurately viewed as associated with cognitive arousal and readiness to vigorously engage the immediate situation; hence it is an index of psychobiological load (for review see McEwen, 1997). Provocative stimuli may have positive as well as negative affective valence; moreover, load does not necessarily correlate with social status, where roles entail differential load. For instance, officers of military Special Forces in action have been found to have higher cortisols than enlisted men (for review see Seeman & McEwen, 1996). In another study, homeless children and middle class schoolboys exhibited equivalently elevated morning cortisols, relative to boys in village and urban squatter families (Panter-Brick and Pollard, Chapter 5). Given the valence-neutral nature of cortisol values, simple comparison of average cortisol in different study groups offers little insight into the bases and meaning of the comparison. Dynamic studies (of the kind described by Flinn, Chapter 4) that examine responses to specific stimuli under known conditions are for this reason much more valuable for understanding variation.

Age-related change

The adrenal undergoes a brief period of postnatal reorganization and establishment of regulatory setpoints, including the emergence of diurnal variation by age 6 months (Onishi *et al.*, 1983; de Zegher *et al.*, 1994; Ramsay & Lewis, 1995). Once the system is up and running, it does not seem to undergo further systematic developmental change over the lifespan (Figure 3.4). No notable age-related change in adrenocortical activity has been identified after the postpartum period, nor have systematic sex differences in circulating cortisol been reported. Nevertheless, the central regulatory hormone, ACTH, exhibits both lower levels in childhood, and sex differences in adulthood, despite lack of concomitant cortisol differences. Sex differences in ACTH output are absent in childhood (Wallace *et al.*, 1991). Both the age and the sex difference suggest differences in ACTH delivery to the adrenal, or in adrenocortical sensitivity or responsiveness to ACTH stimulation. Although aging has not been linked to changes in circulating levels of cortisol, increased ratios of cortisol to adrenal androgens have been observed in the cerebrospinal fluid of children (age 3–8 years) and the elderly (> 60 years), who may thus be more vulnerable to neurolytic effects of stress (Sapolsky, 1992, Guazzo *et al.*, 1996).

Circadian patterns

Pulsatility of cortisol release is established early and persists through life, driven by brain activity via its regulatory hormone, ACTH. The 24-hour pattern of ACTH exhibits a burst-like secretion mode (40 pulses/24 h), wherein amplitude but not frequency modulation creates the circadian rhythm in circulating ACTH (Figure 3.5) (Veldhuis *et al.*, 1990, Wallace *et al.*, 1991). Circadian variation in cortisol roughly follows that of ACTH, although cortisol falls throughout the latter of two time periods of greatest ACTH amplitude (0600–1200 and 1200–1800 h). Diurnal circulating

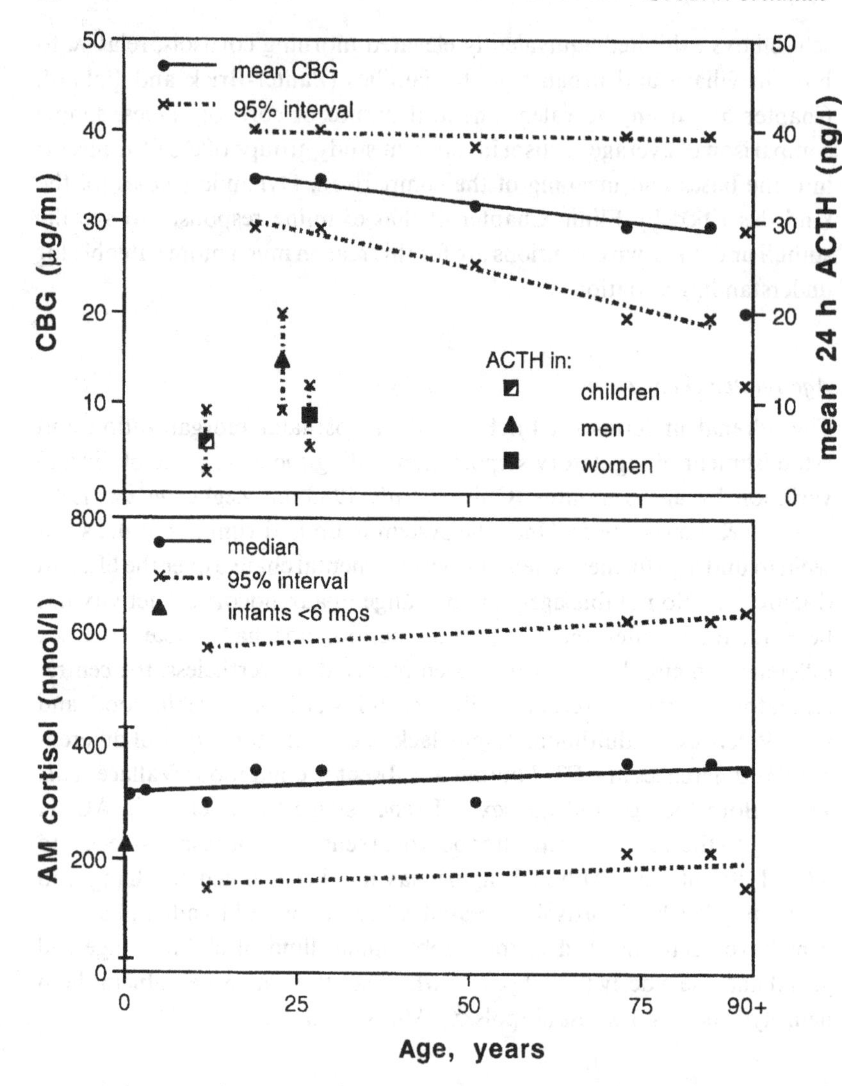

Figure 3.4. Upper panel: circulating plasma CBG concentrations (left axis) by age and mean 24-hour serum ACTH values in children (mean age 11.3 years) and adults (mean age 25 years) of both sexes (with 95% confidence intervals). Slope by age for CBG insignificant (note high age-specific variance), but age and sex differences in ACTH, despite lack of concomitant cortisol correlates, indicate regulatory differences. Lower panel: morning blood cortisol values by age, both sexes (with confidence intervals). Slope by age not significant, but infants under 6 months show differences in cortisol output that indicate a developmental period. (CBG, ACTH: Guazzo *et al.*, 1996; Horrocks *et al.*, 1990; Tietz *et al.*, 1992; Wallace *et al.*, 1991. Cortisol: Knuttson *et al.*, 1997; Tietz *et al.*, 1992.)

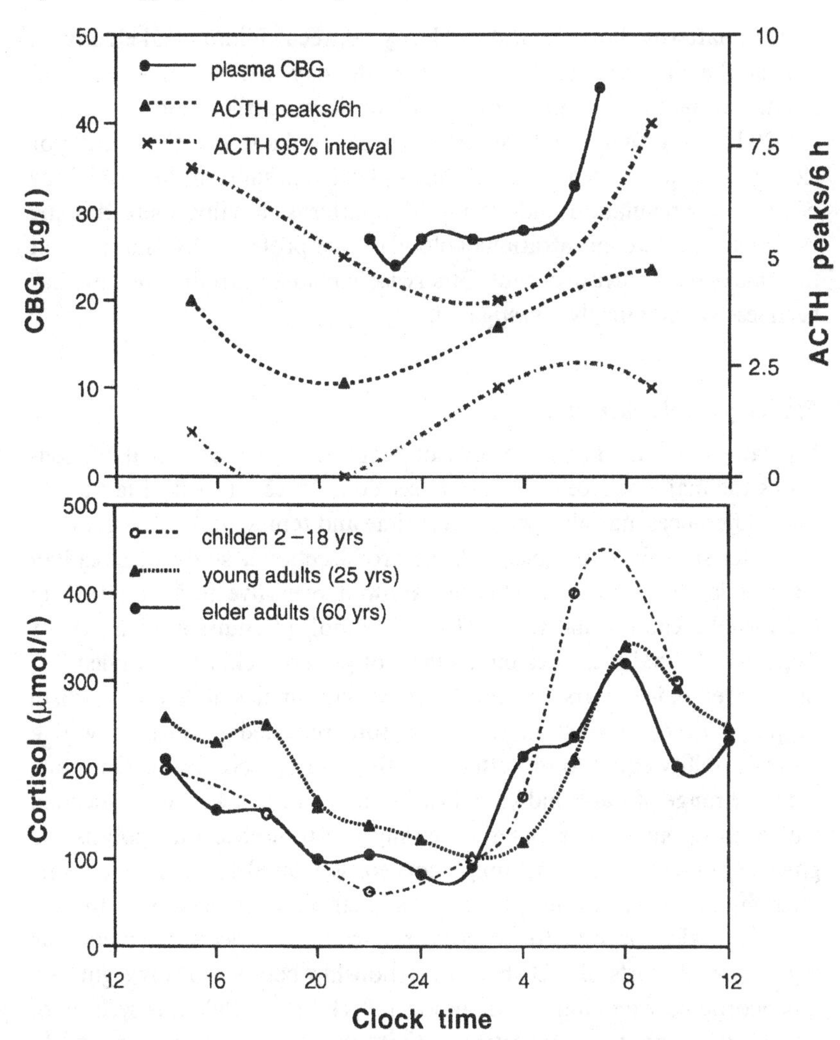

Figure 3.5. Upper panel: nocturnal concentrations of plasma CBG (left axis), and mean number of ACTH pulses per 6-hour time interval (with confidence intervals) in subjects monitored continuously over 24 hours. Lower panel: diurnal variation in serum cortisol concentrations in children, young adults, and elder adults of both sexes, based on continuous sampling. Sex differences are absent at any age, so both sexes are pooled. (Data sources: CBG, ACTH: de Bellis *et al.*, 1996; Horrocks *et al.*, 1990. Cortisol: Knuttson *et al.*, 1997; Sherman *et al.*, 1985.)

cortisol patterns show greatest similarity between children and elderly, in that adults show slower declines across the day, with a later (~0400 h) nocturnal nadir. Children, by contrast, have earlier cortisol nadirs (~2200 h) and higher and earlier nocturnal peaks than either adults or elderly. The age differences have been implicated in shifts in chronobiology of attention regulation and sleep–wake patterns. Finally, there is some evidence that the concentration of blood carrier protein, CBG, increases at the time of peak cortisol output. Otherwise, circadian variation in CBG has been less systematically investigated.

Effects of developmental factors

Virtually all studies involving cortisol document large individual differences as the major source of variation (see confidence intervals, Figure 3.4). Such differences may also persist over time and represent stable individual characteristics. For instance, a large cross-sectional study of circadian cortisol levels in Swedish children showed over five fold variation in individual mean diurnal values (100–510 μmol/l) (Knutsson *et al.*, 1997). Repeated diurnal measures on a subset of pubertal children, studied 2–7 times over 0.5–8 years, showed high longitudinal stability of cortisol output; indeed, production estimates from repeated profiles showed a reproducibility equal to or better than the assay precision. Furthermore, the wide range of stable individual variation in cortisol was not associated with anthropometric correlates, which suggests little effect on lypolysis and glucose homeostasis, even though cortisol is related to their regulation. Such findings may be due to differences in target tissue sensitivity, though they may also be due to genetic organizational effects on metabolic regulation. Reports of a U-shaped relationship between birthweight and glucocorticoid excretion in childhood (Clark *et al.*, 1996), as well as of persistently higher adrenocortical activity in rhesus monkeys exposed to prenatal stress (Clarke *et al.*, 1994), support the latter possibility.

3.6.2 Adrenal androgens

The role of the adrenals in life history merits attention. Adrenal androgens-androstenedione (Δ4), dehydroepiandrosterone (DHEA) and its sulfate (DHEAS) – are weak androgens produced by the zona reticularis of the adrenal cortex in large quantities that, from puberty, exceed those of all other steroids. Functions of adrenal androgens remain poorly understood (Ebeling & Kolvisto, 1994). Their regulation is likewise incompletely characterized; ACTH, the pituitary regulatory hormone for cortisol, is partly involved but does not account for the distinctive, patterned changes in output across the life course (Parker, 1991a; Phillips, 1996). This pattern is particularly evident in DHEAS, which is the quantitatively dominant steroid hormone in the body (Figure 3.6).

Age-related change

Plasma concentrations of DHEAS (Figure 3.6) are high at birth, decline rapidly over the first year of life to the low and slightly rising levels in early childhood, and show an increase in slope between ages 6–8 years. This shift occurs slightly later in boys than girls. The increased adrenal androgen output of mid-childhood, known as adrenarche, is likely not linked to timing of puberty; adrenarche occurs at similar ages in very late as in early maturing populations. At puberty, adrenal androgen output escalates sharply and climbs steadily until it reaches peak values in the early–mid 20s, after which it declines linearly with age. These age-related changes in DHEAS may either have functional implications on their own, related to direct effects of DHEAS, or they may indirectly index aging processes reflected in DHEAS. Alternatively, DHEAS may serve a background or buffering role that moderates the impact of variation in other steroid hormones, such as glucocorticoids.

Circadian patterns

The long half-life and high concentrations of DHEAS lead to an absence of pulsatile or diurnal variation in circulating levels of this hormone, although a precursor, DHEA, does show circadian periodicity. This property makes DHEAS a particularly easy hormone to characterize because sampling does not require controlling for time of day, activity patterns, pulsatility, or other physiological or psychobehavioral variables.

Effects of developmental factors

These tantalizing points about adrenal function in human development and aging have prompted inclusion of their measurement in nearly all my comparative studies of child development. These comparative studies have begun to uncover strong population differences in lifetime trajectories of DHEAS (Worthman *et al.*, 1997*a*). DHEAS rises more slowly in later than earlier maturing populations (Figure 3.4), so that comparing populations

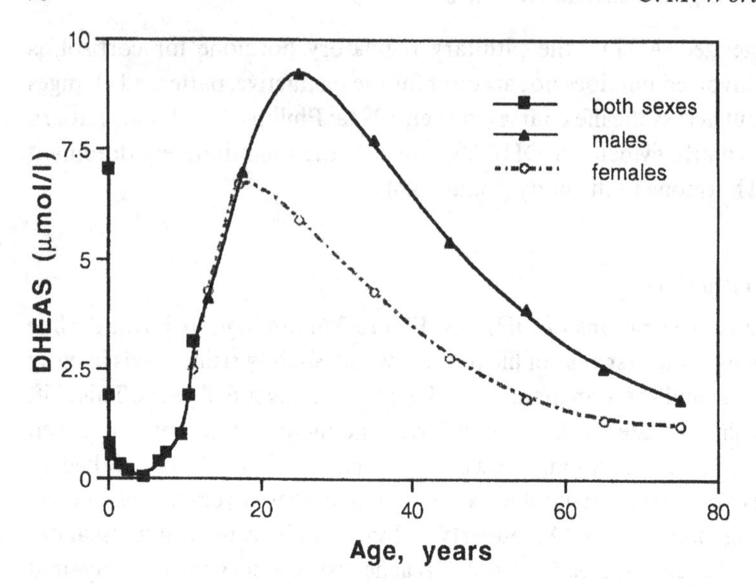

Figure 3.6. Lifespan trajectory of DHEAS in serum, with data for both sexes combined until differences emerge in puberty. Data source: Orentreich *et al.*, 1984.

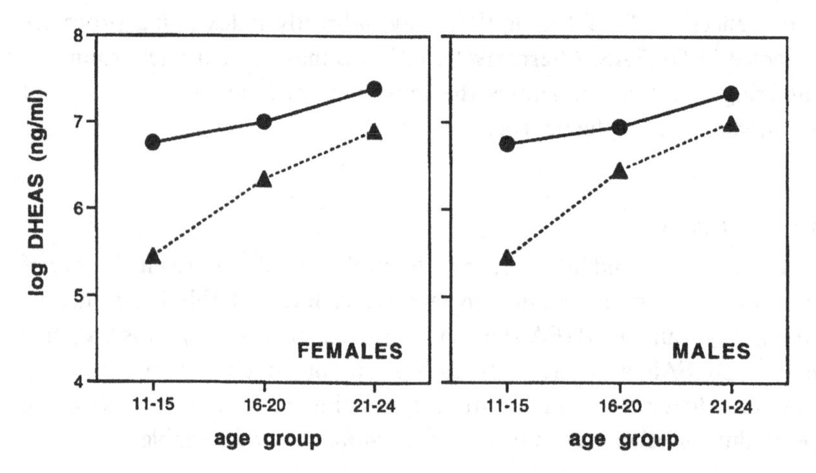

Figure 3.7. Serum DHEAS by age in two later-maturing Bundi populations, one urban, one rural, in whom median age at menarche at the time of study (1983–4) was 17.23 and 15.78 years, respectively. (See also Zemel *et al.*, 1993.)

in the second decade is confounded unless one controls for maturational status. Data from Zemel's study of pubertal maturation in rural and urban Bundi of Papua New Guinea are plotted on Figure 3.7, in which DHEAS values exhibit a different urban-rural contrast by maturational status and age, although there is no such difference in early adulthood (Zemel *et al.*,

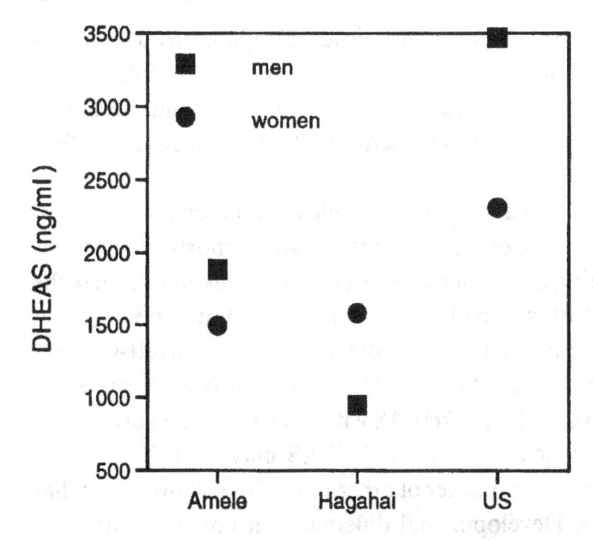

Figure 3.8. Comparison of plasma DHEAS in young adults ages 20 through 24 years from two populations in Papua New Guinea, the fringe highland Hagahai (n = 18) and the subcoastal Amele (n = 36), contrasted with mean values for 13-year-old North American boys (n = 454).

1993). The question arises whether populations differ in peak DHEAS levels achieved in early adulthood and whether, like bone mass and risk for osteoporosis, peak DHEAS predicts later survivorship and health risk. Figure 3.8 displays mean concentrations in young adults ages 20 through 24 years in two Papua New Guinea populations, one (Amele) with relatively good health, nutrition, and life expectancy (Jenkins *et al.*, 1984), another (Hagahai; Jenkins, 1987, Jenkins *et al.*, 1989) with poor health, nutrition, and life expectancy, contrasted with North American teenagers (McDade *et al.*, 1995). Note that 13-year-olds in the North Carolina epidemiological study of puberty already have average values over 1500 ng/dl in both sexes and markedly exceed those in both populations of young adult Papua New Guineans. Unsurprisingly, later maturing populations achieve peak adult values somewhat later in their early 20s than do the Western youth characterized in Figure 3.6 (see also Mavoungou *et al.*, 1986). Note that, moreover, young adult Hagahai show a significant, reversed sex difference. The population difference in values for men is striking, and corresponds to the hierarchy of life expectancy.

Group differences in adrenal androgen profiles may be due to absolute differences in output and/or metabolic variation. A recent report contrasting Peruvian children ages 6–15 years at low and high altitude found both apparently delayed adrenarche at high altitude compared to low, and a high DHEA/androstenedione ratio suggestive of greater DHEA conversion to Δ4, perhaps through greater relative Δ5–Δ4-steroid isomerase (3β-hydroxysteroid dehydrogenase-isomerase) activity (Goñez *et al.*, 1993; Parker, 1991*b*).

The relationships of adrenal androgens to life expectancy are likely not straightforward. Follow-up of populations showing short-term associations of DHEAS with cardiovascular mortality risk has diminished the estimate of their effect on survival risk for men, and shown no effect on women (Barrett-Conner *et al.*, 1986, Goodman-Gruen & Barrett-Conner, 1996). The antiglucocorticoid effects of DHEAS have been adduced to explain the association of reduced DHEAS with neurological disorder such as Alzheimer's (Leblhuber *et al.*, 1993). DHEAS may both have direct health effects, and simply be a marker of other physiological processes that influence health status. Developmental differences in adrenal androgens are associated with strongly contrasted profiles in aging that do not directly correlate with adult life expectancy (Worthman *et al.*, 1997 *a,b*). Little or no decline of DHEAS was found for either sex between ages 25 and 75 years in high altitude populations of Bolivia and Tibet, or between ages 25 and 55 years in the Amele population mentioned above. Characterization and interpretation of variation in adrenal androgen output across the life course in populations with divergent ecology, nutrition, health, and morbidity patterns will likely provide insights into the concommitants of DHEAS variability.

3.7 Thyrotropic axis and central metabolic regulation

The hypothalamo-pituitary–thyroid axis (HTA) comprises the major pathway mediating the crucial process of metabolic regulation. The absolute dependence on thyroid function for adequate brain development leads to disastrous developmental effects from congenital hypothyroidism due to gestational dietary or genetic metabolic factors (Hetzel *et al.*, 1990). After the critical period of brain development through infancy, inadequate thyroid function contributes less to irreversible neurological damage, but pervasively associates with poor cognitive functioning and reduced motor performance (Tiwari *et al.*, 1996). The primary cause of hypothyroidism worldwide is iodine deficiency attributable to insufficient dietary iodine and compounded at times by naturally occurring goitrogens in food or water (Delange, 1994). Hence, most developed countries have national neonatal screening programs to detect congenital hypothyroidism, and public health programs worldwide undertake food fortification with iodine when possible. That one billion people are considered at risk for iodine deficiency disorders (IDD) indicates the extent of this problem (Delange, 1994).

Age-related change

The high risk for and functional impact of hypothyroidism has invested the construction of reference ranges for normal thyroid (euthyroid) functioning with considerable importance. Such reference values have proven elusive both for methodological reasons and because of difficulty in identifying a definitive marker of euthyroid status. The pituitary hormone, thyroid stimulating hormone (TSH), yields too crude a measure of actual thyroid function, while total serum levels of the principal thyroid hormone, thyroxine, provide little information about unbound, bioavailable levels because 92–94% is tied up by circulating binding proteins that themselves change with age (Nelson *et al.*, 1993; Wiedemann *et al.*, 1993). As the metabolically available form, free thyroxine (T4) is viewed as an optimal marker of thyroid function. This hormone exhibits a complex pattern of age-related change over a narrow range of variation across the lifespan (Figure 3.9) that is thought to indicate maturational shifts in pituitary and thyroid regulation (Nelson *et al.*, 1993). Following a neonatal TSH and T4 surge, no further systematic age change in T4 has been reliably observed (Wiedemann *et al.*, 1993). Sex differences have not been systematically seen at any age. Reports of elevated TSH (see Hesse *et al.*, 1994) with age are based on populations unscreened for health status. Reports on well-screened healthy aged populations, by contrast, indicate TSH does decline steadily from late infancy throughout adulhood, reflecting diminishing pituitary response to hypothalamic releasing hormone (van Coevorden *et al.*, 1989; Runnels *et al.*, 1991; Mariotti *et al.*, 1993).

Circadian patterns

TSH exhibits pulsatility and diurnal variation with an early morning acrophase and afternoon nadir: nocturnal TSH averages 62% higher than daytime TSH (Figure 3.10). Evening rises in TSH are associated with increased pulse frequency, while sleep itself dampens pulse amplitude (Brabant *et al.*, 1990; Loche *et al.*, 1994). Circadian rhythmicity in TSH emerges after the first month of life (Loche *et al.*,1994) and is established by ages 5–6 months (Mantagos *et al.*, 1992). Characteristics of the nocturnal

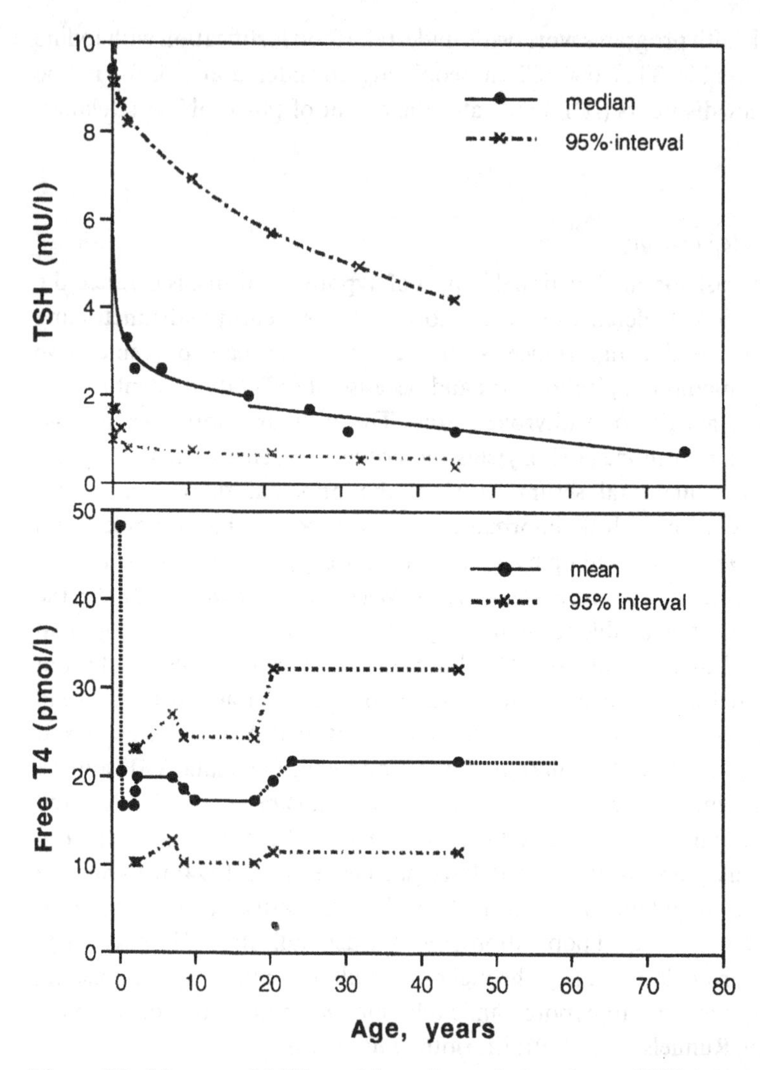

Figure 3.9. Means and 95% confidence intervals for plasma TSH (upper panel) and free T4 (lower panel) across the lifespan, both sexes combined. (Data sources as follows: TSH: Brabant *et al.*, 1990; Mariotti *et al.*, 1993; Van Coevorden *et al.*, 1991; Wiedemann *et al.*, 1993. Free T4: Mariotti *et al.*, 1993; Nelson *et al.*, 1993; Wiedemann *et al.*, 1993.)

TSH surge are otherwise uncorrelated with age, sex, or pubertal status. In studies with 24-hour sampling, no correlation of circulating thyroid hormones with TSH has been reported (Brabant *et al.*, 1990; Chan *et al.*, 1978); rather, the nadir for thyroxine occurs in the early morning hours and shows circadian variation that is nearly the reverse of that for TSH.

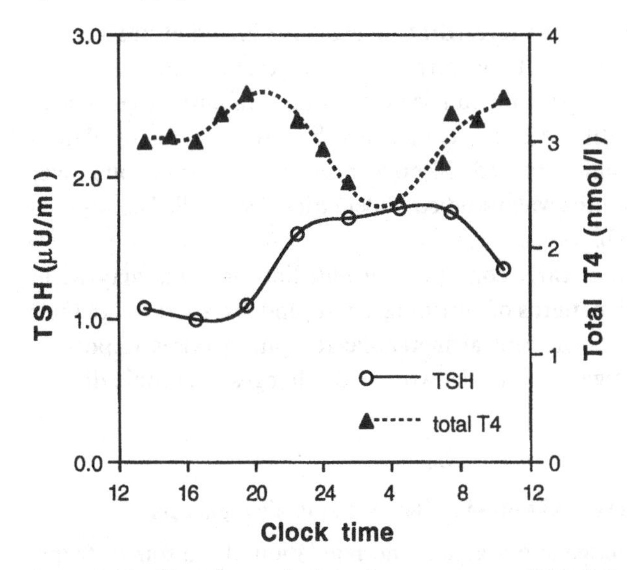

Figure 3.10. Circadian periodicity in mean serum TSH (left axis) and total thyroxine (right axis) from 24 h continuous monitoring. Data for free thyroxine unavailable. Data from Brabant *et al.*, 1990; Loche *et al.*, 1994; van Coevorden *et al.*, 1989.

Effects of developmental factors

Maternal, fetal, and postnatal hypothyroidism all play pathogenic roles in IDD, but their relative importance and long-term organizational impact on the thyrotropic axis have not been fully established (Delange, 1994). Although wide variation in thyroid sufficiency has been reported in apparently euthyroid individuals, organizational effects of fetal or postnatal conditions on normal infants have scarcely been explored. One study of thyroid function in middle-aged British women showed that birthweight correlated with adult TSH and inversely with adult T4 in women who had been bottle fed but not those exclusively breastfed as infants (Phillips *et al.*, 1993). Furthermore, adult T4 was related to duration of exclusive breastfeeding. Breast milk contains numerous immunological, trophic, and other endocrine and nutrient constituents, including thyroid hormones (Grosvenor *et al.*, 1993). Thus, breast feeding may buffer infants from effects of acute transient hypothyroidism to which they are peculiarly prone, particularly where iodine nutrition is even faintly compromised (Delange, 1994). Thyroid responsivity to pituitary stimulation may furthermore relate to degree of breastfeeding, suggesting a long-term organizational effect. Stress may also play a role in development and function of the thyrotropic axis: glucocorticoids exert negative feedback action on TSH secretion at a suprapituitary level (Brabant *et al.*, 1987), so that stress exerts a damping effect on metabolic turnover. Once established, individual differences appear robust. Individuals demonstrate stability of distinctive circadian and pulsatile patterns, as indicated by strong cross-correlation among three to four repeated 24-hour measures within adults over intervals up to 6 months when studied under strictly controlled conditions (Brabant *et al.*, 1990).

The possibility that early conditions of nutrition or stress may wield long-term effects on patterns of nutrition, health, and disease through their organizational effects on central metabolic regulation raises important questions for investigation into the worldwide changes in chronic disease prevalence.

3.8 Somatotropic axis and semi-distributed metabolic regulation

Advances in knowledge about action and regulation of the somatotropic axis (STA) have revealed a highly complex system that integrates growth and somatic maintenance with energetics and metabolic processes. Furthermore, target tissues for this system are dispersed widely rather than localized into glands or organs, and temporal course of functional goals ranges from days (cell turnover) to decades (achievement of adult height and shape). Hence, activity of the STA exhibits acute sensitivity to nutritional status and energy demands, while it also exhibits marked, if limited, ability to adjust growth rates (catch-up or catch-down) in a target-seeking manner. Moreover, recognition of the regulatory and temporal micro-achitecture of growth as an episodic process has opened new windows onto proximal mechanisms of growth and growth regulation (Lampl *et al.*, 1992). The interconnections of growth with nutritional and health status appear to act as important adaptive mechanisms for allocating resources between the demands of growth and maintenance, where we know that mortality risk drives such life history trade-offs.

Intense clinical and popular interest in achieving desired adult heights has led to close scrutiny of this axis, but again most of the fine-grained data are available for healthy, well-nourished members of western populations. Exploration of growth processes and regulation under different socio-ecological circumstances with divergent nutrition, health, and stress profiles will likely remain an important field for investigation well into the future. Meanwhile, what we do know about how the STA works raises formidable challenges for adequate investigation in the field. For instance, as will be shown below, GH release is so strongly pulsatile and sleep entrained, that only integrated 24-hour measures accurately reflect the overall output at a given age. In addition to such physiological complexities, multiple conditions such as infectious status and malnutrition (Wan Nazaimoon *et al.*, 1996), or micronutrient sufficiency (Favier, 1992), powerfully affect axis activity.

Age-related change

The somatotropic axis plays important roles concerning not only growth in the pre-adult years, but also in body maintenance across the lifespan. These dual functions find expression in the lifespan profiles of components of the axis. Strictly speaking, growth hormone does not cause growth *per se*; rather, under appropriate metabolic and local conditions, it stimulates local and hepatic production of tropic hormones (e.g., IGF-I). GH-induced IGF-I, however, requires insulin (Erfurth *et al.*, 1996), which illustrates the condition dependency of action by this axis. Trophic hormones, in turn, act to promote cell division under appropriate cellular and peri-cellular conditions. At the same time, presence of at least four binding proteins for IGF complicates interpretation of circulating levels of tropic hormones. The quantitatively predominant form, IGFBP-3, apparently modulates the bioavailability and bioactivity of IGF-I (Jorgenson *et al.*, 1990). Circulating levels are dependent on age, body mass index (BMI), GH, and IGF-I.

Three principal functional components of the somatotropic axis – GH for pituitary regulation, IGF-I for target tissue response, and IGFBP-3 for the bound circulating reservoir – are represented by age in Figures 3.11 and 3.12. Figure 3.11 shows trajectories for the first two decades only, to highlight axis activity over the postnatal growth period. Mean 24-hour GH is known only from mid-childhood onward, because of the high participant burden imposed by 24-hour cannulation studies required to obtain a realistic estimate of mean GH. Sex differences in activity of the axis appear slight over infancy and childhood, but become significant with pubertal onset: IGF-I differs by sex across all stages of puberty (Juul *et al.*, 1994*a*). During this period, age itself exerts an effect on IGF-I that is independent of puberty and varies among pubertal stages. The latter observation suggests absolute maturational change in IGF-I regulation at these ages (Juul *et al.*, 1994*a*). That maximum levels of IGF-I and IGFBP-3 occur 2 years later than peak height velocity further drives home the lesson that growth is not simply and directly a function of somatotropic axis activity (Juul *et al.*, 1995). For instance, gonadal hormones have synergistic effects on growth at puberty (Brook *et al.*, 1988).

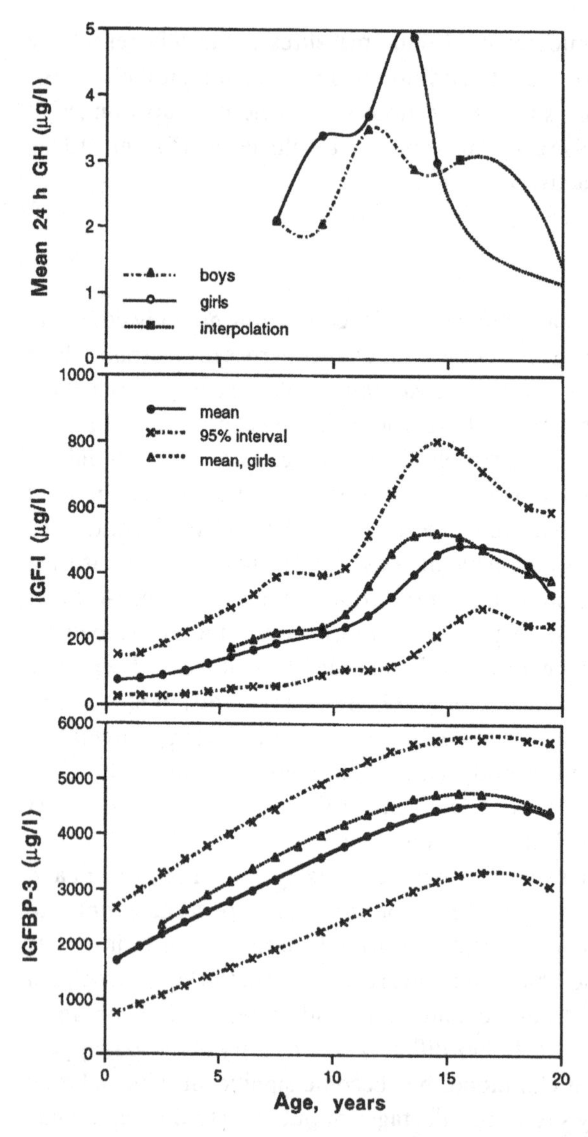

Figure 3.11. Activity during the first two decades of three principal components of the somatotropic axis. Top panel: mean 24 h serum GH, based on continuous sampling. Middle panel: mean and 95% confidence intervals for cross-sectional serum IGF-I, boys and girls combined until age 6 years. Girls and boys shown separately when sex differences become pronounced, after age 5 years; the separate dotted line is for girls and the continuing solid line, for boys. Bottom panel: mean and 95% confidence intervals for serum IGFBP-3, with the sexes shown separately from age 3 years onwards. The continuing solid line represents boys and the dotted one, girls. Sources of data: 24 h GH: van Coevorden *et al.*, 1991; Ho *et al.*, 1987; Mendlewicz *et al.*, 1985. IGF-I: Juul *et al.*, 1994*a*; Hesse *et al.*, 1994; Kelly *et al.*, 1993. IGFBP-3: Juul *et al.*, 1994*b*, 1995; Baum *et al.*, 1996; Martha *et al.*, 1993)

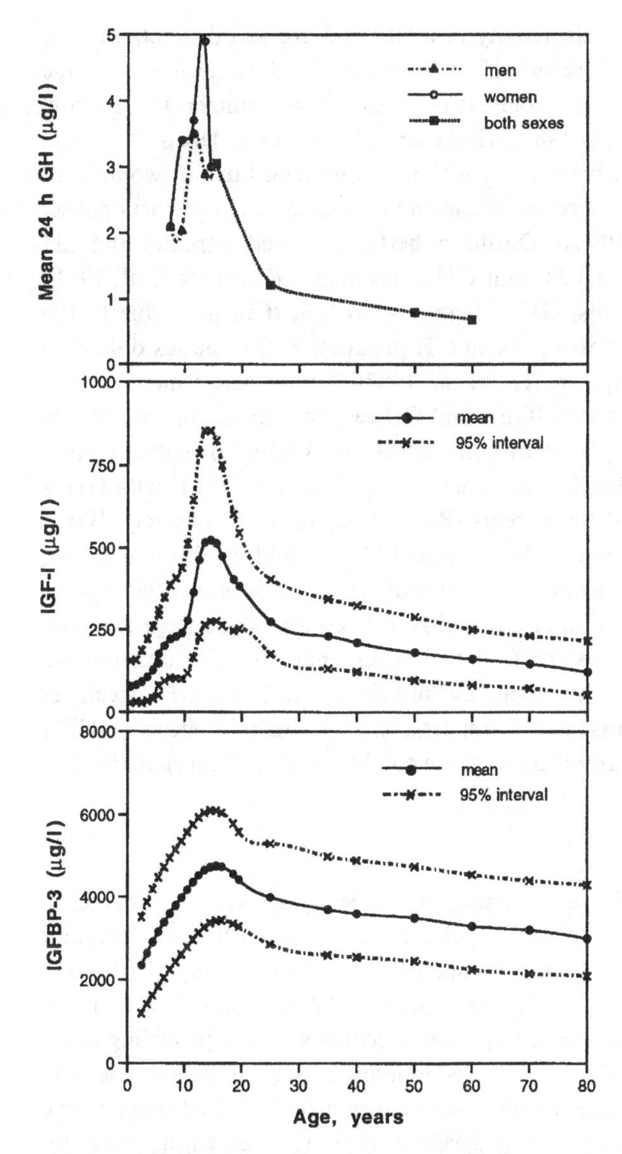

Figure 3.12. Lifespan profiles of three principal components of the somatotropic axis. Top panel: mean 24 h serum GH, sexes combined after age 16 years. Extrapolated curve drawn for gap in available data between ages 16 and 22 years. Middle panel: mean and 95% confidence intervals for IGF-I. The curve to age 20 years is based on data for females. Bottom panel: serum IGFBP-3, mean and 95% confidence intervals across the lifespan. The curve to age 20 years is based on data for females. (Data sources as for Figure 3.11.)

Sex differences in axis activity in adulthood are less distinct, and are influenced by several factors. First, 24-hour GH does decline with age (Figure 3.12), being lower in older (57–76 years) than younger (18–33 years) individuals (Ho *et al.*, 1987). However, 24-hour GH depends on age, physical fitness, and body composition among men but not women, an effect attributable to estrogen effects on GH regulation in premenopausal women (Veldhuis, 1996*a*). During puberty, increased estradiol and LH correlate with increased 24-hour GH levels in girls (Wennick *et al.*, 1991). Among younger adults, GH is higher in women than men, due to the stimulatory effects of estrogens on GH production, but the sex difference vanishes after menopause (Ho *et al.*, 1987). Among adult men, BMI is negatively associated with IGF-I and GH response to stimulation (Baum *et al.*, 1996), and body fat is inversely related to 24-hour GH concentration (Weltman *et al.*, 1994). The same negative association of BMI with GH is seen in children and adolescents (Rose *et al.*, 1991). Conversely, IGF-I shows strong positive correlation with BMI in children through adults, though not in the very old (Hartman *et al.*, 1992; Hesse *et al.*, 1994; Harris *et al.*, 1997). Dissociation of GH and IGF-I responses to energetic stores, reflected in BMI, likely stems from their divergent roles – GH as metabolic, IGF-I as trophic – when energy sufficiency declines: GH organizes metabolic adjustments to caloric shortfalls, while anabolic effects of IGF-I are costly and undesirable under these conditions (Hartman *et al.*, 1992).

Circadian patterns

Diurnal activity of the somatotropic axis varies by age and sex. Characteristics of pulsatility and circadian release are significant because temporal patterns of GH and tropic hormone delivery can markedly affect tissue response, in terms of quality and quantity (Veldhuis, 1996*a*). Clinical interest in evaluating somatotopic axis adequacy and in providing treatment for growth insufficiency has led to intense focus on the brain activity reflected in GH release, to the near exclusion of IGF-I or other tropic hormones. Infants exhibit high levels of tonic GH secretion. Pulse frequency and amplitude decrease as the infant matures, which brings down tonic levels. Prepubertal children exhibit a pulsatile pattern of GH release, with increased pulse amplitude in sleep (Ho *et al.*, 1987). Changes in GH pulsatility at puberty likely organize the accelerated growth characteristic of this period: at puberty, nocturnal GH pulse amplitude, but not frequency, increases to a peak at pubertal stages 3–4 in girls and 3 in boys (Rose *et al.*, 1991). Nocturnal GH is negatively related to BMI, though the effect is significant only in girls, among whom the correlation strengthens between early and late puberty.

Men and women differ markedly in the GH circadian rhythm of pulsatility, having strongly and modestly circadian release patterns, respectively (Veldhuis, 1996*a*). Men have one or few large GH spikes during sleep, specifically slow wave sleep; women have more daytime pulses, and sleep-associated pulses account for a minority of daily GH release (Van Cauter and Plat, 1996). GH pulse frequency and duration, but not amplitude, decline with age in men (Figure 3.13) (Veldhuis *et al.*, 1996).

Serum concentrations of IGF-I fluctuate in a circadian manner lagged in time from stimulatory pulses from the pituitary, reaching nocturnal peaks in the early morning hours (Figure 3.14). Notably, levels of the dominant serum binding protein, IGFBP-3 are temporally stable and evidence no apparent circadian variation (Jorgenson *et al.*, 1990), which implies that diurnal variation in IGF-I drives availability of its free form.

Effects of developmental factors

Malnourished or fasted individuals show increased GH and decreased IGF-I secretion. Dissociation of the response by these hormones to caloric stress is attributable perhaps to the mediating role of GH in metabolic adaptation to hunger, contrasted with growth-decelerating effects from reduced synthesis of IGF-I (Hartman *et al.*, 1992). Furthermore, caloric and micronutrient insufficiency may often operate synergistically. For instance, Malaysian children experiencing malnutrition and iodine deficiency exhibited not only low levels of IGF-I and IGFBP-1, but also correlation of these with T4 (Wan Nazaimoon *et al.*, 1996). The expected correlation of IGF-I and IGFBP-3 with height for age in childhood also persisted through puberty, despite hypothyroid status and low levels of both moieties. At puberty, gonadal steroids stimulate rises in GH, IGF-I, and IGBP-3 (Argente *et al.*, 1993; Kerrigan & Rogol, 1992). Such observations suggest that cross-axis interactions strongly influence activity of the STA (Mauras *et al.*, 1996), so that variations in timing of puberty will be reflected in variation in STA hormone profiles by age.

Restitution of nutritional deficits generally abolishes the endocrine profile associated with energetic stress, but whether there are developmentally sensitive periods at which such stress exerts a lasting impact on growth and metabolic regulatory processes remains unresolved. In samples of British and Indian children, birthweight was found to be negatively correlated to current height, weight, and IGF-I (Fall *et al.*, 1995). The

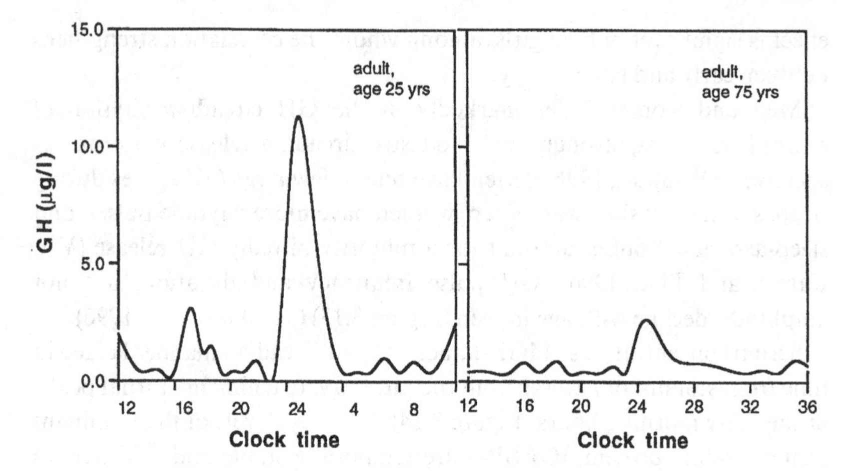

Figure 3.13. Pulsatile and circadian patterns of serum GH in younger and older adults. (Data from van Coevorden *et al.*, 1991.)

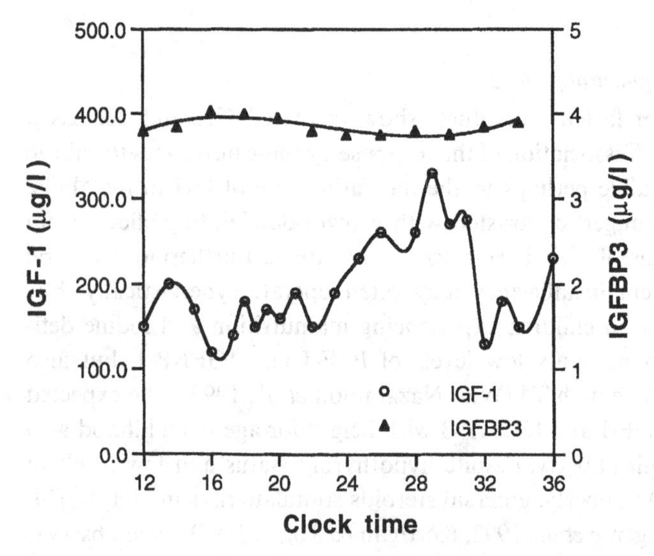

Figure 3.14. Circadian pattern of serum IGF-I in young men around age 21 years and IGFBP-3 in boys and girls around age 15 years. (Data sources: Jorgensen *et al.*, 1990; Stratakis *et al.*, 1996.)

investigators viewed the results as consistent with reprogramming of the somatotropic axis from gestational undernutrition, and suggested that high IGF-I may mediate linkages of reduced fetal growth to adult hypertension.

Marked, consistent individual differences in endocrine profiles of the STA have been noted, though these may be due to genetic or constitutional as much as to environmental or developmental effects. Growth hormone binding protein, for instance, varied over a 12-fold range among pubertal boys, but varied more narrowly, averaging a three-fold range, within individuals (Martha *et al.*, 1993). Moreover, boys evaluated three to five times over 8–16 months and adults observed twice over a year showed reproducible 24-hour GH patterns in which inter- strongly exceeded intra-individual variation (Martha *et al.*, 1993). Genetics may play a role in individual differences, although not in a simple fashion: thus, STA function did not differ in young Greek men with normal or tall stature, a statural difference assumed to be due to genetic factors (Stratakis *et al.*, 1996). Individual differences may be related to differences in body composition and its known effects on GH secretion.

Growth is recognized as a powerful marker of health and nutritional status that predicts life expectancy and functional outcomes in adulthood. Early insults or suboptimal conditions – particularly in gestation and under age 2 years, but also through childhood – consistently predict losses in adult height, and thus the factors that drive height deficits are viewed as significant public health targets (Schroeder & Brown, 1994). Nonetheless, height, in and of itself, does not mediate the observed variation in adult well-being; it is simply a marker, for being taller has no absolute biological virtue on its own. The proximal physiological mechanisms that do organize developmental health risks and are reflected in relationships between height and well-being remain uncertain: processes of growth and metabolic regulation, and physiological mediators of resource allocation among various life history parameters of growth, maintenance, and reproduction must be involved. Comparative study of endocrine regulation across these domains should help us unravel the association of height and health to more deeply understand the determinants of variation in human well-being.

3.9 Bases of variation

3.9.1 Secular trends and population variation

As discussed at the outset, sources of variation and secular trends in growth and maturation rates have been heavily documented, and fetal, infant, and child growth indices inform public health policy and resource allocation worldwide. Yet the physiological causes and consequences underlying these important outcomes are just beginning to be explored. Human physiology is assumed to be generic and universal: findings that LH regulates T and that T is the principal hormone of male reproductive function and dimorphic morphology are assumed to apply to everyone, everywhere. And indeed, increasing data from non-western settings confirm that physiological fundamentals do conform to this presumption. Yet other findings (see also Panter-Brick and Pollard, Chapter 5; Ellison, Chapter 6 and Whitten, Chapter 7) suggest that, within this pattern of universality, resides a core of quantitative endocrine variation that may have qualitative functional implications.

Work by my laboratory with various co-investigators will be used here to illustrate current approaches to comparative endocrine study of development, focusing on the HPG axis and on DHEAS. Hormones – gonadotropins and gonadal steroids – act to initiate puberty and drive its progression, and can be used to study reproductive maturation in both sexes. The adrenal androgens also exhibit a developmental profile with distinctive shifts in circulating levels from mid-childhood through early adulthood, and thus provide another window onto maturational progression over this period. Our comparative endocrine studies of puberty have shown that populations with later maturation do enter puberty later than early maturing ones, but they also go through it more slowly. In other words, pubertal onset is not so delayed as their late ages at menarche or peak height velocity would suggest. For instance, among Kikuyu of Kenya where median age at menarche was 15.9 years, onset of puberty based on endocrine or morphological markers occurred in girls at median age 13.0 (Worthman, 1986, 1993). Hence, Kikuyu girls took an average 2.9 years to go from pubertal onset to menarche, contrasted with 2.3 years for a closely studied sample of British girls (Marshall & Tanner, 1969).

Quality of environment, as it determines nutritional status and morbidity in children, has been implicated as the major source of variation in rates of child growth and maturation (Bielicki, 1986). Thus, class or other markers of socioeconomic status have been consistently linked with differences in growth (Mascie-Taylor, 1991). Nevertheless, environmental quality can differ sharply not only by socio-economic status but also within households, particularly in relation to sex-differentiated child care practices. Sex differences in timing of pubertal onset vary across populations (Worthman, in press), and likely reflect sex-differentiated environments of rearing. For instance, based on endocrine or morphological markers, it was found that Kikuyu boys entered puberty either 6 months ahead of, or at the same time as, girls (Worthman, 1993). By contrast, girls in the British sample mentioned above entered puberty 6 months ahead of boys, a pattern commonly observed in Western populations with good health, nutrition, and gender-egalitarian care. At the other extreme, boys experience pubertal onset over 2 years later than girls among Hadza of Tanzania (Worthman, in press). Timing of pubertal onset, rate of pubertal progression, and sex differences in these, all provide endocrine markers of differentiated developmental processes in relation to environmental variation.

Developmental variation also corresponds to adult variation. The two New Guinean populations mentioned earlier, Amele and Hagahai, differ not only in environmental quality, health, and life expectancy, but also in maturation rates. In comparison to Amele, Hagahai have extensive growth stunting, late pubertal onset, and blunted and delayed peak height velocity. Reproductive function of young men ages 20 through 24 contrast sharply. Amele have twice the concentration of T (675 ± 28 ng/dl, mean $\pm$ s.e.) and LH (7.6 ± 0.49 IU/l) as do Hagahai (320 ± 37 ng/dl and 4.07 ± 0.80 IU/l, respectively); these values place many Hagahai men at the very end of the normal range for North Americans or Amele. Interpretation of these data hinges on the low normal LH observed among Hagahai, because it indicates central downregulation of testicular function, not peripheral refractoriness or suppression. In this regard, the very high values of LH apparently required among Amele to achieve levels comparable to Western peers stand out. These populations differ, therefore, in key aspects of male reproductive function, and in the lifetime exposure to testosterone. Such observations of relatively high levels of gonadal steroids in Western compared to non-Western men concur with comparisons of ovarian function among women. Implications for reproductive cancer risk have been raised (see Ellison, Chapter 6, and Whitten, Chapter 7), but other functional, psychobehavioral and health trade-offs associated with such large population differences in gonadal steroid exposure for men and women have scarcely been considered.

Between-population quantitative variation in hormone profiles occurs not only in the reproductive axis, but also for DHEAS (see Figures 3.7 and 3.8, above), and such variation likely occurs for other endocrine axes as well. The question requires further study on variation in normative rather than pathogenic terms. By illustration, worldwide scrutiny of thyroid function has understandably focused on identification and etiology of hypothyroidism rather than population variation in metabolic regulation, and work on the STA in various populations has concentrated on bases for small stature in specific populations, or on the effects of malnutrition on growth. The above survey of lifespan function of five endocrine axes provides a basis on which to consider how these population differences may arise developmentally, as summarized in Table 3.1. First, most axes show pulsatility and circadian rhythmicity which both increase the information content of regulatory signals, and renders the axis vulnerable to acute effects of affective (e.g., stress), behavioral (e.g., workload, sleep patterns), and physical condition (e.g., illness, caloric shortfall). Given the technical difficulty of intensive, round-the-clock sampling, knowledge about intra- and cross-population variation in pulsatile and circadian patterns remains limited. Most research tries to control for these factors, rather than to study them directly. DHEAS is an informative exception by having no ultradian variation, which may contribute to its value as a metabolic buffer reflecting neuroregulatory setpoints that establish allocation biases among life history trade-offs.

Second, all axes for which data are available show intense functional shifts indicating regulatory maturation in the postnatal and early infancy period. Notably, cortisol evidences no further patterned developmental change across the lifespan. The thyrotropic and somatotropic axes appear to show similarly swift postnatal establishment of regulatory setpoints, whereas the HPG axis in females and DHEAS in both sexes show a more prolonged lag to stabilization. The postnatal, and not only the fetal, period may represent an interesting sensitive period in which real-world conditions are sampled for the first time. Whether there is variation in postnatal endocrine trajectories, and whether these are responsive to effects of nutrition, morbidity, childcare practices, or cultural practices such as circumcision are questions that merit exploration.

Third, various parameters leap out as relevant for comparative endocrine study. These include not just the amount of circulating hormone, but also levels of its binding protein, timing of functional shifts, and slopes of change with age. Fourth, endocrine axes cannot be assumed to vary independently of each other; indeed, we should assume the converse. The adrenal, gonadal, thyroid, and tropic axes all exhibit extensive regulatory interrelationships (e.g., Brabant *et al.*, 1987; Veldhuis *et al.*, 1996; Erfurth *et al.*, 1996; Mauras *et al.*, 1996; Wan Nazaimoon *et al.*, 1996) and these, I would argue, represent the endocrine organization of life history strategy required to translate that strategy into an individual developmental life span. The bases of such translation could comprise facultative and organizational adjustment of regulatory set-points within and across axes. Multi-axis study will be needed to test the concept of cross-axis regulatory set-points and to consider how trade-offs in life history allocation are effected physiologically and affected by context or experience. Again, one must note that the bulk of available data are from epidemiologically and nutritionally privileged Western populations, and that regulatory trade-offs may shift strongly in more constrained settings. The lifespan summaries of functional parameters for several axes presented in this chapter should provide a basis from which to launch such a comparative effort.

3.9.2 Intrapopulation variation

Compounding population variation is large within-population variation in endocrine profiles. Clinical research, and the data discussed in previous sections, document the wide range of endocrine output that appears generally compatible with normal functioning. By illustration, after the neonatal period, individual variation in TSH and T4 at any age far outstrip developmentally organized changes over the lifespan (Figure 3.9). My own early work on !Kung San with Konner and Shostak vividly revealed this difference in what might be termed "endocrine economy". Our 10-day study of !Kung hunters showed similar diurnal patterns and hunt-associated shifts in T values, but actual baselines and dynamic ranges over which individuals responded varied markedly. Degree of diurnal and day-to-day variation in T also differed among men. Figure 3.15 shows morning and evening values over the study period from a man with what one might call a "florid" endocrine economy, on the left, and, on the right, a man with a modest one. What developmental course and functional difference do these variations in endocrine economy represent? Likewise, in the North Carolina study of puberty, enormous individual variation is seen in the level of endocrine stimulation required to attain pubertal onset and progression (McDade *et al.*, 1995). Furthermore, in virtually all realms of endocrine research, large amounts of individual variation lurk in the standard errors of population means. In terms of hormone levels required to exert a physiological effect, for some a "little dab will do it," whereas for others only a large measure will do.

One axis included in the present analysis, the HPA, participates in acute cognitive, behavioral, and physiological responses to ongoing experience; hence, its dynamic response properties shape the impact of experience (van Eck *et al.*, 1996). Reciprocally, experience can affect response properties of the HPA. Response properties, while they emerge early in life and thereafter show little or no lifespan change on the population level (Figure 3.4), exhibit early organizational and later facultative sensitivity to experience on the individual level (Seeman & McEwan, 1996). Prenatal exposure to maternal HPA stress responses permanently enhance postnatal reactivity (Clark *et al.*, 1994; Wadwha *et al.*, 1996). Cortisol reactivity, or the threshold, magnitude and duration of HPA response to experience, is related to both early traumatic experience and constitutional or heritable

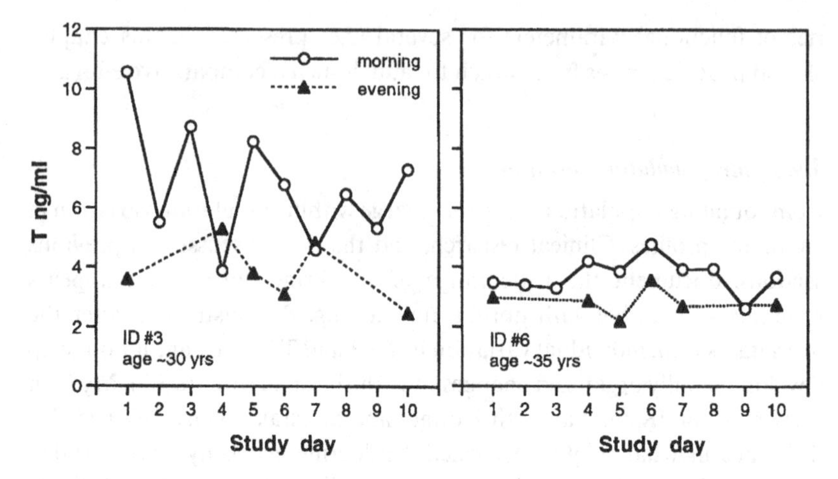

Figure 3.15. Serum testosterone profiles for two !Kung San men, based on morning and evening sampling over 10 consecutive days. Direct observation and report document that both men engaged in similar activity, in camp and hunting, during the study period.

factors (Higley *et al.*, 1991, 1993; Suomi, 1991; Berman *et al.*, 1994). Individual differences in HPA reactivity have increasingly been linked to differences in risk-taking behavior, immune function, health indices, attachment, and social relationships (Stansbury & Gunnar, 1994; Boyce *et al.*, 1995*a,b*; Gunnar *et al.*, 1997). These effects are consistent with a role for the HPA in adjusting life history allocation biases toward immediate physiological and cognitive–behavioral coping, and away from long-term investment in maintenance. Social and cultural factors that determine kinds and qualities of experiences encountered in the pre-adult years thus also influence health and psychosocial well-being indirectly, through experiential effects on the HPA (Seeman & McEwen, 1996). This chain of relationships provides an elegant instance of interactions among hormones, health, and behavior that arise through developmental organization of life history.

3.10 Implications for health

The question is not only if and how individuals arrive at different developmental trajectories, but also whether the diverse outcomes have meaningful functional implications. That is, we are not interested in just any human variation, we are interested mainly in that variation which predicts or corresponds to significant differences in outcomes – in function, quality of life, or fitness. These are precisely epidemiological questions, and the statistical tools and conceptual logic of epidemiology are highly relevant to human biology. Attention to risk and to mortality or disease outcomes is closely congruent with current life history concerns, with mortality rates, and temporally dissociated correlations. Epidemiologists concern themselves with the problem of discerning meaningful from significant variation, which likewise concerns us. Moreover, their methods provide statistical means to deal with a range of contextual variables in a much more nuanced way than much of human biology research does. Like human biology, epidemiology lacks stong models of development, although the questions epidemiology addresses are inherently developmental (Costello & Angold, 1995).

So far, I have emphasized that human development both reflects patterns of environmental quality and physiologically organizes life history trade-offs through the endocrine–neuroendocrine system. Here I consider how, epidemiologically, human development also predicts differential outcomes in terms of risk to adult function, disease, and death. Widespread interest by human biologists, pediatricians, and public health scientists in growth and timing of puberty is due less to their intrinsic biological value, than to their value as markers of differential well-being that predict future morbidity and mortality. Essentially, a marker represents a kind of "bioassay" for quality of life, embodying all the processes that are too complex, inaccessible, or temporal to either predict or to study directly. Child growth and maturation rate are two of the most powerful of recognized biological markers for early effects on long-term outcomes, for they are sensitive indicators of nutrition, physical health, and even psychological stress that also predict life expectancy, later health, and cognitive function (Pollitt *et al.*, 1995; Martorell *et al.*, 1996). Although the pathways to variation in child growth status are not fully understood, the study of causes and consequences of growth variation has provided a platform for teasing out linkages between culture and variation in well-being, as well as for understanding growth itself.

This important work on growth has not considered *why*, in ultimate terms of organic design, growth and health are so closely linked. The present analysis addresses that question, albeit in preliminary form. The model for developmental epidemiology is presented in Figure 3.16, in a framework that aims to integrate evolutionary, life history concepts (upper box) with individual ontogeny and life course (organismic level, lower boxes), by examining how the rubber of organic design meets the road of everyday life. In this model, features of the social and physical environment set up differential resources and challenges to the organism that, on a remote evolutionary scale, represent selection pressure in action while, on the proximal individual level, they inform ongoing processes that mediate ontogeny and determine outcomes. Selection pressures set up by environmental affordances and challenges push the organism to effect trade-offs in resource allocation that are effected through physiological, behavioral, and psychological adjustments on the processual level. Such adjustments, however, also drive the direction of ontogeny (e.g., learning, differentiation) and function (e.g., regulatory setpoints) that, in turn, invoke trade-offs among life history parameters (growth, reproduction, maintenance) constrained by the allocation rule. Allocation trade-offs – for instance, reproducing earlier by growing less, or short-changing development for enhancing immediate survival – are instantiated in differential outcomes, in terms of health, longevity, fertility, or even cognitive capacities. Finally, these differential life history or life course outcomes define differential fitness.

Additionally, linkage between individual and evolutionary levels transpires through design features of the organism (lower left, Figure 3.16), namely, the evolved form and manner in which ontogeny and function happen through real time, in dynamic relationship to or coaction with context. Organisms do not "decide" to trade off growth and health, say; they possess myriad functional design features that, severally and in combination, probabilistically drive physiological processes generating individual response in a particular way to particular circumstances at a particular time. The present analysis has pursued the notion of an endocrine architecture of life history in some detail for several axes but even so, many levels of design and ontogenetic differentiation (e.g., setpoints, or size, types, and distributions of receptor populations) remain. The overall point has nonetheless been provisionally established: evolution and life history are intimately related through individual ontogeny. Evolutionary processes are comprised of lifespan developmental ones, but the study of evolution has come to be abstracted from the proximal pragmatics of everyday life, from the actual life of the individual, and vice versa.

We are now in a position to reconsider the question: why is growth linked to health? Growth and maturation are valuable markers of well-being precisely because they are organized by the endocrine regulatory systems. The link to health is forged because these same systems, in turn, determine allocation biases and life history trade-offs, as well as drive the functional capacities, vulnerabilities, and resiliencies that shape adult psychobehavioral competence, health, and mortality risk. Secular trends to

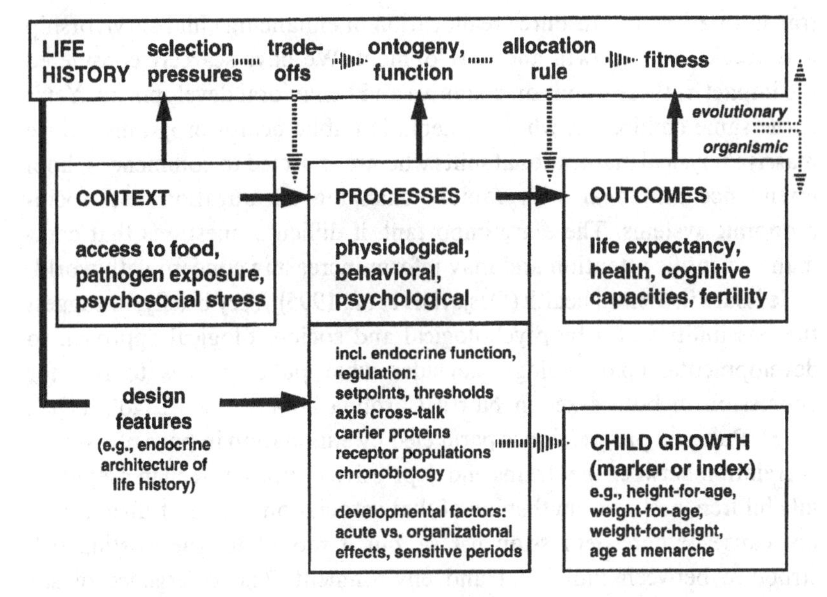

Figure 3.16. Relationships of life history parameters to individual ontogeny and function, focusing on mediating endocrine factors, as determinants of adult outcomes in health, survival, and competence (on the organismic level), as well as fitness (on the evolutionary level).

increased growth and accelerated mortality occur because of facultative shifts in developmental organization and function of endocrine-neuroendocrine axes. Epidemiological identification of secular trends and individual differences associated with variation in function or health therefore rests on these underlying endocrine processes that organize risk structure by driving function and functional trade-offs in the short and long term. This insight concerning the endocrine and developmental, or time-lagged, bases of risk underscores the need for comparative developmental research on endocrine regulation and function. The chronic conditions – diabetes, hypertension, cancer, obesity – currently undergoing global increases in prevalence all have a developmental course, and the recognition is growing that focus on adult behavior will not suffice to explain these trends (Barker, 1997*a*, *b*; Ellison, Chapter 6; Whitten, Chapter 7).

In sum, the contention is that epidemiology of development informs the epidemiology of adult illness and well-being. A further pressing issue in developmental epidemiology remains, namely, the relationship of psychosocial and behavioral outcomes to developmental variation. Human biologists investigate the conditions that affect human growth and maturation rates, and document the enormous worldwide impact of social transformation and resource reallocation in enhancing child survivorship and accelerating growth and development. We have scarcely considered the impact of these trends on cognitive and behavioral development. Yet it is the same children in whom systematic public health programs induce massive physical maturational shifts who are expected to commence school attendance and deal with rapidly changing socialization and socio-economic systems. These are important, if difficult, questions that command scientific attention and may inform increasing concern with world-wide issues in mental health (Desjarlais *et al.*, 1995). Very briefly, to suggest the possibilities of a biopsychological and socio-ecological approach to developmental epidemiology, consider what puberty does to risk for depression in both sexes in North Carolina adolescents (Angold *et al.*, 1998). Pubertal progression is paralleled by elimination in boys of a strong correlation between life strains and depressive symptoms seen in prepubertal children, and a strengthening of that association in girls. Puberty does not cause or cure depression *per se*, but it modulates the existing risk structure between individual and environment. The emergence of sex differences in rates of depression over puberty represents the largest developmental effect in psychiatric epidemiology for Western populations (for review see in Angold & Worthman, 1993). What, then, do worldwide changes in pubertal timing imply for mental health risk? Furthermore, given the connections of stress and distress with immunosuppression (Ader *et al.*, 1991), what does variation in mental well-being imply for physical health?

3.11 Conclusions

This chapter concludes with a summary overview of the present argument in relation to the theme of hormones, health, and behavior. The case was made for a developmental lifespan approach based on life history trade-offs for allocation of time and resources among growth, maintenance, and reproduction. Hormones, it was suggested, instantiate the architecture of life history. Hormones effect life history strategy and facultatively establish trade-offs among competing needs and demands in the face of existing conditions experienced by the developing individual. Such trade-offs are reflected in hormone-mediated phenomena such as delayed growth, accelerated maturation, or upregulated adult reproductive function. They are furthermore reflected in cross-axis effects in adrenal, gonadal, thyroid, and somatotopic activity across the lifespan. Temporal dimensions of endocrine adjustment allow distinctions between acute and organizational effects; the first concerns moment-to-moment responses to physiological or cognitive load, while the latter concerns permanent adjustment of regulatory response set-points that define the dynamic properties of the system. Furthermore, the endocrine and neuroendocrine systems are integrated functionally, as evidenced in the brain-driven, hypothalamus-mediated components of the HPG, HPA, HPT, and STA axes. Intimate endocrine–neuroendocrine interconnections forge significant links of brain, cognition, and affect with physical function and health, and vice versa.

Behavior plays multiple roles in this model. Culturally and socially patterned conditions and experiences constitute the environment of rearing and of adult function. Class structure, subsistence patterns, and childcare practices exemplify salient structural conditions probabilistically affecting individual experience. Such macro-level conditions define exposures to challenge, while they also affect individual vulnerability or resilience to challenge by determining the distribution of needed resources, environmental insults, and benefits. On this level, worldwide patterns of economic change, public health interventions, and macro- and household-level demographic shifts organize the micro-level conditions under which people grow up and function, and it is these changes to which the secular trends to increased growth and accelerated maturation have been attributed. Concurrently, such changes transform the conditions of everyday adult life and construct the field on which endocrine organization underlying secular trends intersects or collides with new cognitive–functional demands in adulthood. Structural transformation of work and social relationships not only define the demands and rewards faced throughout life, but they also define the bases of perceived psychosocial as well as physical competence and normative performance. On the level of individual behavior, lifestyle shifts in patterns of activity, food consumption, values, and perceptions exert short- and long-term effects on well-being via, in part, their endocrine impact.

Finally, health is itself a complex phenomenon. Most crudely, health can be gauged by reference to mortality or morbidity, but defining morbidity proves to be difficult because the bases for evaluating ill-being must be established. Some measure of functional impairment may help to define dysfunction. Yet again, local definitions of and demands for physical and psychobehavioral competence may represent the most salient gauge of morbidity, and lead to a need to define health less in absolute than in local terms. If psychobehavioral competence importantly defines well-being, then the usual tendency to conflate physical health with well-being is undermined by the recognition that mental well-being defines health as well. Furthermore, short-term gains in perceived well-being may be bought at the expense of long-term costs to health and survival; a life history approach emphasizes that these trade-offs inevitably occur (Hill & Hurtado, 1995). We may look forward to an exciting period in which the bases of differential health and human development are reconceptualized and investigated in the light of biocultural, life history models that integrate evolutionary, cultural–structural, cognitive–behavioral, and endocrine–neuroendocrine levels of analysis.

3.12 References

Ader, R., Felten, D., & Cohen, N. (1991). *Psychoneuroimmunology*. Orlando, FL: Academic Press.

Angold, A., & Worthman, C. (1993). Puberty onset of gender differences in rates of depression: a developmental, epidemiologic and neuroendocrine perspective. *Journal of Affective Disorders*, **29**, 145–58.

Angold, A., Costello, A., & Worthman, C. (1998). Puberty and depression: the roles of age, pubertal status, and pubertal timing. *Psychological Medicine*, **28**, 51–61.

Apter, D., Butzow, T., Laughlin, G., & Yen, S. (1993). Gonadotropin-releasing hormone pulse generator activity during pubertal transition in girls: Pulsatile and diurnal patterns of circulating gonadotropins. *Journal of Clinical Endocrinology and Metabolism*, **76**(4), 940–9.

Apter, D., Cacciatore, B., Alfthan, H., & Stenman, H. (1989). Serum luteinizing hormone concentrations increase 100-fold in females from 7 years of age to adulthood, as measured by time-resolved immunofluorometric assay. *Journal of Clinical Endocrinology and Metabolism*, **68**(1), 53–7.

Argente, J., Barrios, V., Pzoz, J., Munoz, M., Hervas, F., Stene, M., & Hernandez, M. (1993). Normative data for insulin-like growth factors (IGFs, IGF-binding proteins, and growth hormone-binding protein) in a healthy Spanish pediatric population: Age- and sex-related changes. *Journal of Clinical Endocrinology and Metabolism*, **77**, 1522–8.

Barker, D. J. (1997*a*). Intra-uterine programming of the adult cardiovascular system. *Current Opinion in Nephrology and Hypertention*, **6**, 106–10.

Barker, D. J. (1997*b*). Fetal nutrition and cardiovascular disease in later life. *British Medical Bulletin*, **53**, 96–108.

Barker, D., Osmond, C., Simmonds, S., & Wield, G. (1993*a*). The relation of small head circumference and thinness at birth to death from cardiovascular disease in adult life. *British Medical Journal*, **306**, 422–6.

Barker, D., Hales, C., Fall, C., Osmond, C., Phipps, K., & Clark, P. (1993*b*). Type 2 (non-insulin-dependent) diabetes mellitus, hypertension and hyperlipidaemia (syndrome X): relation to reduced fetal growth. *Diabetologia*, **36**, 62–7.

Barker, D., Gluckman, P., Godfrey, K., Harding, J., & Owens, J. (1993*c*). Fetal nutrition and cardiovascular disease in adult life. *Lancet*, **341**, 938–41.

Barrett-Connor, E., Khaw, K., & Yen, S. (1986). A prospective study of dehydroepieandrosterone sulfate, mortality, and cardiovascular disease. *New England Journal of Medicine*, **315**, 1519–24.

Baum, H., Biller, B., Katznelson, L., Oppenheim, D., Clemmons, D., Cannistraro, K., Schoenfeld, D., Best, S., & Klibanski, A. (1996). Assessment of growth hormone (GH) secretion in men with adult-onset GH deficiency compared with that in normal men – a clinical research center study. *Journal of Clinical Endocrinology and Metabolism*, **81**, 84–92.

Belanger, A., Locong, A., Noel, C., Cusan, L., Dupont, A., Prevost, J., Caron, S., & Sevigny, J. (1989). Influence of diet on plasma steroid and sex plasma binding globulin levels in adult men. *Journal of Steroid Biochemistry*, **32**, 829–33.

Belgorosky, A., & Rivrola, M. (1986). Progressive decrease in serum sex hormone-binding globulin from infancy to late prepuberty in boys. *Journal of Clinical and Endocrinological Metabolism*, **63**, 510–12.

Belgorosky, A., & Rivarola, M. (1987). Changes in serum sex hormone-binding globulin and in serum non-sex hormone-binding globulin-bound testosterone during prepuberty in boys. *Journal of Steroid Biochemistry*, **27**, 291–5.

Belgorosky, A., & Rivarola, M. (1988). Progressive increase in nonsex hormone-binding globulin-bound testosterone and estradiol from infancy to late prepuberty in girls. *Journal of Clinical Endocrinology and Metabolism*, **67**, 234–7.

Berman, C., Rasmussen, K., & Suomi, S. (1994). Responses of free-ranging rhesus monkeys to a natural form of social separation. I. Parallels with mother-infant separation in captivity. *Child Development*, **65**, 1028–41.

Bielicki, T. (1986). Physical growth as a measure of the economic well-being of populations: The twentieth century. In *Human Growth: A Comprehensive Treatise*, ed. F. Falkner & J. Tanner, 2nd edn, New York: Plenum.

Bolton, N., Tapanainen, J., Koiviston, M., & Vihko, R. (1989). Circulating sex hormone-binding globulin and testosterone in newborns and infants. *Clinical Endocrinology*, **31**, 201–7.

Boyce, W., Adams, S., Tschann, J., Cohen, F., Wara, D., & Gunnar, M. (1995*a*). Adrenocortical and behavioral predictors of immune responses to starting school. *Pediatric Research*, **38**, 1009–17.

Boyce, W., Chesney, M., Alkon, A., Tschann, J., Adams, S., Chesterman, B., Cohen, F., Kaiser, P., Folkman, S., & Wara, D. (1995*b*). Psychologic reactivity to stress and childhood respiratory illnesses: Results of two prospective studies. *Psychosomatic Medicine*, **57**, 411–22.

Boyd, R., & Richardson, P. (1985). *Culture and the Evolutionary Process*. Chicago: University of Chicago Press.

Brabant, G., Brabant, A., Ranft, A., Ocran, K., Kohrle, J., Hesch, R., & Von Zur Muhlen, A. (1987). Circadian and pulsatile thyrotropin secretion in euthyroid man under the influence of thyroid hormone and glucocorticoid administration. *Journal of Clinical Endocrinology and Metabolism*, **65**, 83–8.

Brabant, G., Prank, K., Ranft, U., Schuermeyer, T., Wagner, T., Hauser, H., Kummer, B., Feistner, H., Hesch, R., & Von Zur Muhlen, A. (1990). Physiological regulation of circadian and pulsatile thyrotropin secretion in normal man and woman. *Journal of Clinical Endocrinology and Metabolism*, **70**, 403–9.

Breen, R. (1996). *Regression Models: Censored, Sample-selected, or Truncated Data*. Thousand Oaks: Sage Publications.

Brook, C., Hindmarch, P., & Stanhope, R. (1988). Growth and growth hormone secretion. *Journal of Endocrinology*, **119**, 179–84.

Burger, H., Yamada, Y., Bangah, M., McCloud, P., & Warne, G. (1991). Serum

gonadotropin, sex steroid, and immunoreactive inhibin levels in the first two years of life. *Journal of Clinical Endocrinology and Metabolism*, **72**, 682–5.
Cairns, R., Gariepy, J-L., & Hood, K. (1990). Development, microevolution, and social behavior. *Psychological Review*, **97**, 49–65.
Campbell, K. (1994). Blood, urine, saliva and dip-sticks: Experiences in Africa, New Guinea, and Boston. *Annals of the New York Academy of Science*, **709**, 312–30.
Cemeroglu, A., Foster, C., Warner, R., Kletter, G., Marshall, J., & Kelch, R. (1996). Comparison of the neuroendocrine control of pubertal maturation in girls and boys with spontaneous puberty and in hypogonadal girls. *Journal of Clinical Endocrinology and Metabolism*, **81**, 4352–7.
Chan, V., Jones, A., Liendo-Ch, P., McNeilly, A., Landon, J., & Besser, J. (1978). The relationship between circadian variations in circulating thyrotrophin, thyroid hormones and prolactin. *Clinical Endocrinology* **9**, 337–49.
Changeux, J-P. (1985). *Neuronal Man*. New York: Pantheon.
Chanoine, J., Rebuffat, E., Kahn, A., Bergmann, P., & Van Vliet, G. (1995). Glucose, growth hormone, cortisol, and insulin responses to glucagon injection in normal infants. *Journal of Clinical Endocrinology and Metabolism*, **80**, 3032–5.
Charnov, E. (1993). *Life History Invariants: Some Explorations of Symmetry in Evolutionary Ecology*. Oxford: Oxford University Press.
Cheung, G., Thornton, J., Kuijper, J., Weigle, D., Cliton, D., & Stteiner, R. (1997). Leptin is a metabolic gate for the onset of puberty in the female rat. *Endocrinology*, **138**, 855–8.
Clark, P., Hindmarsh, P., Shiell, A., Law, C., Honour, J., & Barker, D. (1996). Size at birth and adrenocortical function in childhood. *Clinical Endocrinology*, **45**, 721–6.
Clarke, A., Wittwer, D., Abbott, D., & Schneider, M. (1994). Long-term effects of prenatal stress on HPA axis activity in juvenile rhesus monkeys. *Developmental Psychobiology*, **27**, 257–69.
Costello, E., & Angold, A. (1995). Developmental epidemiology. In *Developmental Psychopathology*, ed. D. Cicchetti & D. Cohen, pp. 23–56. New York: John Wiley.
Cunningham, S., Loughlin, T., Culliton, M., & McKenna, T. (1985). The relationship between sex steroids and sex-hormone-binding globulin in plasma in physiological and pathological conditions. *Annals of Clinical Biochemistry*, **22**, 489–97.
de Bellis, M., Dahl, R., Perel, J., Birmaher, B., al-Shabbout, M., Williamson, D., Nelson, B., & Ryan, N. (1996). Noctural ACTH, cortisol, growth hormone, and prolactin secretion in prepubertal depression. *Journal of the American Academy of Child and Adolescent Psychiatry*, **35**, 1130–8.
de Zegher, F., Vanhole, C., Van den Berghe, G., Devlieger, H., Eggermont, E., & Veldhuis, J. (1994). Properties of thyroid-stimulating hormone and cortisol secretion by the human newborn on the day of birth. *Journal of Clinical Endocrinology and Metabolism*, **79**, 576–81.
Delange, F. (1994). The disorders induced by iodine deficiency. *Thyroid*, **4**, 107–28.
Dennison, E., Fall, C., Cooper, C., & Barker, D. (1997). Prenatal factors influencing long-term outcome. Hormone Research, **40**, 25–9.
Desjarlais, R., Eisenberg, L., Good, B., & Kleinman, A. (1995). *World Mental Health: Problems and Priorities in Low-income Countries*. New York: Oxford University Press.

Dubos, R. (1966). Man and his environment – Biomedical knowledge and social action. *Perspectives in Biology and Medicine*, **9**, 523–36.
Durham, W. (1991). *Coevolution: Genes, Culture, and Human Diversity*. Stanford, CA: Stanford University Press.
Ebeling, P., & Koivisto, V. (1994). Physiological importance of dehydroepiandrosterone. *The Lancet*, **343**, 1479–81.
Edelman, G. (1987). *Neural Darwinism: The Theory of Neuronal Group Selection*. New York: Basic Books.
Ellison, P. (1988). Human salivary steroids: Methodological considerations and applications in physical anthrology. *Yearbook of Physical Anthropology*, **31**, 115–42.
Ellison, P. (1993). Measurement of salivary progesterone. *Annals of the New York Academy of Science*, **694**, 161–76.
Ellison, P., Panter-Brick, C., Lipson, S., & O'Rourke, M. (1993). The ecological context of human ovarian function. *Human Reproduction*, **8**, 2248–58.
Erfurth, E., Hagmar, L., Saaf, M., & Hall, K. (1996). Serum levels of insulin-like growth factor I and insulin-like growth factor-binding protein 1 correlate with serum free testosterone and sex hormone binding globulin levels in healthy young and middle-aged men. *Clinical Endocrinology*, **44**, 659–64.
Eveleth, P., & Tanner, J. (1990). *Worldwide Variation in Human Growth*. New York: Cambridge University Press.
Falkner, F., & Tanner, J. (1985). *Human Growth: A Comprehensive Treatise*. New York: Plenum.
Fall, C., Pandit, A., Law, C., Yajnik, C., Clark, P., Breier, B., Osmond, C., Shiell, A., Gluckman, P., & Barker, D. (1995). Size at birth and plasma insulin-like growth factor-1 concentrations. *Archives of Diseases of Childhood*, **73**, 287–93.
Favier, A. (1992). Hormonal effects of zinc on growth in children. *Biological Trace Element Research*, **32**, 383–98.
Field, A., Colditz, G., Willett, W., Longcope, C., & McKinlay, J. (1994). The relation of smoking, age, relative weight, and dietary intake to serum adrenal steroids, sex hormones, and sex hormone-binding globulin in middle-aged men. *Journal of Clinical Endocrinology and Metabolism*, **79**, 1310–16.
Finch, C., & Rose, M. (1995). Hormones and the physiological architecture of life history evolution. *The Quarterly Review of Biology*, **70**, 1–52.
Genazzani, A., Petraglia, F., Sgarbi, L., Montanini, V., Hartmann, B., Surico, N., Biolcati, A., Volpe, A., & Genazzani, A. (1997). Difference of LH and FSH secretory characteristics and degree of concordance between postmenopausal and aging women. *Maturitas: Journal of the Climacteric and Postmenopause*, **26**, 133–8.
Ghai, K., & Rosenfield, R. (1994). Maturation of the normal pituitary-testicular axis, as assessed by gonadotropin-releasing hormone agonist challenge. *Journal of Clinical Endocrinology and Metabolism*, **78**, 1336–40.
Gilbert, S., Opitz, J., & Raff, R. (1996). Resynthesizing evolutionary and developmental biology. *Developmental Biology*, **173**, 357–72.
Gõnez, C., Villena, A., & Gonzales, G. (1993). Serum levels of adrenal androgens up to adrenarche in Peruvian children living at sea level and at high altitude. *Journal of Endocrinology*, **136**, 517–23.
Goodman-Gruen, D., & Barrett-Connor, E. (1996). A prospective study of sex hormone-binding globulin and fatal cardiovascular disease in Rancho Bernardo men and women. *Journal of Clinical Endocrinology and Metabolism*, **81**, 2999–3003.

Gottlieb, G. (1991). Experimential canalization of behavioral development: theory. *Developmental Psychology*, **27**, 4–13.
Gould, S. (1977). *Ontogeny and Phylogeny*. Cambridge: MA: Harvard University Press.
Grosvenor, C., Picciano, M., & Baumrucker, C. (1993). Hormones and growth factors in milk. *Endocrine Reviews*, **14**, 710–28.
Guazzo, E., Kirkpatrick, P., Goodyer, I., Shiers, H., & Herbert, J. (1996). Cortisol, dehydroepiandrosterone (DHEA), and DHEA sulfate in the cerebrospinal fluid of man: Relation to blood levels and the effects of age. *Journal of Clinical Endocrinology and Metabolism*, **81**, 3951–60.
Gunnar, M. R., Tout, K., de Haan, M., Pierce, S., & Stansbury, K. (1997). Temperament, social competence, and adrenocortical activity in preschoolers. *Developmental Psychobiology*, **31**, 65–85.
Hales, C., Barker, D., Clark, P., Cos, L., Fall, C., Osmond, C., & Winter, P. (1991). Fetal and infant growth and impaired glucose tolerance at age 64. *British Medical Journal*, **303**, 1019–22.
Harris, T., Kiel, D., Roubenoff, R., Langlois, J., Hannan, M., Havlik, R., & Wilson, P. (1997). Association of insulin-like growth factor-I with body composition, weight history, and past health behaviors in the very old: The Framingham Heart Study. *Journal of the American Geriatric Society*, **45**, 133–9.
Hartman, M., Veldhuis, J., Johnson, M., Lee, M., Alberti, K., Samojlik, E., & Thorner, M. (1992). Augmented growth hormone (GH) secretory burst frequency and amplitude mediate enhanced GH secretion during a two-day fast in normal men. *Journal of Clinical Endocrinology and Metabolism*, **74**, 757–65.
Hesse, V., Jahreis, G., Schambach, H., Vogel, H., Vilser, C., Seewald, H., Borner, A., & Deichl, A. (1994). Insulin-like growth factor I correlations to changes of the hormonal status in puberty and age. *Experimental and Clinical Endocrinology*, **102**, 289–98.
Hetzel, B., Potter, B., & Dulberg, E. (1990). The iodine deficiency disorders: nature, pathogenesis, and epidemiology. In *Aspects of Some Vitamins, Minerals, and Enzymes in Health and Disease*, ed. G. Bourne, pp. 59–112. Basel: Karger.
Higley, J., Suomi, S., & Linnoila, M. (1991). CSF monoamine metabolite concentrations vary according to age, rearing, and sex, and are influenced by the stressor of social separation in rhesus monkeys. *Pharmacology*, **103**, 551–6.
Higley, J., Thompson, W., Champoux, M., Goldman, D., Hasert, M., Kraemer, G., Scanland, J., Suomi, S., & Linnoila, M. (1993). Paternal and maternal genetic and environmental contributions to cerebrospinal fluid monoamine metabolites in rhesus monkeys (*Macaca mulatta*). *Archives of General Psychiatry*, **50**, 615–23.
Hill, K., & Hurtado, A. (1995). *Ache Life History: The Ecology and Demography of a Foraging People*. New York: Aldine de Gruyter.
Ho, K., Evans, W., Blizzard, R., Veldhuis, J., Merriam, G., Samojlik, E., Furlanetto, R., Rogol, A., Kaiser, D., & Thorner, M. (1987). Effects of sex and age on the 24-hour profile of growth hormone secretion in man: Importance of endogenous estradiol concentrations. *Journal of Clinical Endocrinology and Metabolism*, **64**, 51–8.
Hofer, M. (1996). On the nature and consequences of early loss. *Psychosomatic*

Medicine, **58**, 570–81.

Horrocks, P., Jones, A., Ratcliffe, W., Holder, G., White, A., Holder, R., Ratcliffe, J., & London, D. (1990). Patterns of ACTH and cortisol pulsatility over twenty-four hours in normal males and females. *Clinical Endocrinology*, **32**, 127–34.

Jenkins, C. (1987). Medical anthropology in the Western Schrader Range, Papua New Guinea. *National Geographic Research*, **3**, 412–30.

Jenkins, C., Dimitrakakis, M., Cook, E., Sanders, R., & Stallman, N. (1989). Culture change and epidemiological patterns among the Hagahai, Papua New Guinea. *Human Ecology*, **17**, 27–57.

Jenkins, C., Orr-Ewing, A., & Heywood, P. (1984). Cultural aspects of early childhood growth and nutrition among the Amele of lowland Papua New Guinea. *Ecology of Food and Nutrition*, **14**, 261–75.

Johnston, T. (1987). Developmental explanation and the ontogeny of birdsong: Nature/nurture redox. *Behavioral and Brain Sciences*, **11**, 617–63.

Jorgensen, J., Blum, W., Møller, N., Ranke, M., & Christiansen, J. (1990). Circadian patterns of serum insulin-like growth factor (IGF) II and IGF binding protein 3 in growth hormone-deficient patients and age- and sex-matched normal subjects. *Acta Endocrinologica* (*Copenh.*), **123**, 257–62.

Juul, A., Bang, P., Hertel, N., Main, K., Dalgaard, P., Jorgensen, K., Muller, J., Hall, K., & Skakkebaek, N. (1994*a*). Serum insulin-like growth factor-I in 1030 healthy children, adolescents, and adults: Relation to age, sex, stage of puberty, testicular size, and body mass index. *Journal of Clinical Endocrinology and Metabolism*, **78**, 744–52.

Juul, A., Dalgaard, P., Blum, W., Bang, P., Hall, K., Michalesen, K., Muller, J., & Skakkebaek, N. (1995). Serum levels of insulin-like growth factor (IGF)-binding protein-3 (IGFBP-3) in healthy, infants, children, and adolescents: the relation of IGF-I, IGF-II, IGFBP-1, IGFBP-2, age, sex, body mass index, and pubertal maturation. *Journal of Clinical Endocrinology and Metabolism*, **80**, 2534–42.

Juul, A., Main, K., Blum, W., Lindholm, J., Ranke, M., & Skakkebæk, N. (1994*b*). The ratio between serum levels of insulin-like growth factor (IGF-1) and the IGF binding proteins (IGFBP-1, 2 and 3) decreases with age in healthy adults and is increased in acromegalic patients. *Clinical Endocrinology*, **41**, 85–93.

Kelly, J., Rajkovic, I., O'Sullivan, A., Sernia, C., & Ho, K. (1993). Effects of different oral oestrogen formulations on insulin-like growth factor-I, growth hormone and growth hormone binding protein in post-menopausal women. *Clinical Endocrinology*, **39**, 561–7.

Kerrigan, J., & Rogol, A. (1992). The impact of gonadal steroid hormone action on growth hormone secretion during childhood and adolescence. *Endocrine Reviews*, **13**, 281–98.

Klein, K., Martha, P., Blizzard, R., Herbst, T., & Rogol, A. (1996). A longitudinal assessment of hormonal and physical alterations during normal puberty in boys II. Estrogen levels as determined by an ultrasensitive bioassay. *Journal of Clinical Endocrinology and Metabolism*, **81**, 3203–7.

Knutsson, U., Dahlgren, J., MArcus, C., Rosberg, S., Bronnegard, M., Stierna, P., & Albertsson-Wikland, K. (1997). Circadian cortisol rhythms in healthy boys and girls: Relationship with age, growth, body composition, and pubertal development. *Journal of Clinical Endocrinology and Metabolism*, **82**, 536–40.

Kwekkeboom, D., deJong, F., van Hemert, A., Vandenbroucke, J., Valkenburg,

H., & Lamberts, S. (1990). Serum gonadotropins and α-subunit decline in aging normal postmenopausal women. *Journal of Clinical Endocrinology and Metabolism*, **70**, 944–50.
Lampl, M., Veldhuis, J., & Johnson, M. (1992). Saltation and stasis: A model of human growth. *Science*, **258**, 801–3.
Landauer, T., & Whiting, J. (1981). Correlates and consequences of stress in infancy. In *Handbook of Cross-Cultural Research in Human Development*, ed. R. Munroe, R. Munroe, & B. Whiting, pp. 355–75. New York: Garland Press.
Law, C., Gordon, G., Shiell, A., Barker, D., & Hales, C. (1995). Thinness at birth and glucose tolerance in seven-year-old children. *Diabetic Medicine*, **12**, 24–9.
Leblhuber, F., Neubauer, C., Peichl, M., Reisecker, F., Steinparz, F., Windhager, E., & Dienstl, E. (1993). Age and sex differences of dehydroepiandrosterone sulfate (DHEAS) and cortisol (CRT) plasma levels in normal controls and Alzheimer's disease (AD). *Psychopharmacology*, **111**, 23–6.
Lee, S., Lenton, E., Sexton, L., & Cooke, I. (1988). The effect of age on the cyclical patterns of plasma LH, FSH, oestradiol and progesterone in women with regular menstrual cycles. *Human Reproduction*, **3**, 851–5.
Loche, S., Cherubini, V., Bartolotta, E., Lampis, A., Carta, D., Tomasi, P., & Pintor, C. (1994). Pulsatile secretion of thyrotropin in children. *Journal of Endocrinological Investigation*, **17**, 189–93.
Mantagos, S., Koulouris, A., Makri, M., & Vagenakis, A. (1992). Development of thyrotropin circadian rhythm in infancy. *Journal of Clinical Endocrinology and Metabolism*, **74**, 71–4.
Mariotti, S., Barbesino, G., Caturegli, P., Bartalena, L., Sansoni, P., Fagnoni, F., Monte, D., Fagiolo, U., Franceschi, C., & Pinchera, A. (1993). Complex alteration of thyroid function in healthy centenarians. *Journal of Clinical Endocrinology and Metabolism*, **77**, 1130–4.
Marshall, W. & Tanner, J. (1969). Variation in pattern of pubertal changes in girls. *Archives of Diseases of Childhood*, 44.
Marshall, J., & M. Griffen. (1993). The role of changing pulse frequency in the regulation of ovulation. *Human Reproduction*, **8**, 57–61.
Martha Jr., P., Rogol, A., Carlsson, L., Gesundheit, N., & Blizzard, R. (1993). A longitudinal assessment of hormonal and physical alterations during normal puberty in boys. I. Serum growth hormone-binding protein. *Journal of Clinical Endocrinology and Metabolism*, **77**, 452–7.
Martorell, R., Ramakrishnan, U., Schroeder, D. G., & Ruel, M. (1996). Reproductive performance and nutrition during childhood. *Nutrition Reviews*, **54**, S15–21.
Mascie-Taylor, C. (1991). Biosocial influences on stature: A review. *Journal of Biosocial Science*, **23**, 113–28.
Mauras, N., Rogol, A., Haymond, M., & Veldhuis, J. (1996). Sex steroids, growth hormone, and insulin-like growth factor-1: Neuroendocrine and metabolic regulation in puberty. *Hormone Research*, **45**, 74–80.
Mavoungou, D., Gass, R., Emane, M., Cooper, R., & Roth-Meyer, C. (1986). Plasma dehydroepiandrosterone, its sulfate, testosterone and FSH during puberty of African children in Gabon. *Journal of Steroid Biochemistry* **24**, 645–51.
Mayer, K. & Tuma, N. (1990). *Event History Analysis in Life Course Research*. Madison, Wisconsin: University of Wisconsin Press.
McDade, T., Angold, A., Costello, E., Stallings, J., & Worthman, C. (1995). Physiologic bases of individual variation in pubertal timing and progression:

Report from the Great Smoky Mountains Study. In *American Journal of Physical Anthropology, Suppl.* **20**, 148.
McEwen, B. (1997). Hormones as regulators of brain development: Life-long effects related to health and disease. *Acta Paediatrica, Suppl.*, **422**, 41–4.
Mendlewicz, J., Linkowski, P., Kerkhofs, M., Desmedt, D., Golstein, J., Copinschi, G., & van Cauter, E. (1985). Diurnal hypersecretion of growth hormone in depression. *Journal of Clinical Endocrinology and Metabolism*, **60**, 505–12.
Morales, A., Holden, J., & Murphy, A. (1992). Pediatric and adolescent gynecologic endocrinology. *Current Opinion in Obstetrics and Gynecology*, **4**, 860–6.
Mulligan, T., Iranmanesh, A., Gheorghiu, S., Godschalk, M., & Veldhuis, J. (1995). Amplified noctural luteinizing hormone (LH) secretory burst frequency with selective attenuation of pulsatile (but not basal) testosterone secretion in healthy aged men: Possible Leydig cell desensitization to endogeonous LH signaling – a clinical research center study. *Journal of Clinical Endocrinology and Metabolism*, **80**, 3025–31.
Nelson, J., Clark, S., Borut, D., Tomei, R., & Carlton, E. (1993). Age-related changes in serum free thyroxine during childhood and adolescence. *Journal of Pediatrics*, **123**, 899–905.
Onishi, S., Miyazawa, G., Nishimura, Y., Sugiyama, S., Yamakawa, T., Inagaki, H., Katoh, T., Itoh, S., & Isobe, K. (1983). Postanatal development of circadian rhythm in serum cortisol levels in children. *Pediatrics*, **72**, 399–404.
Orentreich, N., Brind, J., Rizer, R., & Vogelman, J. (1984). Age changes and sex differences in serum dehydroepiandrosterone sulfate concentrations throughout adulthood. *Journal of Clinical Endocrinology and Metabolism*, **59**, 551–5.
Oyama, S. (1985). *The Ontogeny of Information: Developmental Systems and Evolution.* Cambridge: Cambridge University Press.
Panter-Brick, C., Worthman, C., Lunn, P., Baker, R., & Todd, A. (1996). Urban, rural, and class differences in biological markers of stress among Nepali children. *American Journal of Human Biology* Suppl., **8**, 126.
Parker, L. (1991a). Control of adrenal androgen secretion. *Endocrinology and Metabolism Clinics of North America*, **20**, 401–21.
Parker, L. (1991*b*). Adrenarche. *Endocrinology and Metabolism Clinics of North America*, **20**, 71–83.
Phillips, D., Barker, D., & Osmond, C. (1993). Infant feeding, fetal growth and adult thyroid function. *Acta Endocrinologica*, **129**, 134–8.
Phillips, D., Barker, D., Hales, C., Hirst, S., & Osmond, C. (1994). Thinness at birth and insulin resistance in adult life. *Diabetologia*, **37**, 150–4.
Phillips, G. (1996). Relationship between serum dehydroepiandrosterone sulfate, androstenedione, and sex hormones in men and women. *European Journal of Endocrinology*, **134**, 201–6.
Politt, E., Gorman, K., Engle, P., Rivera, J., & Martorell, R. (1995). Nutrition in early life and the fulfillment of intellectual potential. *Journal of Nutrition*, **125**, 1111S–18S.
Raff, R. (1996). *The Shape of Life: Genes, Development, and the Evolution of Animal Form.* Chicago: Chicago University Press.
Ramsay, D., & Lewis, M. (1995). The effects of birth condition on infants' cortisol response to stress. *Pediatrics*, **95**, 546–9.
Reame, N., Kelch, R., Beitins, I., Yu, M-Y., Zawacki, C., & Padmanabhan, V. (1996). Age effects on follicle-stimulating hormone and pulsatile luteinizing

hormone secretion across the menstrual cycle of premenopausal women. *Journal of Clinical Endocrinology and Metabolism*, **81**, 1512–18.

Rivera, J. A., Martorell, R., Ruel, M. T., Habicht, J. P., Haas, J. D. (1995). Nutritional supplementation during the preschool years influences body size and composition of Guatemalan adolescents. *Journal of Nutrition*, **125**, 1068S–77S.

Rose, S., Giovanna, M., Barnes, K., Kamp, G., Uriarte, M., Ross, J., Cassoria, F., & Cutler, G. Jr. (1991). Spontaneous growth hormone secretion increases during puberty in normal girls and boys. *Journal of Clinical Endocrinology and Metabolism*, **73**, 428–35.

Runnels, B., Garry, P., Hunt, W., & Standefer, J. (1991). Thyroid function in a healthy elderly population: Implications for clinical evaluation. *Journal of Gerontology: Biological Sciences*, **46**, B39–44.

Sapolsky, R. M. (1992). *Stress, the Aging Brain, and the Mechanisms of Neuron Death*. Cambridge, MA: MIT Press.

Schroeder, D. & Brown, K. (1994). Nutritional status as a predictor of child survival: Summarizing the association and quantifying its global impact. *Bulletin of the World Health Organization*, **72**, 569–79.

Seeman, T., & McEwen, B. (1996). Impact of social environment characteristics on neuroendocrine regulation. *Psychosomatic Medicine*, **58**, 459–71.

Sherman, B., Wysham, C., & Pfohl, B. (1985). Age-related changes in the circadian rhythm of plasma cortisol in man. *Journal of Clinical Endocrinology and Metabolism*, **61**, 439–43.

Stansbury, K. & Gunnar, M. R. (1994). Adrenocortical activity and emotion regulation. *Monographs of the Society for Research in Child Development*, **59**, 108–34.

Stearns, S. (1992). *The Evolution of Life Histories*. New York: Oxford University Press.

Stearns, S., & Koella, J. (1986). The evolution of phenotypic plasticity in life-history traits: Predictions for norms of reaction for age- and size-at-maturity. *Evolution*, **40**, 893–913.

Stehling, O., Doring, H., Ertl, J., Preibisch, G., & Schmidt, I. (1996). Leptin reduces juvenile fat stores by altering the circadian cycle of energy expenditure. *American Journal of Physiology*, **271**, R1770–4.

Sterelny, K., Smith, K., & Dickison, M. (1996). The extended replicator. *Biology and Philosophy*, **11**, 377–403.

Stratakis, C., Mastorakos, G., Magiakou, M., Papavasiliou, E., Panitsa-Faflia, C., Georgiadis, E., & Batrinos, M. (1996). 24-hour secretion of growth hormone (GH), insulin-like growth factors-I and -II (IGF-I,-II), prolactin (PRL) and thyrotropin (TSH) in young adults of normal and tall stature. *Endocrine Research*, **22**, 261–76.

Suomi, S. (1991). Early stress and adult emotional reactivity in rhesus monkeys. In *The Childhood Environment and Adult Disease*, ed. G. Bock & J. Whelan. Chichester, NY: Wiley.

Tietz, N., Shuey, D., & Wekstein, D. (1992). Laboratory values in fit aging individuals – sexagenarians through centenarians. *Clinical Chemistry*, **38**, 1167–85.

Tiwari, B., Godbole, M., Chattopadhyay, N., Mandal, A., & Mithal, A. (1996). Learning disabilities and poor motivation to achieve due to prolonged iodine deficiency. *American Journal of Clinical Nutrition*, **63**, 782–6.

Tse, W., Hindmarsh, P., & Brook, C. (1989). The infancy–childhood–puberty

model of growth: Clinical aspects. *Acta Paediatrica Scandinavica, Suppl.*, **356**, 38–43.
van Cauter, E., & Plat, L. (1996). Session II: Spontaneous growth hormone secretion and the diagnosis of growth hormone deficiency; physiology of growth hormone secretion during sleep. *Journal of Pediatrics*, **128**, 32S–7S.
van Coevorden, A., Laurent, E., Decoster, C., Kerkhofs, M., Neve, P., Van Cauter, E., & Mockel, J. (1989). Decreased basal and stimulated thyrotropin secretion in healthy elderly men. *Journal of Clinical Endocrinology and Metabolism*, **69**, 177–85.
van Coevorden, A., Mockel, A., Laurent, E., Kerkhofs, M., L'Hermite-Baleriaux, M., Decoster, C., Neve, P., & Van Cauter, E. (1991). Neuroendocrine rhythms and sleep in aging men. *American Journal of Physiology*, **260**, E651–61.
van Eck, M., Berkhof, H., Nicolson, N., & Sulon, J. (1996). The effects of perceived stress, traits, mood states, and stressful daily events on salivary cortisol. *Psychosomatic Medicine*, **58**, 447–58.
Veldhuis, J. (1996*a*). Gender differences in secretory activity of the human somatotropic (growth hormone) axis. *European Journal of Endocrinology*, **134**, 287–95.
Veldhuis, J. (1996*b*). Neuroendocrine mechanisms mediating awakening of the human gonadotropic axis in puberty. *Pediatric Nephrology*, **10**, 304–17.
Veldhuis, J., Iranmanesh, A., Johnson, M., & Lizarralde, G. (1990). Amplitude, but not frequency, modulation of adrenocorticotropin secretory bursts gives rise to the nyctohemeral rhythm of the corticotropic axis in man. *Journal of Clinical Endocrinology and Metabolism*, **71**, 452–63.
Veldhuis, J., Urban, R., Lizarralde, G., Johnson, M., & Iranmanesh, A. (1992). Attenuation of luteinizing hormone secretory burst amplitude as a proximate basis for the hypoandrogenism of healthy aging in men. *Journal of Clinical Endocrinology and Metabolism*, **75**, 707–13.
Vermeulen, A., Kaufman, J., & Giagulli, V. (1996). Influence of some biological indexes on sex hormone-binding globulin and androgen levels in aging or obese males. *Journal of Clinical Endocrinology and Metabolism*, **81**, 1821–6.
Wadhwa, P., Dunkel-Schetter, C., Chicz-DeMet, A., Porto, M., & Sandman, C. (1996). Prenatal psychosocial factors and the neuroendocrine axis in human pregnancy. *Psychosomatic Medicine*, **58**, 432–46.
Wallace, W., Crowne, E., Shalet, S., Moore, C., Gibson, S., Littley, M., & White, A. (1991). Episodic ACTH and cortisol secretion in normal children. *Clinical Endocrinology*, **34**, 215–21.
Wan Nazaimoon, W., Osman, A., Wu, L., & Khalid, B. (1996). Effects of iodine deficiency on insulin-like growth factor-I, insulin-like growth factor-binding protein-3 levels and height attainment in malnourished children. *Clinical Endocrinology*, **45**, 79–83.
Weltman, A., Weltman, J., Hartman, M., Abbott, R., Rogol, A., Evans, W., & Veldhuis, J. (1994). Relationship between age, percentage body fat, fitness, and 24-hour growth hormone release in healthy young adults: Effects of gender. *Journal of Clinical Endocrinology*, **78**, 543–8.
Wennick, J., Delemarre-van de Waal, H., Schoemaker, R., Blaauw, G., van den Braken, C., & Schoemaker, J. (1991). Growth hormone secretion patterns in relation to LH and estradiol secretion throughout normal female puberty. *Acta Endocrinologica (Copenh.)*, **124**, 129–85.
Wiedemann, G., Jonetz-Mentzel, L., & Panse, R. (1993). Establishment of

reference ranges for thyrotropin, triiodothyronine, thyroxine and free thyroxine in neonates, infants, children and adolescents. *European Journal of Clinical Chemistry and Clinical Biochemistry*, **31**, 277–8.
World Bank. 1993. *World Development Report* 1993: *Investing in Health*. New York: Oxford University Press.
Worthman, C. M., (1986). Developmental dysynchrony as normative experience: Kikuyu adolescents. In *School-Age Pregnancy and Parenthood*, ed. J. B. Lancaster & B. Hamburg, pp. 95–112. New York: Aldine.
Worthman, C. (1993). Bio-cultural interactions in human development. In *Juvenile Primates: Life History, Development and Behavior*, ed. M. Pereira & L. Fairbanks, pp. 339–58. Oxford: Oxford University Press.
Worthman, C. (1998). Evolutionary perspectives on the onset of puberty. In *Evolutionary Medicine*, ed. W. Trevathan, J. McKenna & E. Smith. New York: Oxford University Press. In press.
Worthman, C. & Stallings, J. (1994). Measurement of gonadotropins in dried blood spots. *Clinical Chemistry*, **40**, 448–53.
Worthman, C. & Stallings, J. (1997). Hormone measures in finger-prick blood spot samples: New field methods for reproductive endocrinology. *American Journal of Physical Anthropology*, **103**, 1–21.
Worthman, C., Beall, C., & Stallings, J. (1997*a*). Population differences in DHEAS across the lifespan: Implications for aging. *American Journal of Human Biology*, **9**, 149.
Worthman, C., Beall, C., & Stallings, J. (1997*b*). Population variation in reproductive function of men. *American Journal of Physical Anthropology*, Suppl., **24**, 246.
Worthman, C., & McDade, T. (1996). Anthropology meets immunology: A developmental-adaptationist model of immunocompetance. *American Journal of Physical Anthropology*, Suppl., **22**, 247–8.
Wu, F., Butler, G., Kelnar, C., Huhtaniemi, Y., & Veldhuis, J. (1996). Ontogeny of pulsatile gonadotropin releasing hormone secretion from midchildhood, through puberty, to adulthood in the human male: A study using deconvolution analysis and an ultrasensitive immunofluorometric assay. *Journal of Clinical Endocrinology and Metabolism*, **81**, 1798–805.
Wu, F., Butler, G., Kelnar, C., Stirling, H., & Huhtaniemi, I. (1991). Patterns of pulsatile luteinizing hormone and follicle-stimulating hormone secretion in prepubertal (midchildhood) boys and girls and patients with idiopathic hypogonadotrophic hypogonadism (Kallmann's syndrome): A study using an ultrasensitive time-resolved immunofluorometric assay. *Journal of Clinical Endocrinology and Metabolism*, **72**, 1229–37.
Zemel, B., Worthman, C., & Jenkins, C. (1993). Differences in endocrine status associated with urban-rural patterns of growth and maturation in Bundi (Gende-speaking) adolescents of Papua New Guinea. In *Urban Health and Ecology in the Third World*, ed. L. Schell, N. Smith, & A. Bilsborough, pp. 38–60. Cambridge: Cambridge University Press.

4

Family environment, stress, and health during childhood

MARK V. FLINN

4.1 Introduction

In a child's world, the family is of paramount importance. Throughout human evolutionary history, parents and close relatives provided calories, protection, and information necessary for survival, growth, health, social success, and eventual reproduction. The human brain is therefore likely to have evolved special sensitivity to interactions among family caretakers, particularly during infancy and early childhood. This special sensitivity for negotiating the psychosocial dynamic of the family environment has important consequences for child development.

Changing, unpredictable environments require adjustment of priorities. Growth, immunity, digestion, and sex are irrelevant when being chased by a predator (Sapolsky, 1994), or when coping with a traumatic social event. Emergencies, large and small, good and bad, perceived by the brain stimulate a variety of neuroendocrine systems. Hundreds of different endogenous chemicals, hormones, neurotransmitters, cytokines, and so forth, are released from secretory glands and cells in response to information received and processed by the central nervous system (CNS). The movement of these chemicals in plasma and other intercellular fluids communicates information among the different cells and tissues, helping the body to respond appropriately to varying environmental demands.

Physiological stress responses affect the allocation of energetic and other somatic resources to different bodily functions via a complex assortment of neuroendocrine mechanisms. Stress hormones help shunt blood, glucose, cells of the immune system, and other resources to cells and tissues necessary for the task at hand. Chronic and traumatic stress can diminish long-term health, evidently because resources are diverted away from important health functions, including cellular repair, immune response, and brain expansion. Stress during childhood may be particularly taxing because of the additional demands of growth and development (Burns & Arnold 1990; Rutter 1991).

During the first few years of life the human brain roughly triples in size via an extraordinary spurt of cell growth, migration, specialization, remodeling, and pruning. The brain consumes almost half of the infant's resting caloric requirements. The thymus and other parts of the immune system undergo a similarly dramatic transformation, preparing defenses against a nearly infinite variety of pathogens, while selecting out responses to the numerous molecular fingerprints of self-tissues. Even under the best of circumstances successful outcome of these complex ontogenies would seem miraculous; the developing child, however, often faces a most imperfect environment. Emotional and physical stressors, such as abuse, neglect, parental divorce, strenuous work, exogenous toxins, inconsistent punishment, infectious disease, and malnutrition, are powerful stimulants of physiological stress response with potential effects on brain and immune development (e.g., Fukunaga *et al.*, 1992; Boyce *et al.*, 1995b; Maccari *et al.*, 1995; McEwen 1995, 1997). Even common, everyday activities may be important. Figure 4.1 illustrates some of the possible developmental pathways among a child's environment, stress perception, hormonal response, and health.

Extensive research on hormonal stress response has been conducted in clinical, experimental, school, and work settings (e.g., Arnold & Carnahan, 1990; Weiner, 1992; Haggerty *et al.*, 1994; Stansbury & Gunnar, 1994). We know relatively little, however, about the causes and effects of stress endocrinology among children in normal everyday ("naturalistic") environments, particularly in non-industrial societies (Panter-Brick, 1998). Investigation of childhood stress and its effects on development has been hampered by the lack of non-invasive techniques for measurement of stress hormones. Frequent collection of plasma samples in non-clinical settings is not feasible. The development of saliva immunoassay techniques (Walker *et al.*, 1978; Okamura *et al.*, 1997), however, presents new opportunities for stress research (e.g., Kirschbaum *et al.*, 1992; Hart *et al.*, 1996). Saliva is relatively easy to collect and store, especially under adverse field conditions faced by anthropologists (Ellison, 1988). Longitudinal monitoring of a child's daily activities, stress hormones, health, and psychological conditions provides a powerful research design for investigating the effects of naturally-occurring stressors on child health. Figure 4.2 illustrates how measures of the stress hormone cortisol may be used to assess children's responses to events over (*a*) hourly, (*b*) daily, and (*c*) yearly temporal periods.

Analyzing hormone levels from saliva is a useful tool for examining the

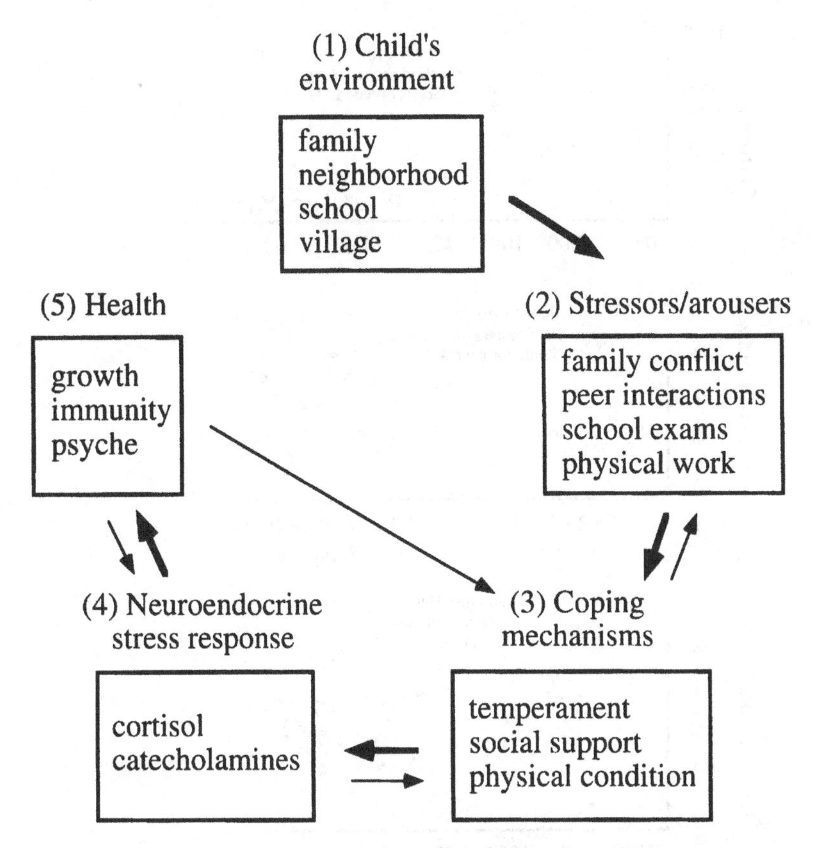

Figure 4.1. Relations among childhood environment, stressors, coping mechanisms, hormonal response, and health. The model suggests a predominantly linear flow from the production of stressors in a child's environment (1) and (2), perceived and modified by coping mechanisms (3), to the stimulation of neuroendocrine stress response (4), with subsequent effects on child health (5).

child's imperfect world and its developmental consequences, especially when accompanied by detailed ethnographic, medical, and psychological information. This chapter discusses how and why stress response may affect child health, and then reviews results from a study of stress hormones, family environment, and health among children living in a rural Caribbean village.

4.2 Stress response mechanisms and theory

Physiological responses to environmental stimuli that are cognitively perceived as "stressful" are modulated by a part of the brain termed the

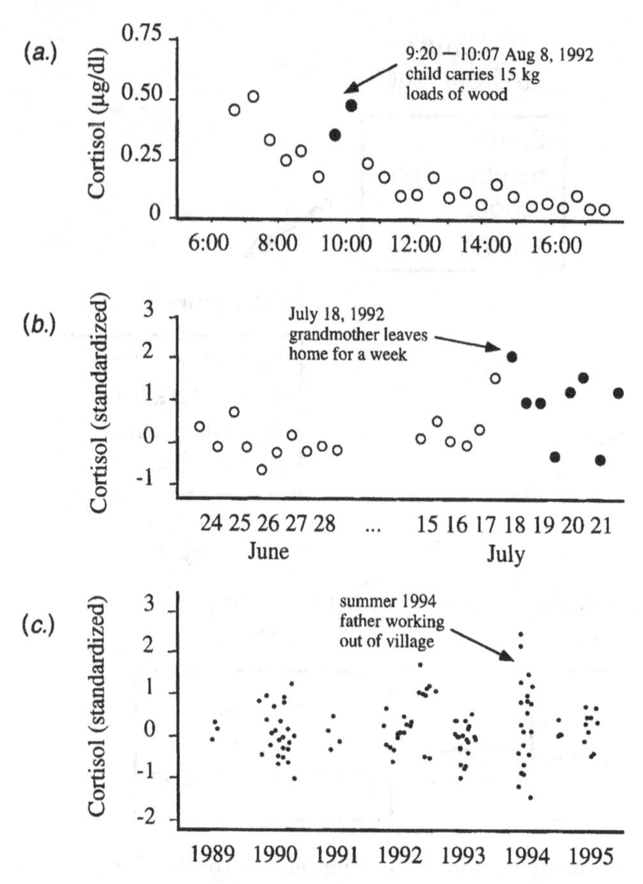

Figure 4.2. Longitudinal monitoring of cortisol levels as a tool for investigating stress response among children in a Caribbean village. (*a*) Hourly sampling of a 12-year-old male demonstrating elevation of cortisol levels associated with carrying heavy loads of wood. (*b*) Twice-daily sampling of a 13-year-old girl demonstrating change in pattern of cortisol levels associated with temporary absence of caretaking grandmother. (*c*) Twice-daily sampling over a 7-year period of a male born in 1985 demonstrating the change in pattern of cortisol levels associated with the absence of his father.

limbic system (amygdala and hippocampus) and basal ganglia which interact with the sympathetic and parasympathetic nervous systems and two neuroendocrine axes: the sympathetic adrenal–medullary system (abbreviated as SAM), and the hypothalamic–anterior pituitary–adrenal cortex system (abbreviated as HPA) via the hypothalamic paraventricular nucleus (Herman *et al.*, 1996). The SAM and HPA systems affect a wide range of physiological functions in concert with other neuroendocrine mechanisms and involve complex feedback regulation. The SAM system controls the catecholamines norepinephrine and epinephrine (adrenaline). The HPA system regulates glucocorticoids, primarily cortisol, which is normally released in seven to fifteen pulses during a 24-hour period (for reviews see Gray, 1987; Sapolsky, 1992a b; Weiner, 1992; McEwen, 1995; Sapolsky, this volume).

Cortisol is a key hormone produced in response to physical and psycho-social stressors. Cortisol modulates a wide range of somatic functions, including: (1) energy release (e.g., stimulation of hepatic gluconeogenesis in concert with glucagon and inhibition of some effects of insulin), (2) immune activity (e.g., control of inflammatory response and the cytokine cascade), (3) mental activity (e.g., alertness, memory, and learning), (4) growth (e.g., inhibition of growth hormone and somatomedins), and (5) reproductive function (e.g., inhibition of gonadal steroids, including testosterone). These complex, multiple effects of cortisol muddle understanding of its adaptive functions. The demands of energy regulation must orchestrate with those of immune function, and so forth. Receptor differentiation (Barrington, 1982) and other mechanisms enable localized targeting (e.g., glucose uptake by active *versus* inactive muscle tissues, neuropeptide directed immune response, modulation by antiglucocorticoids, and cell-specific interactions with macrophages) of the above general physiological effects (e.g., Calandra *et al.*, 1995; Dallman *et al.*, 1995; Vacchio *et al.*, 1996; Orchinik *et al.*, 1997). Cortisol regulation allows the body to respond to changing environmental conditions by preparing for *specific* short-term demands (Mason, 1971; Munck *et al.*, 1984).

These temporary beneficial effects of glucocorticoid stress response, however, are not without costs. Persistent activation of the HPA system is associated with immune deficiency, cognitive impairment, inhibited growth, delayed sexual maturity, damage to the hippocampus, and psychological maladjustment (Ader *et al.*, 1991; Glaser and Kiecolt-Glaser, 1994; Dunn, 1995; Sapolsky, 1996; McEwen & Magarinos, 1997). Chronic stress may diminish cellular energy (Sapolsky, 1991) and produce complications for autoimmune protection (Munck & Guyre, 1991; Buckingham *et al.*, 1996). Stressful life events, such as divorce, death of a family member, change of residence, or loss of a job, are associated with infectious disease and other health problems during adulthood (House *et al.*, 1988; Kaplan, 1991; Cohen *et al.*, 1993; Herbert & Cohen, 1993; Maier *et al.*, 1994).

Current psychosocial stress research suggests that cortisol response is stimulated by uncertainty that is perceived as significant and for which behavioral responses will have unknown effects (Fredrikson *et al.*, 1985;

Levine, 1993; Kirschbaum & Hellhammer, 1994). In a child's world, important events are going to happen, the child does not know how to react, but is highly motivated to figure out what should be done. Cortisol release is associated with unpredictable, uncontrollable events that require full alert readiness and mental anticipation. In appropriate circumstances, temporary moderate increases in stress hormones (and associated neurotransmitters such as dopamine) may enhance mental activity for short periods in localized areas and prime memory storage, hence improving cognitive processes for responding to social challenges (see Eysenck, 1982; Martignoni *et al.*, 1992; McEwen and Sapolsky, 1995; McGaugh *et al.*, 1996; Joels *et al.*, 1997). Mental processes unneccesary for appropriate response may be inhibited, perhaps to reduce external and internal "noise" (Servan-Schreiber *et al.*, 1990; see Newcomer *et al.*, 1994; Kirschbaum *et al.*, 1996; Lupien *et al.*, 1997). In addition, changes in the blood–brain barrier during psychosocial stress may produce autoimmune complications in the CNS requiring downregulation of immune function and protection from elevated levels of catecholamines.

Experimental studies that expose subjects to temporal stressors such as public speaking or parachute jumping reliably elevate stress hormones. Relations between cortisol production and emotional distress in natural settings, however, are difficult to assess because of temporal and inter-individual variation in HPA response (Tennes & Mason, 1982; Dabbs & Hopper, 1990; Kagan, 1992; Pollard, 1995; Nachmias *et al.*, 1996; Walker, 1996). Habituation may occur to repeated events for which a child or adult acquires an effective mental model. Apparently "stressful" job environments may not stimulate increased levels of stress hormones if individuals have adjusted to them (Walsh *et al.*, 1997). Expressions of behavioral distress (e.g., crying) among children are not reliably associated with elevated cortisol (Gunnar, 1992), and phobic individuals exhibit only moderate rises in cortisol during clinical phobic episodes (Nesse *et al.*, 1985). Attenuation and below-normal levels of cortisol may follow a day or more after emotionally charged events. Chronically stressed children may develop abnormal cortisol response, possibly via changes in binding globulin levels, and/or reduced affinity or density of glucocorticoid, CRH, and vasopressin receptors in the brain (see Cohen *et al.*, 1989; De Kloet, 1991; Fuchs & Flugge, 1995). Early experience, such as perinatal stimulation of rats (Meaney *et al.*, 1991; Takahashi, 1992), prenatal stress of rhesus macaques (Schneider *et al.*, 1992; Clarke, 1993: Coe *et al.*, 1992), prenatal exposure to cocaine among humans (Karmel & Gardner, 1996), maternal–infant attachment among humans (Spangler & Grossmann, 1993), and sexual abuse among humans (De Bellis *et al.*, 1994), may permanently alter HPA response (Insel 1990). And personality may affect HPA response (and vice versa), because children with inhibited temperaments tend to have higher cortisol levels than extroverted children (Kagan *et al.*, 1988; see Suomi, 1991; Gunnar, 1994; Higley & Suomi, 1996; Nachmias *et al.*, 1996).

Further complications arise from interaction between HPA stress response and a wide variety of other neuroendocrine activities, including modulation of catecholamines (SAM), melatonin, testosterone, serotonin, β-endorphins, cytokines, and enkephalins (Axelrod & Reisine, 1984; De Kloet, 1991; Sapolsky, 1990b, 1992b; Scapagnini, 1992; Saphier *et al.*, 1994). Changes in cortisol for energy allocation and modulation of immune function may be confused with effects of psychosocial stress. Cortisol may be a co-factor priming oxytocin and vasopressin intracerebral binding sites that are associated with familial attachment in mammals (Insel & Shapiro, 1992; Winslow *et al.*, 1993; Corter & Fleming, 1995), and hence may influence distress involving caretaker–child relationships. Other components of the HPA axis, such as corticotropin releasing hormone (CRH) and melanocyte stimulating hormone, have additional stress-related effects that are distinct from cortisol. Finally, a variety of hormones and other endogenous chemicals, including "antiglucocorticoids", mediate specific actions of cortisol. Concurrent monitoring of all these neuroendocrine activities would provide important information about stress response, but is not possible in a non-clinical setting with current techniques.

Relations between stress-induced cortisol response and immunosuppression are perhaps even more complex and enigmatic (Besedovsky & del Ray, 1991; Kapcala *et al.*, 1995; Biondi & Zannino, 1997). Stress is associated with a variety of illness, including infectious disease, reactivation of latent herpesvirus, cancer, and cardiovascular problems. The wide range of health effects of stress suggest that a number of immune mechanisms are involved. Cortisol influences many functions of lymphocytes, macrophages, and leukocytes, and as with energy use, may direct their movement to specific locations and even modulate apoptosis (Costas *et al.*, 1996). Cortisol also inhibits the production of some cytokines (e.g., interleukin 1) and mediates several components of the inflammatory response. In concert with the sympathetic system (SAM), which generally down-modulates lymphocyte and monocyte functions, HPA stress response affects all of the major components of the immune system. The effects of neuroendocrine stress response are not all inhibitory, and may involve temporary upregulation and/or localized enhancement of some immune functions (Jefferies, 1990). Teasing apart the effects of stress response on different parts of immune function (e.g., macrophage activity, neutrophilia, suppressor T cells, helper T cells, cytotoxic T cells, NK cells, B cells, and production of IgA, IgE, and IgM) in temporal context (e.g., before, during, 10 minutes after, 1 hour after, 1 day after, 1 week after, 1 year after, etc., a particular stress event) may help resolve some of the apparent paradoxes.

Munck *et al.* (1984) hypothesize that some effects of stress on health may be incidental consequences of the multiple regulatory functions of cortisol. As cortisol modulates (1) immune response, including protection from autoimmune reactions, (2) mental processes (e.g., energy allocation to the CNS and enhancement of neural circuits vital to flight–fight and some types of memory retention), and (3) protective responses to damaging effects of catecholamines, immunity could be inadvertently turned off during psycho-social stress. Alternatively, the increased permeability and flow rate across the blood-brain barrier during stress may require greater protection of the CNS from autoimmune and catecholamine effects.

Sapolsky (1990*a*, 1994) suggests that stress response involves an optimal allocation problem. Energy resources are diverted to mental and other short-term (stress emergency) functions, at cost to long-term functions of growth, development, and immunity. Under normal conditions of temporary stress, there would be little effect on health. Indeed, there may be brief enhancement and directed trafficking of immune function. Persistent stress and associated hyper- or hypo-cortisolemia, however, may result in pathological immunosuppression and depletion of energy reserves.

Assessment of relations among psychosocial stressors, hormonal stress response, and health is complex, requiring (1) longitudinal monitoring of social environment, emotional states, hormone levels, immune measures, and health, (2) control of extraneous effects from physical activity, circadian rhythms, and food consumption, (3) knowledge of individual differences in temperament, experience, and perception, and (4) awareness of cultural context. Anthropological research that integrates human biology and ethnography is particularly well suited to these demands (e.g., Dressler, 1996; Panter-Brick, 1998). Physiological and medical assessment in concert with ethnography and coresidence with children and their families in anthropological study populations can provide intimate, prospective, naturalistic information that is not feasible to collect in clinical studies.

The following sections of this chapter review a 10-year study of childhood stress, family environment, and health in a rural village on the east coast of Dominica. Data from this study indicate that abnormal hormone response profiles and frequent illness are associated with household composition and unstable marital relationships of parents/caretakers. Preliminary analyses of salivary immune measures suggest that stress has complex effects on immune function. These results suggest that family environment has important effects on childhood psychosocial stress and illness.

4.3 Studying childhood stress in a rural Caribbean village

In this study, longitudinal monitoring was used to identify associations between health and psychosocial stressors. Data analyses examined both long term (ten years) and short term (day-to-day, hour-by-hour) associations among cortisol levels, family composition, socio-economic conditions, behavioral activities, events, temperament, growth, medical history, immune measures, and illness.

4.3.1 Variables and measures used in the study

Physiological stress response was assessed by radioimmunoassay (RIA) of cortisol levels in saliva. Analyses included mean values, variation, and day-to-day and hour-by-hour profiles of standardized (circadian time control) cortisol data. *Family composition* was assessed by age, sex, genealogical relationship, and number of individuals in the caretaking household. *Socio-economic conditions* included household income, material possessions, land ownership, occupations, and educational attainment. *Caretaking attention* was assessed by (1) frequencies and types of behavioral interaction, (2) informant ratings of caretaking that children received, and (3) informant interviews. *Personality and temperament* were assessed by (1) culturally appropriate versions of psychological instruments (Buss & Plomin, 1984; Goldberg, 1992, 1993), (2) informant (peers, parents, teachers, neighbors) interview, and (3) behavioral observation. *Immune response* was assessed by RIA of neopterin and interleukins 1 and 8, turbidimetric immunoassay of secretory-immunoglobulin A, and microparticle enzyme immunoassay of microglobulin $\beta 2$ from saliva samples. *Health* was assessed by (1) observed type, frequency, and severity of medical problems (diarrhea, influenza, common cold, asthma, abrasions, rashes, etc.), (2) informant (parents, teachers, neighbors) ratings, (3) medical records, (4) growth (standard anthropometric measures, including height, weight, and skinfolds) patterns, and (5) physical examination by a medical doctor. The primary measure of health used here is *percentage of days ill*, the proportion of days that a child was observed (directly by researchers) with common benign temporary infectious disease (89% were common-cold upper respiratory tract infections with nasal discharge, cough, or myalgia, e.g., rhinovirus, adenovirus, parainfluenza, and influenza; 6% were diarrheal; 5% were miscellaneous indeterminate, e.g., febrile without other symptoms). *Daily activities* and *emotional states* were assessed from (1) caretaker and child self-report questionnaires, and (2) systematic behavioral observation (focal follow and instantaneous scan sampling). Multiple sources of information were cross-checked to assess reliability (Bernard *et al.*, 1984).

The primary focus of the following sections of this chapter is on relations among: cortisol levels, percentage of days ill, immune function, and family composition. Caretaking attention, socioeconomic conditions, temperament, daily activities, and emotional states are analyzed as secondary or control variables.

4.3.2 The study village

"Bwa Mawego" is a rural village located on the east coast of Dominica. About 750 residents live in 200 structures/households that are loosely clumped into five "hamlets" or neighborhoods. The population is of mixed African, Carib, and European descent. The village is isolated because it sits at the dead-end of a rough road passable by small trucks except for occasional periods during the rainy season. Part-time residence is common, with many individuals emigrating for temporary work to other parts of Dominica, the USA, or Canada. Most residents cultivate bananas and/or bay leaves as cash crops, and plantains, dasheen, and a variety of fruits and vegetables as subsistence crops. Fish are caught by free-diving with spear-guns and from small boats (hand-built wooden "canoes" of Carib design) using lines and nets. Land is communally "owned" by kin groups, but parcelled for long-term individual use.

Most village houses are strung close together along roads and tracks. Older homes are constructed of wooden planks and shingles hewn by hand from local forest trees; concrete block and galvanized roofing are more popular today. Most houses have one or two sleeping rooms, with the kitchen and toilet as outbuildings. Children usually sleep together on foam or rag mats. Wealthier households typically have "parlors" with sitting furniture. Electricity became available in 1988; during the summer of 1995 about 70% of homes had "current," 41% had telephones, 11% had refrigerators, and 7% had televisions. Water is obtained from streams, spring catchments, and run-off from roofs.

The village of Bwa Mawego is appropriate for the study of relations among psychosocial stress and child health for the following reasons: (1) there is substantial variability among individuals in the factors under study (socio-economic conditions, family environments, cortisol response, and health), (2) the village and housing are relatively open, hence behavior is easily observable, (3) kin tend to reside locally, (4) the number of economic variables is reduced relative to urban areas, (5) the language and culture are familiar to the investigator, (6) there are useful medical records, and (7) local residents welcome the research and are most helpful.

The study involved 264 individuals aged 2 months to 18 years residing in 82 households. This is a nearly complete sample (> 98%) of all children living in four of the five village hamlets during the period of fieldwork. More than 30 months of field research were conducted over a 10-year period (1988 to 1997).

4.3.3 Collecting saliva and assaying hormones

Information on socio-economic conditions, household environment, care-taking attention, temperament, and health was collected by standard ethnographic techniques including interviews, behavioral scans, participant observation, and questionnaire instruments. These methods are described in more detail in other publications (Flinn, 1988; Quinlan, 1995).

Data on physiological stress response are derived from RIA (see below) of cortisol from saliva samples. Salivary cortisol is a reliable measure of adrenal cortical function (Kirschbaum *et al.*, 1992). Saliva is easier, safer, and less expensive to collect than blood or urine, and is the fluid of choice for temporal measures of cortisol and a number of other hormones (Ellison, 1988; Dabbs, 1990, 1992) and immune products (e.g., secretory-IgA).

The cost of materials (polystyrene test tubes, disposable pipettes, plastic cups, tube labels, preservative solution of sodium azide in double-distilled water, stimulant such as chewing gum, plastic storage bags, parafilm for wrapping/sealing test tubes, and rubber bands for bundling test tubes) for collecting saliva is about US$ 0.30 per sample. Absorbant cotton collection devices that are placed in the mouth may be used in situations where subjects would be uncomfortable expectorating into plastic cups, but they add cost (about US$ 1.00 per sample). Additional materials (repetitive syringe dispenser for pipetting sodium azide solution into test tubes, foam test tube racks, test tube labels and data collection sheet printer, and storage box) add start-up costs of about US$ 500, but are optional and necessary only if large numbers (1000 or more) of saliva samples are to be collected. The cost of laboratory immunoassay processing varies considerably (approximately US$ 2.00 to 20.00 per sample).

In the study, saliva was collected by two routines. The primary routine was a twice-daily collection in which an anthropologist and research assistant walked set routes from house to house, once in the morning (5:30am – 9:00am), and once in the afternoon (3:00pm – 6:30pm). Most (16 652 of 22 438) saliva samples were collected this way. A secondary routine used a "focal follow" technique in which hourly saliva samples were collected (3424 of 22 438) from (1) older children from dawn until early afternoon or evening, or (2) infants from dawn until early afternoon. The hourly collection protocol provides much richer information; however, it is more invasive and costly in terms of collection labor. Saliva samples from some parents and other caretakers were collected at the same time as their children. As the majority of saliva samples were collected at the child's household during periods when family interactions were common (early morning and late afternoon), the effects of family environment on cortisol response may be accentuated.

The saliva collection protocol was as follows. First, children rinsed their mouths with fresh water. At this point, they were checked for oral bleeding. Both food and blood contamination may affect the integrity of samples (Ellison, 1988). Next, children were given one-quarter to half a stick of Wrigley's gum (spearmint) to stimulate saliva production. After chewing the gum for about 1 minute, saliva was deposited in disposable plastic cups for about 3 minutes. For infants, saliva was collected by swabbing with cotton rolls (Turner, 1995). Approximately 4 ml of saliva was pipetted into labeled (name, date, time, number) polystyrene test tubes and preserved using sodium azide and refrigeration. Analysis of cortisol levels requires precise information on *time of collection, time of waking up from sleep*, and *individual sleep schedule* because there is a circadian pattern to cortisol release (Fredrikson *et al.*, 1985; Van Cauter 1990). Hourly samples were taken for finer-grained analysis of individual differences in temporal fluctuation in cortisol levels. Daily activity, emotional state, and health questionnaires were administered concomitant with saliva collection (see below).

Children readily acclimated to the collection procedure. Shy or inhibited children in particular tended to have cortisol levels that were higher than normal for the first few days of saliva collection. Multiple samples (more than 50 days of morning and afternoon samples over several years) from each child provide much better information about stress response than a single collection design. A minimum of 15 samples per individual is recommended (see Coste *et al.*, 1994).

Cortisol levels in saliva samples were determined by standard RIA techniques: (1) Saliva was centrifuged, (2) 200 μl of saliva was pipetted into receiver tubes from DPC ^{125}I cortisol solid phase radioimmunoassay kits (each sample was duplicated), (3) radioactive label was added to the tubes, which were then vortexed, placed in 37°C water baths for 45 minutes, aspirated (twice), and run through a gamma counter. Data from the counter were analyzed with a statistical program (StatLIA). All samples were analyzed at the Ligand Assay Laboratory at the University of Michigan Hospitals.

Cortisol release follows a circadian pattern, with highest levels around waking up in the morning, diminishing to low levels during evening hours and a nadir just before or during sleep. For example, mean cortisol levels when children have been awake for 10 minutes is about 0.6 μg/dl, for 2 hours 0.2 μg/dl, and for 8 hours 0.07 μg/dl. To control for time (circadian) effects, raw cortisol data were standardized by 5-minute time intervals from wake-up time. Use of precise wake-up time is critically important for studies of cortisol because small differences among individuals in their sleep schedules (e.g., 6:30am *versus* 7:00am wake-up) result in large differences in morning cortisol levels. For each 5-minute time interval from wake-up time, mean values and standard deviations were computed. Standardized values were generated as time-controlled measures of cortisol response as with the following hypothetical example: the mean cortisol level of children that have been awake for 60 minutes = 0.3 μg/dl with standard deviation of 0.1 μg/dl. A cortisol measure of 0.4 μg/dl from this time period would have a standardized value of 1.0, i.e., 0.4 μg/dl is one standard deviation (0.1 μg/dl) from the average (mean) value of 0.3 μg/dl. This procedure allows comparison of cortisol values from saliva collected at different times. All cortisol data presented in this chapter are time standardized with the exception of Figure 4.2a.

During saliva collection, children and their caretakers were asked a series of questions concerning what activities the child did that day, and how the child felt during these activities. This self-report and caretaker-report information on daily activities was compared with behavioral observation data. Health evaluations were also conducted daily during saliva collection. If illness was indicated, body temperature was checked using an oral thermometer. Blood pressure was measured once a week.

Most individuals exhibit a slight rise (about 0.01 μg/dl at lunch, or 13%) in cortisol levels after eating a meal or drinking caffeinated beverages. This post-prandial rise is most significant for the mid-day meal. Underweight children (two standard deviations below NCHS weight-for-age references, Hamill *et al.*, 1997) had greater (27% versus 9%) post-prandial cortisol elevations than their heavier peers. Data were not adjusted for eating and caffeine intake because very few samples were taken during lunch time, the effect is small, and the occurrence of eating and caffeine intake is presumed random with regard to hypotheses tested. Individuals commonly have small elevations in cortisol levels during mid-day, usually in association with potential minor stressors (Holl *et al.*, 1984).

Most individuals show a rise in cortisol levels during and shortly after intensive physical exertion (e.g., carrying heavy loads). Physical exertion involving social interaction, such as competitive sports, is associated with more substantial elevation of cortisol levels, particularly for males. Data are not adjusted for physical activity because only a small proportion of samples were collected when children were physically exerted, it was difficult to determine how exerted children were, and such activities are presumed random with regard to hypotheses tested. As general activity levels are associated with cortisol, this presents a confounding effect. Some children may have increased cortisol response to psychosocial events when they are healthy, active, and have abundant energetic resources, compared to when they are inactive and have depleted energy reserves.

4.4 Stress response and family environment

Human infants and juveniles cannot survive, let alone develop effective social skills, without assistance from parents or other caretakers. Relationships within the caretaking household are essential for the developing child (Bornstein, 1995). Composition of the family or caretaking household may have important effects on child development. For example, in western cultures, children with divorced parents may experience more emotional tension or "stress" than children living in a stable two-parent family (Wallerstein, 1983; Pearlin & Turner, 1987; Emery, 1988). Family stress may subsequently affect health and psychological well-being (Gottman & Katz, 1989; Arnold & Carnahan, 1990), mediated in part by hormonal processes (Tennes, 1982).

Associations between average cortisol levels of children and household composition are presented in Figure 4.3. Children living with distant relatives, step-fathers and half-siblings (step-father has children by the step-child's mother), or single parents without kin support, had higher average levels of cortisol than children living with both parents, single

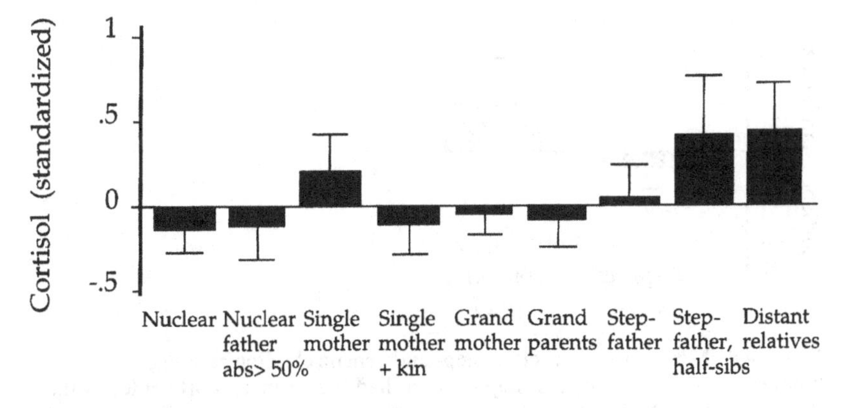

Figure 4.3. Household composition and cortisol. Vertical bars represent 95% confidence intervals (1.96 SE). Sample sizes (*N* of children, *N* of cortisol saliva assays) are: 89, 6905; 28, 2234; 30, 2296; 31, 2581; 32, 2645; 16, 1341; 5, 279; 24, 1870; 9, 482. (Figure adapted from Flinn & England, 1997.)

mothers with kin support, or grandparents. A further example is provided by comparison of step and genetic children residing in the same households (Figure 4.4). Step-children had higher average cortisol levels than their half-siblings residing in the same household who were genetic offspring of both parents.

Several caveats need emphasis. First, not all children in difficult family environments have elevated cortisol levels. Second, household composition is not a uniform indicator of family environment. Some single-mother households, for example, appear more stable, affectionate, and supportive than some two-parent households. Parenting "style" may be an important mediator (MacDonald, 1992). Third, children appear differentially sensitive to specific aspects of their caretaking environments, reflecting temperamental and other individual differences.

Consideration of these factors, however, does not invalidate the general association between household composition and child cortisol levels. There are several possible reasons underlying this result. Children in difficult caretaking environments may experience chronic stress resulting in moderate-high levels of cortisol (i.e., a child has cortisol levels that are above average day after day). They may experience more acute stressors that substantially raise cortisol for short periods of time. They may experience more frequent stressful events (e.g., parental chastisement or marital quarreling (see Wilson *et al.*, 1980; Daly & Wilson, 1988; Flinn,

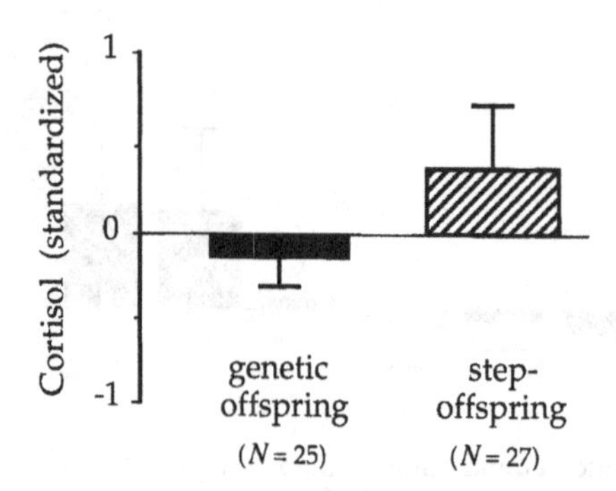

Figure 4.4. Mean cortisol levels of step- and genetic children residing in the same household. In 38 of 43 dyads, step children had higher mean cortisol levels than their co-resident half-siblings who are genetic offspring of both resident parents. Average age of step-children is 11.3 years, genetic children 8.4 years. 95% confidence intervals are shown by vertical lines. (Figure adapted from Flinn & England, 1997.)

1988; Finkelhor & Dzuiba-Leatherman, 1994)) that temporarily raise cortisol. There may be a lack of parental consolation or reconciliation, and they may have inadequate coping abilities, perhaps resulting from difficult experiences in early development. The following case examples present temporal analyses of family relations and cortisol levels that illustrate some of the above possibilities.

"Jenny" was a 12-year-old girl who lived with her grandparents, aunt, and uncle. Her mother had lived in Guadeloupe for the past 10 years. At 9:17am on 17 July 1994, I observed the following events: "Wayonne," a 6-year-old male cousin that was visiting for the week, threw a stone at Jenny, who was sweeping in front of the house. She responded by scolding Wayonne, who pouted and retreated behind a mango tree. Wayonne found a mango pit, and lobbed it towards Jenny, but missed and hit a dress hanging on a clothesline, marking it with a streak of red dirt. Jenny ran to Wayonne, and struck him on the legs with her broom. He began to cry, arousing the interest of "granny Ninee", who emerged from the cooking room asking what happened, and upon hearing the story, scolded Jenny for "beating" Wayonne. Jenny argued that she was in the right, but granny Ninee would not hear of it, and sent her into the house. Jenny appeared frustrated, but looked down and kept quiet despite a quivering lip.

Jenny's cortisol levels were substantially elevated that afternoon, followed by subnormal levels the next day (a possible recovery period?). Three days after the incident she reported feeling ill and had a runny nose and oral temperature of 99.9°F (Figure 4.5).

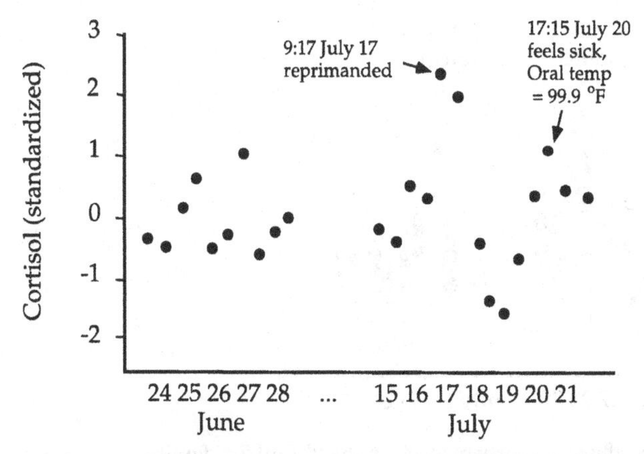

Figure 4.5. Morning and afternoon cortisol levels of "Jenny" during summer 1994. Late June cortisol levels are normal, but after being reprimanded by her grandmother on the morning of July 17, she has elevated cortisol levels for one day, followed by depressed cortisol levels for two days. Jenny exhibits symptoms of an upper respiratory infection with slight fever (common cold, probably rhinovirus) on the afternoon of July 20. (Figure adapted from Flinn & England, 1997.)

On 28 June 1992, a serious marital conflict erupted in the "Franklin" household. "Amanda" was a 34-year-old mother of six children, five of whom (ages 2, 3, 5, 8, and 14 years) were living with her and their father/step-father, "Pierre Franklin". Amanda was angry with Pierre for spending money on rum. Pierre was vexed with Amanda for "shaming" him in front of his friends. He left the village for several weeks, staying with a relative in town. His three genetic children (ages 2, 3, and 5 years) showed abnormal cortisol levels for a prolonged period following their father's departure (Figure 4.6). Children usually habituated to most stressful events, but absence of a parent often resulted in abnormal patterns of elevated and/or subnormal cortisol levels. Following the return of their father, the Franklin children's cortisol levels resumed a more normal profile. Children living in families with high levels of marital conflict (observed and reported serious quarreling, fighting, or residence absence) were more likely to have abnormal cortisol profiles than children living in more amiable families.

The events in children's lives that were associated with elevated cortisol are not always traumatic or even "negative". Eating meals, hard physical work, routine competitive play such as cricket, basketball, and "king of the mountain" on ocean rocks, and return of a family member that was temporarily absent (e.g., father returning from a job in town for the

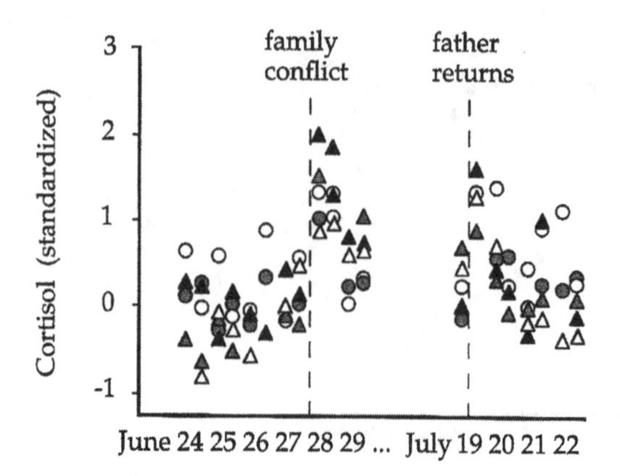

Figure 4.6. Marital conflict and cortisol levels in the "Franklin" family. Three genetic children (2, 3, and 5-year-old males) are represented by △▲▲ triangles and two step-children (8 and 14-year-old females) are represented by ○● dots respectively. Cortisol levels of three genetic children are normal before the conflict, rise during the conflict and during father's absence, briefly rise upon his return, and return to normal (lower) levels. The younger of the two step-children has a pattern of abnormally high cortisol, although her levels are moderate during step-father's absence. The older step daughter has a similar, but more normal pattern of cortisol levels.
(Figure adapted from Flinn & England, 1997.)

weekend) were associated with temporary moderate increases (about 10–100%) in cortisol among healthy children. These moderate stressors usually had rapid attenuation (< 1 hour) of cortisol levels (some stressors had characteristic temporal "signatures" of cortisol level and duration).

High stress events (cortisol increases from 100% to 2000%), however, most commonly involved trauma from family conflict or change (Flinn & England, 1995, 1997; Flinn *et al.*, 1996). Punishment, quarreling, and residence change substantially increased cortisol levels, whereas calm affectionate contact was associated with diminished (– 10%, – 50%) cortisol levels. In all, 19.2% of all cortisol values that were more than two standard deviations above mean levels (i.e., indicative of substantial stress) were temporally associated with traumatic family events (residence change of child or parent/caretaker, punishment, "shame", serious quarreling, and/or fighting) within a 24-hour period. Also, 42.1% of traumatic family events were temporally associated with substantially elevated cortisol (i.e., at least one of the saliva samples collected within 24-hours was > 2 standard deviation above mean levels).

It is important to note that there was considerable variability among children in cortisol response to family disturbances. Not all individuals had detectable changes in cortisol levels associated with family trauma. Some children had significantly elevated cortisol levels during some episodes of family trauma but not during others. Cortisol response is not a simple or uniform phenomenon. Numerous factors, including preceding events, habituation, specific individual histories, nutritional status, illness, context, and temperament, might affect how children respond to particular situations.

Nonetheless, traumatic family events were associated with elevated cortisol levels for all ages of children more than any other factor that we examined. These results suggest that family interactions were a critical psycho-social stressor in most children's lives, although the sample collection during periods of intense family interaction (early morning and late afternoon) may have exaggerated this association.

Although elevated cortisol levels are associated with traumatic events such as family conflict, long-term stress may result in diminished cortisol response. In some cases chronically stressed children had blunted response to physical activities that normally evoked cortisol elevation. Similarly, some shy or inhibited children had little or no cortisol response to social activities (from which they often withdrew) that stimulated cortisol elevations in more outgoing children. Comparison of cortisol levels during "non-stressful" periods (no reported or observed crying, punishment, anxiety, residence change, family conflict, or health problem during 24-hour period before saliva collection) indicates a striking reduction and, in many cases, reversal of the family environment-stress association (Flinn & England, 1995). Chronically stressed children sometimes had subnormal cortisol levels when they were not in stressful situations. For example, cortisol levels immediately after school (walking home from school) and during non-competitive play were lower among some chronically stressed children. Some chronically stressed children appeared socially "tough" or withdrawn, and exhibited little or no arousal to the novelty of the first few days of the saliva collection procedure.

Some children exhibited longitudinal change in cortisol profiles concomitant with an improved family environment, but others did not. Unlike their responses to many other potential stressors, most children did not seem to habituate readily to family trauma. Some parental actions may lack predictability and controllability necessary for development of actions or perceptions that reliably alleviate stress response (Garmezy, 1983; Hinde & Stevenson-Hinde, 1987; Rutter, 1991; Hetherington & Blechman, 1996).

Longitudinal analysis of caretaking histories indicates that children may have "sensitive periods" for development of stress response. Children with severe caretaking problems during infancy (neglect, parental alcoholism, and/or maternal absence), and/or growth disruptions (weight loss of > 10% of body weight) during the first 2 years of life, and/or high frequencies of minor physical abnormalities at birth usually exhibit one of two distinct cortisol profiles: (1) unusually low basal cortisol levels with occasional high spikes, or (2) chronically high cortisol levels. The first type (low basal, high spikes) is associated with hostility and antisocial behavior (e.g., theft, running away from home) and is more common among males. The second type (chronically high) is associated with anxiety and withdrawal behavior and is more common among females. These profiles are suggestive of diminished glucocorticoid regulatory function of the hippocampus (see Gray, 1987; Sapolsky, 1991; Yehuda *et al.*, 1991; Fuchs & Flugge, 1995).

4.5 Stress, illness, and immune function

A large and convincing research literature confirms commonsense intuition that psychosocial stress affects health (Black, 1994; Glaser & Kiecolt-Glaser, 1994; Ader *et al.*, 1995). Retrospective studies indicate that traumatic life events, such as divorce or death of a close relative, are associated with subsequent health problems such as cancer or cardiovascular disease (Rabkin & Struening, 1976). Clinical studies indicate that individuals with stressful lives are more susceptible to the common cold (Mason *et al.*, 1979; Cohen *et al.*, 1991; Evans & Pitt, 1994; Stone *et al.*, 1992; Boyce *et al.*, 1995*a*).

Stress response may deplete cellular energy and immune reserves (perhaps involving protection from autoimmunity). Although cortisol may provide short-term benefits, the body needs to replenish energy reserves to provide for immunity, growth, and other functions. Hence chronic stress and high average cortisol levels may be predicted to associate with frequency of illness. Data suggest that chronically stressed children with high cortisol levels tend to be ill more frequently than their counterparts (Figure 4.7), and that mean levels of cortisol are associated with frequency of illness (Flinn & England, 1997).

Temporal patterns of cortisol and observed stressful events also are associated with increased risk of illness, as anecdotally illustrated in the above example of Jenny (Figure 4.5), in which illness follows a high stress event. Children in Bwa Mawego have a nearly twofold increased risk of illness for several days following naturally occurring high-stress events (Figure 4.8).

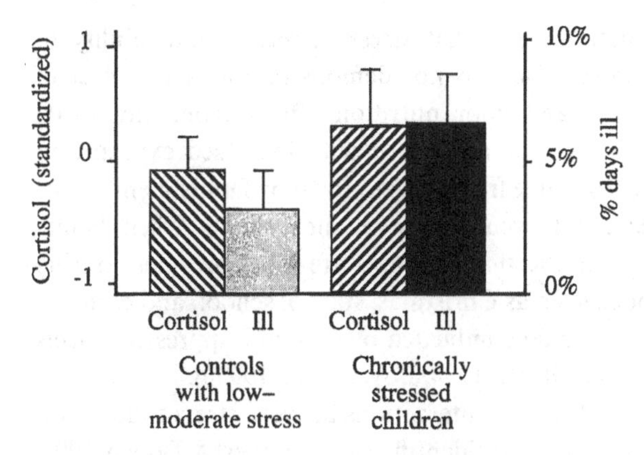

Figure 4.7. Chronic stress, cortisol levels, and frequency of illness. Chronically stressed children (N = 36) had higher average cortisol levels and were ill more often than controls (N = 141). Chronically stressed children were defined as having two or more major risk factors, including: parental conflict, mild abuse or neglect, high frequency (upper 10%) of reported daily stressors, inhibited or anxious temperament, parental alcoholism, low (bottom 10%) peer friendship ranking, and reported anti-social (theft, fighting, or runaway) behaviors.

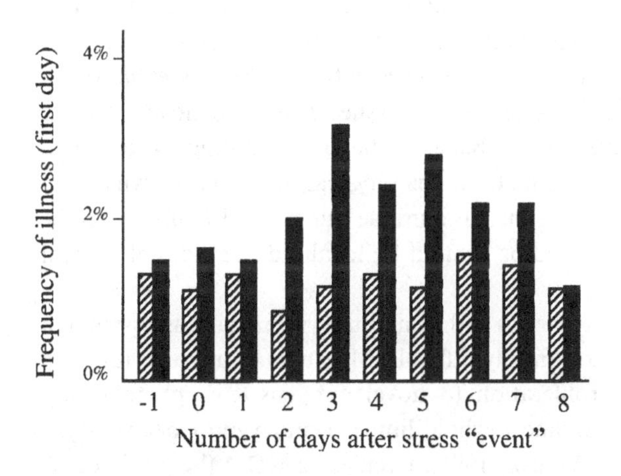

Figure 4.8. Temporal association between naturally occurring stress events and frequency of illness. Children had higher frequencies of illness for 3–5 days following a stress event (illustrated by solid black bars), than when no stress event had occurred (diagonally striped bars). The stress event was an observed or reported stressor that was accompanied within an 8-hour period by an elevated cortisol level of more than two standard deviations above mean levels.

These prospective data suggest that stress increases vulnerability to infectious disease; however, they do not demonstrate a direct effect of cortisol. Sleep disruption and poor nutrition often accompany social trauma. Stressful events may be associated with increased exposure to pathogens, resulting for example from trips to town by family members or residence changes. Stressful events may be more likely when family members are ill. Common infectious diseases are more prevalent during stressful seasonal periods such as Christmas, start of school, and carnival. A more direct causality would be indicated by immunosuppressive effects of stress (Glaser & Kiecolt-Glaser, 1994; Boyce *et al.*, 1995*b*).

Numerous (hundreds) of specific interactions between stress endocrinology and immune function have been identified (e.g., Herbert & Cohen, 1993; Dunn, 1995; Biondi & Zannino, 1997). Indeed, so many different and complex mechanisms appear to be involved in psycho-neuro-immunological (PNI) interactions that a general explanation remains elusive. It seems paradoxical that an organism would suppress an immune response during periods of stress when exposure and vulnerability to pathogens may be high.

Several non-exclusive, complementary hypotheses appear feasible. Allocation of energy to "emergency" demands may favor diversion from immunity (Sapolsky, 1994). Overreactive defensive responses to stress can result in autoimmunity; anti-inflammatory effects of glucocorticoid stress hormones may be protective of some types of tissues (Munck *et al.*, 1984). The possibility of damage to peripheral tissues generating novel antigens (e.g., collagen in joints, nerve tissue in the brain) during exposure to stressors, such as disease and strenuous physical or mental activity, may require particular suppression of immune function. Finally, immune trafficking may be enhanced or focused by localized overrides of general suppressive effects.

The complexity and dynamics of the immune system make assessment of immune function extraordinarily difficult. Measures of immune function include lymphocyte proliferation to novel antigens (e.g., phytohemagglutinin, concanavalin A, and keyhold limpet hemocyanin); delayed-type hypersensitivity (Shell-Duncan, 1995); circulating CD-4 (helper T cells), CD-8 (suppressor/cytotoxic T cells) and CD-15/56 (natural killer) cells; and levels of immunoglobulins (Ig) and cytokines (e.g., interleukins). A battery or panel of different measures provides a broad assessment of immune function, and allows for examination of possible mechanism-specific effects of glucocorticoid stress response on immunity. Blood samples are required for many of these measures, however, making non-invasive study of immune function problematic.

Table 4.1. *Associations between measures of immune function, stress, and illness conditions among children in a Caribbean village*

	Average cortisol	Chronically stressed vs. controls when both healthy	Chronically stressed vs. controls when both ill	Post-stress	During illness
s-Immunoglobulin A	n.s.	–	n.s.	n.s.	+
Interleukin 1	n.s.	n.s.	n.s.	n.s.	+
Interleukin 8	n.s.	n.s.	+	n.s.	n.s.
Neopterin	n.s.	n.s.	n.s.	–	n.s.
Microglobulin β_2	n.s.	n.s.	n.s.	–	n.s.

Note: Significant elevations of immune measures from saliva samples are indicated by +, significant decreases by −, and non-significant associations by n.s. Data indicate: (1) no significant association between a child's mean cortisol levels and any of the five immune measures ($R^2 = -0.18, 0.04, -0.07, -0.01, -0.02$); (2) chronically stressed children have lower s-immunoglobulin A levels than controls, when healthy (two-tailed *t*-tests, $p < 0.05$); (3) chronically stressed childen have lower interleukin 8 levels than do controls, when sick (two-tailed *t*-tests, $p < 0.05$); (4) children have lower neopterin and microglobulin β_2 levels for at least 2 days after a stress event (two-tailed *t*-tests, $p < 0.05$); and (5) children have higher s-immunoglobulin A and interleukin 1 levels when they are sick (two-tailed *t*-tests, $p < 0.05$).

I have begun an exploratory investigation of several components of immune function among children in Bwa Mawego using saliva samples. Levels of neopterin and microgloblin β_2 were examined as indicators of cell-mediated (cytotoxic T cell) activity, secretory-immunoglobulin-A as an indicator of humoral (antibody) response, interleukin 8 as a measure of non-specific (neutrophil recruitment) immunity, and interleukin 1 as an indicator of general immune activation (cytokine cascade). Preliminary analyses of these data suggest that psychosocial stress may have different effects on these components of immune function (Table 4.1).

Apparent differences between immune functioning of chronically stressed children and their counterparts included levels of secretory-immunoglobulin A (s-IgA) and interleukin 8. These measures changed in response to stress events (Figure 4.9). Interleukin 8 levels of controls were more elevated during illness than were those of chronically stressed children. During periods of good health, s-IgA levels of the controls were higher than those of chronically stressed children (see also Scanlan *et al.*, 1987). Neopterin levels were lower for several days after stress events. I do not know if these differences in immune parameters affect morbidity for

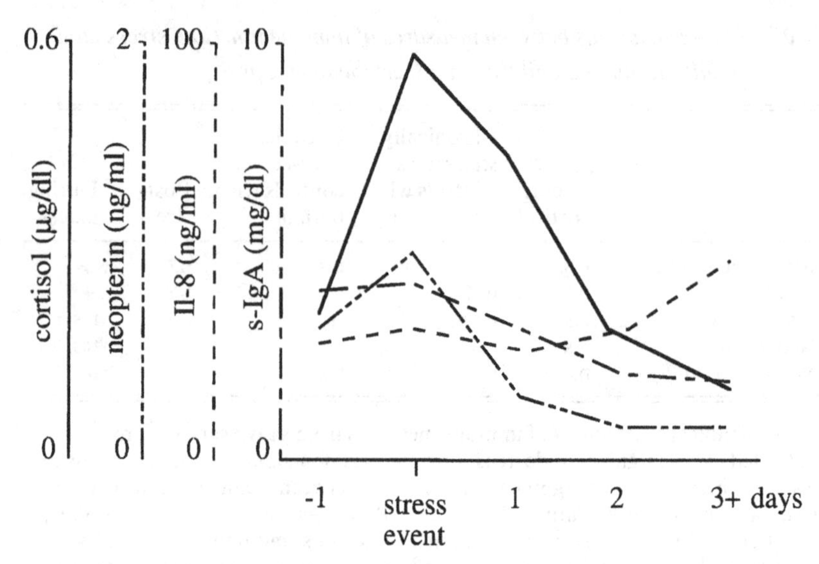

Figure 4.9. Temporal changes in salivary cortisol, neopterin, interleukin 8 (Il-8), and s-immunoglobulin A (s-IgA) levels of a 12-year-old male over a 5-day period (July 25–29, 1994). Elevated cortisol levels were associated with a fight with a peer ("stress event"); subsequent moderate changes occurred in neopterin, interleukin-8, and s-immunoglobulin A levels.

any specific pathogen, but they are suggestive of possible links among psychosocial stress, immune function, and illness. Longitudinal analyses of associations among these immune measures and stressful events, cortisol levels, and illness may provide better understanding of such connections, and are currently being conducted.

4.6 Summary and concluding remarks

Glucocorticoid stress response may be viewed as an adaptive mechanism that allocates somatic resources to different bodily functions, including immunity, growth, muscle action, and cognition (Maier *et al.*, 1994; Sapolsky, 1994; McEwen, 1995). Understanding how stress response affects these processes is important, because of consequences for health and psychological development. Release of cortisol and other stress hormones in response to traumatic family events may modulate energy and mental activity to resolve perceived psychosocial problems, but may diminish immunity and other health functions.

The objective of the long-term ethnographic study in Bwa Mawego is to monitor children's social and physical environment, behavioral activities, health, mental perceptions, and physiological states in a naturalistic setting so as to better understand relations among family environment, stress responses, and health. Analyses of data indicate that children living in households with intensive, stable caretaking usually had moderate cortisol levels (Figure 4.3), low frequency of illness (Figure 4.7), and appropriate immune response (Table 4.1). Children living in households with non-intensive, unstable caretaking were more likely to have abnormal (usually high and variable, but sometimes low) cortisol levels. Traumatic family events were associated temporally with elevated cortisol levels (e.g., Figure 4.5). Some children with caretaking and growth problems during infancy had unusual cortisol profiles. These associations indicate that family environment was a significant source of stress and illness risk for children living in Bwa Mawego. The variability of stress response, however, suggests a complex mix of each child's perceptions, neuroendocrinology, temperament, developmental history, and specific context.

Relations between family environment and cortisol stress response appear to result from a combination of factors, including frequency of traumatic events, frequency of positive "affectionate" interactions, frequency of negative interactions such as irrational punishment, frequency of residence change, security of "attachment", development of coping abilities, and availability or intensity of caretaking attention. Probably the most important correlate of household composition that affects childhood stress is maternal care. Mothers in socially "secure" households (i.e., permanent amiable coresidence with mate and/or other kin) appeared more able and more motivated to provide physical, social, and psychological care for their children. Mothers without mate or kin support were likely to exert effort attracting potential mates, and may have viewed dependent children as impediments to this. Hence co-residence of father may provide not only direct benefits from paternal care, but also may affect maternal care (Lamb *et al.*, 1987; Draper & Harpending, 1988; Lancaster, 1989; Belsky *et al.*, 1991; Flinn, 1992; Hurtado & Hill, 1992; Daly & Wilson, 1995). Young mothers without mate support usually relied extensively upon their parents or other kin for help with childcare.

Children born and raised in household environments in which mothers have little or no mate or kin support were at greatest risk for abnormal cortisol profiles and associated health problems. As socio-economic conditions influence family environment, they have consequences for child health that extend beyond direct material effects. Health in turn may affect an individual's social and economic opportunities, and a cycle of poor health and low socio-economic conditions may be perpetuated generation after generation.

Results of this study suggest that naturally occurring psychosocial events associated with stress hormones have significant short- and long-term effects on child health. This finding is consistent with a large body of clinical and retrospective studies that indicate "stress" has negative effects on health. The remarkable sensitivity of the human CNS to psychosocial stressors and consequent release of cortisol with its apparent health costs presents an unresolved evolutionary paradox. We do not have satisfying explanations for how natural selection might have favored these complex associations among energy regulation, immune function, and cognition, and we have only begun to explore the ontogenetic responses of this system to different social environments faced by the developing child.

4.7 References

Ader, R., Felten, D. L., & Cohen, N., eds. (1991). *Psychoneuroimmunology*. San Diego: Academic Press.

Ader, R., Cohen, N., & Felten, D. (1995). Psychoneuroimmunology: Interactions between the nervous system and the immune system. *Lancet*, **345**, 99–103.

Arnold, L. E., & Carnahan, J. A. (1990). Child divorce stress. In L. E. Arnold, ed., *Childhood Stress*, pp. 374–403. Wiley: New York.

Axelrod, J., & Reisine, T. D. (1984). Stress hormones: Their interaction and regulation. *Science*, **224**, 452–9.

Barrington, E. J. (1982). Evolutionary and comparative aspects of gut and brain peptides. *British Medical Bulletin*, **38**, 227–32.

Belsky, J., Steinberg, L., & Draper, P. (1991). Childhood experience, interpersonal development, and reproductive strategy: An evolutionary theory of socialization. *Child Development*, **62**, 647–70.

Bernard, H. R., Killworth, P. D., Kronenfeld, D., & Sailer, L. (1984). The problem of informant accuracy: The validity of retrospective data. *Annual Review of Anthropology*, **13**, 495–517.

Besedovsky, H. O., & del Rey, A. (1991). Feedback interactions between immunological cells and the hypothalamus–pituitary–adrenal axis. *Netherlands Journal of Medicine*, **39**, 274–80.

Black, P. H. (1994). Central nervous system–immune system interactions: Psychoneuroendocrinology of stress and its immune consequences. *Antimicrobial Agents and Chemotherapy*, **38**, 1–6.

Biondi M., & Zannino, L. G. (1997). Psychological stress, neuroimmunomodulation, and susceptibility to infectious diseases in animals and man: a review. *Psychotherapy and Psychosomatics*, **66**, 3–26.

Bornstein, M. H. ed. (1995). *Handbook of Parenting*. Mahwah, NJ: Lawrence Erlbaum Associates.

Boyce, W. T., Chesney, M., Alkon, A., Tschann, J. M., Adams, S., Chesterman, B., Cohen, F., Kaiser, P., Folkman, S., & Wara, D. (1995*a*). Psychobiologic reactivity to stress and childhood respiratory illnesses: Results of two

prospective studies. *Psychosomatic Medicine*, **57**, 411–22.
Boyce, W. T., Adams, S., Tschann, J. M., Cohen, F., Wara, D., & Gunnar, M. R. (1995*b*). Adrenocortical and behavioral predictors of immune responses to starting school. *Pediatric Research*, **38**, 1009–17.
Buckingham, J. C., Loxley, H. D., Christian, H. C., & Philip, J. G. (1996). Activation of the HPA axis by immune insults: roles and interactions of cytokines, eicosanoids, glucocorticoids. *Pharmacology, Biochemistry and Behavior*, **54**, 285–98.
Burns, E. M., & Arnold, L. E. (1990). Biological aspects of stress: Effects on the developing brain. In *Childhood Stress*, ed. Eugene Arnold, pp. 73–107. New York: Wiley.
Buss, A. H., & Plomin, R. (1984). *Temperament: Early Developing Personality Traits*. Hillsdale, New Jersey: Lawrence Erlbaum Associates.
Calandra, T., Bernhagen, J., Metz, C. N., Spiegel, L. A., Bacher, M., Donnelly, T., Cerami, A., & Bucala, R. (1995). MIF as a glucocorticoid-induced modulator of cytokine production. *Nature*, **377**, 68–71.
Clarke, A. S. (1993). Social rearing effects on HPA axis activity over early development and in response to stress in rhesus monkeys. *Developmental Psychobiology*, **26**, 433–46.
Coe, C. L., Lubach, G., Schneider, M. L., Dierschke, D. J., & Ershler, W. B. (1992). Early rearing conditions alter immune responses in the developing infant primate. *Pediatrics*, **90**, 505–9.
Cohen, P., Velez, C. N., Brook, J., & Smith, J. (1989). Mechanisms of the relation between perinatal problems, early childhood illness, and psychopathology in late childhood and adolescence. *Child Development*, **60**, 701–9.
Cohen, S. (1988). Psychosocial models of the role of social support in the etiology of physical disease. *Health Psychology*, **7**, 269–97.
Cohen, S., Tyrrell, D. A., & Smith, A. P. (1991). Psychological stress and susceptibility to the common cold. *New England Journal of Medicine*, **325**, 606–12.
Cohen, S., Tyrrell, D. A., & Smith, A. P. (1993). Negative life events, perceived stress, negative effect, and susceptibility to the common cold. *Journal of Personality and Social Psychology*, **64**, 131–40.
Corter, C. M. & Fleming, A. S. (1995). Psychobiology of maternal behavior in human beings. In *Handbook of Parenting*, Vol. 2; *Biology and Ecology of Parenting*, ed. M. H. Bornstein, pp. 87–116. Mahwak, NJ: Lawrence Erlbaum Associates, Inc.
Costas, M., Trapp, T., Pereda, M. P., Sauer, J., Rupprecht, R., Nahmod, V. E., Reul, J. M., Holsboer, F., & Arzt, E. (1996). Molecular and functional evidence for *in vitro* cytokine enhancement of human and murine target cell sensitivity to glucocorticoids. TNF-alpha priming increases glucocorticoid inhibition of TNF-alpha-induced cytotoxicity/apoptosis. *Journal of Clinical Investigation*, **98**, 1409–16.
Coste, J., Strauch, G., Letrait, M., & Bertagna, X. (1994). Reliability of hormonal levels for assessing the hypothalamic–pituitary–adrenocortical system in clinical pharmacology. *British Journal of Clinical Pharmacology*, **38**, 474–9.
Dabbs, J. M. Jr. (1990). Salivary testosterone measurements: Reliability across hours, days, and weeks. *Physiology and Behavior*, **48**, 83–6.
Dabbs, J. M. Jr. (1992). Testosterone measurements in social and clinical psychology. *Journal of Social and Clinical Psychology*, **11**, 302–21.
Dabbs, J. M. Jr., & Hopper, C. H. (1990). Cortisol, arousal, and personality in

two groups of normal men. *Personality and Individual Differences*, **11**, 931–5.

Dallman, M. F., Akana, S. F., Strack, A. M., Hanson, E. S., & Sebastian, R. J. (1995). The neural network that regulates energy balance is responsive to glucocorticoids and insulin and also regulates HPA axis responsivity at a site proximal to CRF neurons. *Annals of the New York Academy of Sciences*, **771**, 730–42.

Daly, M., & Wilson, M. (1988). Evolutionary social psychology and family homicide. *Science*, **242**, 519–24.

Daly, M., & Wilson, M. (1995). Discriminative parental solicitude and the relevance of evolutionary models to the analysis of motivational systems. In *The Cognitive Neurosciences*, ed. M. S. Gazzaniga, pp. 1269–86. Cambridge, MIT Press.

De Bellis, M., Chrousos, G. P., Dorn, L. D., Burke, L., Helmers, K., Kling, M. A., Trickett, P. K., & Putnam, F. W. (1994). Hypothalamic–pituitary–adrenal axis dysregulation in sexually abused girls. *Journal of Clinical Endocrinology and Metabolism*, **78**, 249–55.

De Kloet, E. R. (1991). Brain corticosteroid receptor balance and homeostatic control. *Frontiers in Neuroendocrinology*, **12**, 95–164.

Draper, P., & Harpending, H. A. (1988). Sociobiological perspective on the development of human reproductive strategies. In *Sociobiological Perspectives on Human Development*, ed. K. MacDonald, pp. 340–72. New York: Springer-Verlag.

Dressler, W. W. (1996). Hypertension in the African American community: Social, cultural, and psychological factors. *Seminars in Nephrology*, **16**, 71–82.

Dunn, A. J. (1995). Interactions between the nervous system and the immune system: implications for psychopharmacology. In *Psychopharmacology: The Fourth Generation of Progress*. ed. F. R. Bloom & D. J. Kupfer. New York: Raven Press.

Ellison, P. (1988). Human salivary steroids: methodological considerations and applications in physical anthropology. *Yearbook of Physical Anthropology*, **31**, 115–42.

Emery, R. E. (1988). *Marriage, divorce, and children's adjustment*. Volume 14, Developmental Clinical Psychology and Psychiatry. SAGE: Newbury Park, CA.

Evans, P., & Pitt, M. (1994). Vulnerability to respiratory infection and the four-day desirability dip: Comments on Stone, Porter and Neale. *British Journal of Medical Psychology*, **67**, 387–9.

Eysenck, M. W. (1982). *Attention and Arousal: Cognition and Performance*. New York: Springer-Verlag.

Finkelhor, D., & Dzuiba-Leatherman, J. (1994). Victimization of children. *American Psychologist*, **49**, 173–83.

Flinn, M. V. (1988). Step and genetic parent/offspring relationships in a Caribbean village. *Ethology and Sociobiology*, **9**, 1–34.

Flinn, M. V. (1992). Paternal care in a Caribbean village. In *Father–Child Relations: Cultural and Biosocial Contexts*, ed. B. Hewlett, pp. 57–84. Hawthorne, NY: Aldine.

Flinn, M. V., & England, B. G. (1995). Family environment and childhood stress. *Current Anthropology* **36**, 854–66.

Flinn, M. V., & England, B. G. (1997). Social economics of childhood

glucocorticoid stress response and health. *American Journal of Physical Anthropology*, **102**, 33–53.
Flinn, M. V., Turner, M. T., Quinlan, R., Decker, S. D., & England, B. G. (1996). Male–female differences in effects of parental absence on glucocorticoid stress response. *Human Nature*, 7, 125–62.
Fredrikson, M., Sundin, ., & Frankenhauser, M. (1985). Cortisol excretion during the defense reaction in humans. *Psychosomatic Medicine*, **47**, 313–19.
Fuchs, E., & Flugge, G. (1995). Modulation of binding sites for corticotropin-releasing hormone by chronic psychosocial stress. *Psychoneuroendocrinology*, **30**, 33–51.
Fukunaga, T., Mizoi, Y., Yamashita, A., Yamada, M., Yamamoto, Y., Tatsuno, Y., & Nishi, K. (1992). Thymus of abused/neglected children. *Forensic Science International*, **53**, 69–79.
Garmezy, N. (1983). Stressors of childhood. In *Stress, Coping, and Development in Children*, ed. N. Garmezy & M. Rutter, pp. 43–83. New York: McGraw-Hill.
Glaser, R., & Kiecolt-Glaser, J. K., eds. (1994). *Handbook of Human Stress and Immunity*. New York: Academic Press.
Goldberg, L. R. (1992). The development of markers for the big-five factor structure. *Psychological Assessment*, **4**, 26–42.
Goldberg, L. R. (1993). The structure of phenotypic personality traits. *American Psychologist*, **48**, 26–34.
Gottman, J. M., & Katz, L. F. (1989). Effects of marital discord on young children's peer interaction and health. *Developmental Psychology*, **25**, 373–81.
Gray, J. A. (1987). *The Psychology of Fear and Stress* (2nd edn.). Cambridge: Cambridge University Press.
Gunnar, M. (1992). Reactivity of the hypothalamic-pituitary-adrenocortical system to stressors in normal infants and children. *Pediatrics*, **90**, 491–7.
Gunnar, M. R. (1994). Psychoendocrine studies of temperament and stress in early childhood: Expanding current models. In *Temperament: Individual Differences at the Interface of Biology and Behavior*. APA science volumes, ed. J. E. Bates & T. D. Wachs, pp. 175–98. Washington, DC: American Psychological Association.
Haggerty, R. J., Sherrod, L. R., Garmezy, N., & Rutter, M. eds., (1994). *Stress, Risk, and Resilience in Children and Adolescents*. Cambridge: Cambridge University Press.
Hamill, P. U. V., Drizd, T. A., Johnson, C. L., Reed, R. B., & Roche, A. F. (1997). *NCHS Growth Curves for Children, Birth-18 Years*. United States: USDHEW.
Hart, J., Gunnar, M. R., & Cicchetti, D. (1996). Altered neuroendocrine activity in maltreated children related to symptoms of depression. *Development and Psychopathology*, **8**, 201–14.
Herbert, T. B., & Cohen, S. (1993). Stress and immunity in humans: A meta-analytic review. *Psychosomatic Medicine*, **55**, 364–79.
Herman, J. P., Prewitt, C. M., & Cullinan, W. E. (1996). Neuronal circuit regulation of the hypothalamo-pituitary-adrenocortical stress axis. *Critical Reviews in Neurobiology*, **10**, 371–94.
Hetherington, E. M., & Blechman, E. A., eds., (1996). *Stress, Coping, and Resiliency in Children and Families*. Mahwah, NJ: Lawrence Erlbaum Associates.
Higley, J. D., & Suomi, S. J. (1996). Effect of reactivity and social competence on individual responses to severe stress in children: Investigations using

nonhuman primates. In *Severe Stress and Mental Disturbance in Children*, ed. C. R. Pfeffer, pp. 3–57. Washington, DC: American Psychiatric Press.
Hinde, R. A. & Stevenson-Hinde, J. (1987). Interpersonal relationships and child development. *Development Review*, **7**, 1–21.
Holl, R., Fehm, H., Voigt, K., & Teller, W. (1984). The "mid-day surge" in plasma cortisol induced by mental stress. *Hormones and Metabolism Research*, **16**, 158–9.
House, J. S., Landis, K. R., & Umberson, D. (1988). Social relationships and health. *Science*, **241**, 540–4.
Hurtado, A. M., & Hill, K. R. (1992). Paternal effect on offspring survivorship among Ache and Hiwi hunter-gatherers: Implications for modeling pair-bond stability. In *Father-Child Relations: Cultural and Biosocial Contexts*, ed. B. Hewlett, pp. 31–55. New York: Aldine De Gruyter.
Insel, T. R. (1990). Long-term neural consequences of stress during development: Is early experience a form of chemical imprinting? In *The Brain and Psychopathology*, ed. B. J. Carroll, New York: Raven Press.
Insel, T. R., & Shapiro, L. (1992). Oxytocin receptors and maternal behavior. In Oxytocin and Maternal, Sexual, and Social Behaviors. *Annals of the New York Academy of Sciences*, ed. C. A. Pederson, J. D. Caldwell, G. F. Jirikowski, & T. R. Insel, vol. 652, pp. 122–41.
Jefferies, W. M. (1990). Cortisol and immunity. *Medical Hypotheses*, **34**, 198–208.
Joels, M., Karten, Y., Hesen, W., & De Kloet, E. R. (1997). Corticosteroid effects on electrical properties of brain cells: Temporal aspects and role of antiglucocorticoids. *Psychoneuroendocrinology*, **22**, 81–6.
Kagan, J. (1992). Behavior, biology, and the meanings of temperamental constructs. *Pediatrics*, **90**, 510–13.
Kagan, J., Resnick, J. S., & Snidman, N. (1988). The biological basis of childhood shyness. *Science*, **240**, 167–71.
Kapcala, L. P., Chautard, T., & Eskay, R. L. (1995). The protective role of the hypothalamic–pituitary–adrenal axis against lethality produced by immune, infectious, and inflammatory stress. *Annals of the New York Academy of Sciences*, **771**, 419–37.
Kaplan, H. B. (1991). Social psychology of the immune system: A conceptual framework and review of the literature. *Social Science and Medicine*, **33**, 909–23.
Karmel, B. Z., & Gardner, J. M. (1996). Prenatal cocaine exposure effects on arousal–modulated attention during the neonatal period. *Developmental Psychobiology*, **29**, 463–80.
Kirschbaum, C., & Hellhammer, D. H. (1994). Salivary cortisol in psychneuroendocrine research: recent developments and applications. *Psychoneuroendocrinology*, **19**, 313–33.
Kirschbaum, C., Read, G. F., & Hellhammer, D. H., eds. (1992). *Assessment of Hormones and Drugs in Saliva in Biobehavioral Research*. Seattle: Hogrefe and Huber.
Kirschbaum, C., Wolf, O. T., May, M., Wippich, W., & Hellhammer, D. H. (1996). Stress and treatment-induced elevations of cortisol levels associated with impaired declarative memory in healthy adults. *Life Sciences*, **58**, 1475–83.
Lamb, M., Pleck, J., Charnov. E., & Levine, J. (1987). A biosocial perspective on paternal behavior and involvement. In *Parenting Across the Lifespan: Biosocial Dimensions*, ed. J. B. Lancaster, J. Altmann, A. Rossi, & L. Sherrod, pp. 111–42. Hawthorne, NY: Aldine de Gruyter.

Lancaster, J. (1989). Evolutionary and cross-cultural perspectives on single-parenthood. In *Sociobiology and the Social Sciences*, ed. R. W. Bell & N. J. Bell, pp. 63–71. Lubbock, Texas: Texas Tech University Press.
Levine, S. (1993). The influence of social factors on the response to stress. *Psychotherapy and Psychosomatics*, **60**, 33–8.
Lupien, S. J., Gaudreau, S., Tchiteya, B. M., Maheu, F., Sharma, S., Nair, N. P., Hauger, R. L., McEwen, B. S., & Meaney, M. J. (1997). Stress-induced declarative memory impairment in healthy elderly subjects: Relationship to cortisol reactivity. *Journal of Clinical Endocrinology and Metabolism*, **82**, 2070–5.
Maccari, S., Piazza, P. V., Kabbaj, M., Barbazanges, A., Simon, H., & Le Moal, M. (1995). Adoption reverses the long-term impairment in glucocorticoid feedback induced by prenatal stress. *Journal of Neuroscience*, **15**, 110–16.
MacDonald, K. (1992). Warmth as a developmental construct: An evolutionary analysis. *Child Development*, **63**, 753–73.
McEwan, B., & Sapolsky, R. (1995). Stress and cognitive function. *Current Opinions in Neurobiology*, **5**, 205–12.
Maier, S. F., Watkins, L. R., & Fleshner, M. (1994). Psychoneuroimmunology: The interface between behavior, brain, and immunity. *American Psychologist*, **49**, 1004–17.
Martignoni, E., Costa, A., Sinforiani, E., Luzzi, A., Chiodini, P., Mauri, M., Bono, G., & Nappi, G. (1992). The brain as a target for adrenocortical steroids: Cognitive implications. *Psychoneuroendocrinology*, **17**, 343–54.
Mason, J. W. (1971). A re-evaluation of the concept of "non-specificity" in stress theory. *Journal of Psychosomatic Research*, **8**, 323–33.
Mason, J. W., Buescher, E. L., Belfer, M. L., Artenstein, M. S., & Mougey, E. H. (1979). A prospective study of corticosteroids and catecholamine levels in relation to viral respiratory illness. *Journal of Human Stress*, **5**, 18–28.
McEwen, B. S. (1995). Stressful experience, brain, and emotions: Developmental, genetic, and hormonal influences. In *The Cognitive Neurosciences*, ed. M. S. Gazzaniga, pp. 1117–35. Cambridge: MIT Press.
McEwen, B. S. (1997). Possible mechanisms for atrophy of the human hippocampus. *Molecular Psychiatry*, **2**, 255–62.
McEwen, B. S., & Magarinos, A. M. (1997). Stress effects on morphology and function of the hippocampus. *Annals of the New York Academy of Sciences*, **821**, 271–84.
McGaugh, J. L., Cahill, L., & Roozendaal, B. (1996). Involvement of the amygdala in memory storage: Interaction with other brain systems. *Proceedings of the National Academy of Sciences, USA*, **93**, 13508–14.
Meaney, M., Mitchell, J., Aitken, D., Bhat Agar, S., Bodnoff, S., Ivy, L., & Sarriev, A. (1991). The effects of neonatal handling on the development of the adrenocortical response to stress: Implications for neuropathology and cognitive deficits later in life. *Psychoneuroendocrinology*, **16**, 85–103.
Munck, A., Guyre, P. M., & Holbrook, N. J. (1984). Physiological functions of glucocorticoids in stress and their relation to pharmacological actions. *Endocrine Reviews*, **5**, 25–44.
Munck, A., & Guyre, P. M. (1991). Glucocorticoids and immune function. In *Psychoneuroimmunology*, ed. R. Ader, D. L. Felten, & N. Cohen (eds.), San Diego: Academic Press.
Nachmias, M., Gunnar, M., Mangelsdorf, S., Parritz, R. H., & Buss, K. (1996). Behavioral inhibition and stress reactivity: The moderating role of

attachment security. *Child Development*, **67**, 508–22.
Nesse, R. M., Curtis, G. C., Thyer, B. A., McCann, D. S., Huber-Smith, M. J., & Knopf, R. F. (1985). Endocrine and cardiovascular responses during phobic anxiety. *Psychosomatic Medicine*, **47**, 320–31.
Newcomer, J. W., Craft, S., Hershey, T., Askins, K., & Bardgett, M. E. (1994). Glucocorticoid-induced impairment in declarative memory performance in adult humans. *Journal of Neuroscience*, **14**, 2047–53.
Okamura, T., Nakajima, Y., Matsuoka, M., & Takamatsu, T. (1997). Study of salivary catecholamines using fully automated column-switching high-performance liquid chromatography. *Journal of Chromatography-Biomedical Applications*, **694**, 305–16.
Orchinik, M., Hastings, N., Witt, D., & McEwen, B. S. (1997). High-affinity binding of corticosterone to mammalian neuronal membranes: Possible role of corticosteroid binding globulin. *Journal of Steroid Biochemistry and Molecular Biology*, **60**, 229–36.
Panter-Brick, C. (1998). Biological anthropology and child health: Context, process and outcome. In *Biosocial Perspectives on Children*, ed. C. Panter-Brick, pp. 66–101. Cambridge: Cambridge University Press.
Pearlin, L. I., & Turner, H. A. (1987). The family as a context of the stress process. In *Stress and Health: Issues in Research Methodology*, ed. S. V. Kasl & C. L. Cooper. New York: John Wiley.
Pollard, T. M. (1995). Use of cortisol as a stress marker: Practical and theoretical problems. *American Journal of Human Biology*, **7**, 265–74.
Quinlan, R. (1995). *Father absence, maternal care, and children's behavior in a rural Caribbean village*. Unpublished MA thesis. Columbia: University of Missouri.
Rabkin, J. G., & Streuning, E. L. (1976). Life events, stress, and illness. *Science*, **194**, 1013–20.
Rutter, M. (1991). Childhood experiences and adult psychosocial functioning. In *The Childhood Environment and Adult Disease*, ed. G. R. Bock & J. Whelan, pp. 189–200. Chichester: John Wiley.
Saphier, D., Welch, J. E., Farrar, G. E., Ngunen, N. Q., Aguado F., Thaller, T. R., & Knight, D. S. (1994). Interactions between serotonin, thyrotropin-releasing hormone and substance P in the CNS regulation of adrenocortical secretion. *Psychoneuroendocrinology*, **19**, 779–97.
Sapolsky, R. M. (1990*a*). Stress in the wild. *Scientific American*, January, 116–23.
Sapolsky, R. M. (1990*b*). Adrenocortical function, social rank, and personality among wild baboons. *Biological Psychiatry*, **28**, 862–78.
Sapolsky, R. M. (1991). Effects of stress and glucocorticoids on hippocampal neuronal survival. In *Stress – Neurobiology and Neuroendocrinology*, ed. M. R. Brown, G. F. Koob, & C. Rivier, pp. 293–322. Dekker, New York.
Sapolsky, R. M. (1992*a*). Neuroendocrinology of the stress-response. In *Behavioral Endocrinology*, ed. J. B. Becker, S. M. Breedlove, & D. Crews, pp. 287–324. Cambridge: MIT Press.
Sapolsky, R. M. (1992*b*). *Stress, the Aging Brain, and the Mechanisms of Neuron Death*. Cambridge, MA: MIT Press.
Sapolsky, R. M. (1994). *Why Zebras Don't Get Ulcers*. New York: W. H. Freeman and Co.
Sapolsky, R. M. (1996). Stress, glucocorticoids, and damage to the nervous system: The current state of confusion. *Stress*, **1**, 1–20.
Scanlan, J. M., Coe C. L., Latts A., & Suomi S. (1987). Effects of age, rearing, and

separation stress on imunoglobulin levels in rhesus monkeys. *American Journal of Primatology*, **13**, 11–22.
Scapagnini, U. (1992). Psychoneuroendocrinoimmunology: The basis for a novel therapeutic approach in aging. *Psychoneuroendocrinology*, **17**, 411–20.
Schneider, M. L., Coe, C. L., & Lubach, G. R. (1992). Endocrine activation mimics the adverse effects of prenatal stress on the neuromotor development of the infant primate, *Developmental Psychobiology*, **25**, 427–39.
Servan-Schreiber, D., Printz, H., & Cohen, S. D. (1990). A network model of catecholamine effects: gain, signal-to-noise ratio, and behavior. *Science*, **249**, 892–5.
Shell-Duncan, B. (1995). Impact of seasonal variation in food availability and disease stress on health status of nomadic Turkana children: A longitudinal analysis of morbidity, immunity, and nutritional status. *American Journal of Human Biology*, **7**, 339–55.
Spangler, G., & Grossmann, K. E. (1993). Biobehavioral organization in securely and insecurely attached infants. *Child Development*, **64**, 1439–50.
Stansbury, K., & Gunnar, M. R. (1994). Adrenocortical activity and emotion regulation. *Monographs of the Society for Research in Child Development*, **59**, 108–34, 250–83.
Stone, A. A. L., Bovbjerg, D. H., Neale, J. M., Napoli, A., Valdimarsdottir, H., Cox, D., Hayden, F. G., & Gwaltney, J. M. Jr. (1992). Development of the common cold symptoms following experimental rhinovirus infection is related to prior stressful life events. *Behavioral Medicine*, **13**, 70–4.
Suomi, S. (1991). Uptight and laid-back monkeys: Individual differences in the response to social challenges. In *Plasticity of Development*, ed. S. E. Brauth, W. S. Hall, & R. J. Dooling, pp. 27–56. Cambridge, MA: MIT Press.
Takahashi, L. K. (1992). Prenatal stress and the expression of stress-induced responses throughout the life span. *Clinical Neuropharmacology*, **15**, 153–4.
Tennes, K. (1982). The role of hormones in mother–infant interaction. In *The Development of Attachment and Affiliative Systems*, ed. R. N. Emde & R. J. Harmon, pp. 75–88. New York: Plenum Press.
Tennes, K., & Mason, J. (1982). Developmental psychoendocrinology: An approach to the study of emotions. In *Measuring Emotions in Infants and Children*, ed. C. Izard, pp. 21–37. London: Cambridge University Press.
Turner, M. T. (1995). *Mother and Infant Cortisol Response in a Dominican Village*. Unpublished MA thesis. Columbia: University of Missouri.
Vacchio, M. S., King, L. B., & Ashwell, J. D. (1996). Regulation of thymocyte development by glucocorticoids. *Behring Institute Mitteilungen*, **97**, 24–31.
Van Cauter, E. (1990). Diurnal and ultradian rhythms in human endocrine function: A mini-review. *Hormone Research*, **34**, 45–53.
Walker, B. R. (1996). Abnormal glucocorticoid activity in subjects with risk factors for cardiovascular disease. *Endocrine Research*, **22**, 701–8.
Walker, R., Riad-Fahmy, D., & Read, G. F. (1978). Adrenal status assessed by direct raedioimmunoassay of cortisol in whole saliva or parotid saliva. *Clinical Chemistry*, **24**, 1460–3.
Wallerstein, J. S. (1983). Children of divorce: Stress and developmental tasks. In N. Garmezy & M. Rutter, eds., *Stress, Coping, and Development in Children*, pp. 265–302. New York: McGraw-Hill.
Walsh, J. J., Wilding, J. M., Eysenck, M. W., & Valentine, J. D. (1997). Neuroticism, locus of control, type A behaviour pattern and occupational stress. *Work and Stress*, **11**, 148–59.

Weiner, H. (1992). *Perturbing the Organism.* Chicago: University of Chicago Press.
Wilson, M. I., Daly, M., & Weghorst, S. J. (1980). Household composition and the risk of child abuse and neglect. *Journal of Biosocial Science*, **12**, 333–40.
Winslow, J. T., Hastings, N., Carter, C. S., Harbaugh, C. R., & Insel, T. R. (1993). A role for central vasopressin in pair bonding in monogamous prairie voles. *Nature*, **365**, 545–8.
Yehuda, R., Giller, E. L., Southwick, S. M., Lowy, M. T., & Mason, J. W. (1991). Hypothalamic-pituitary-adrenal dysfunction in posttraumatic stress disorder. *Biological Psychiatry*, **30**, 1031–48.

5

Work and hormonal variation in subsistence and industrial contexts

CATHERINE PANTER-BRICK AND
TESSA M. POLLARD

5.1 Introduction

Working behavior varies considerably across socio-ecological contexts, and over the lifespan. Nearly all humans spend a vast proportion of their time engaged in work-related activities and their form plays a large role in determining variation in health outcomes both between and within populations. This variation is usually measured in terms of a limited range of indicators, focusing on those easiest to obtain for a wide range of communities. Thus rates of mortality, morbidity, fertility, work capacity, growth and nutritional status are currently the gold standards for appraising the relative outcomes of different behavioral patterns, including outcomes of working behavior. Also important are the more elusive measures of well-being and the less obvious consequences of behaviors for day-to-day maintenance and long-term health. Endocrine data offer real potential for providing insight into the impact of working behavior on health and well-being.

Current research utilizing population data on endocrine profiles has examined both the energetic and psychosocial aspects of working behavior, linking endocrine changes with important health outcomes such as reproduction and risk of disease (Figure 5.1). In these studies, "work" is conceived both as an *activity* (exacting significant physical or mental effort) and as a particular *microenvironment* (with social interactions or perceived demands different from those of home).

The gonadal (sex) hormones, progesterone and testosterone in particular, have been studied mostly with respect to their association with *energetic* stress and other metabolic loads. The observed variation in levels of these hormones has a demonstrable impact on reproductive capacity (fecundity) and actual fertility, in both Western industrial and Third World

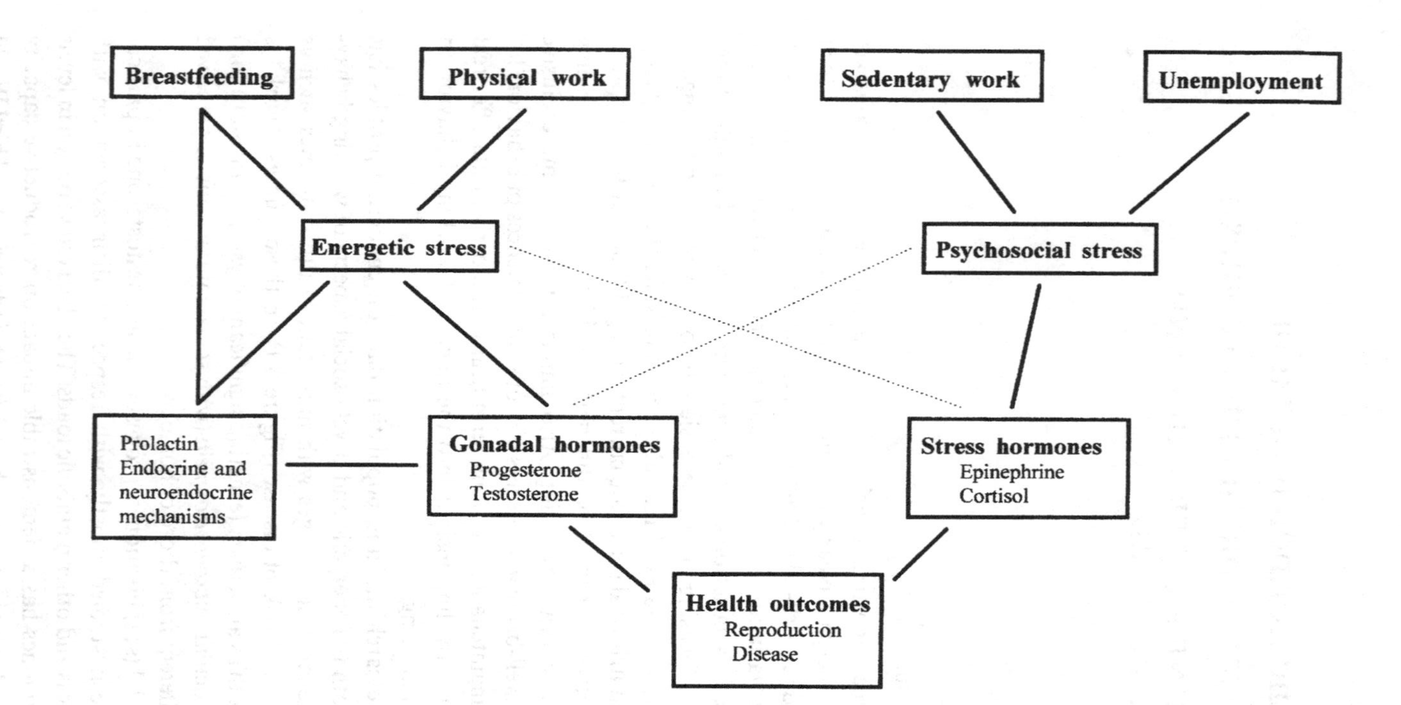

Figure 5.1. A schema representing the salient aspects of working behavior, hormonal pathways, and health outcomes discussed in this chapter.

subsistence contexts. Their associations with psychosocial stress are less well documented.

The so-called "stress hormones", epinephrine (adrenaline) and cortisol, have long been associated with working life and perceived *psychosocial* stress, as well as with physical activity. Altered output of such hormones can be chronically maintained, and have profound long-term consequences for health. For example, it is thought that chronically elevated levels of epinephrine are likely to increase risk of development of the precursors of cardiovascular disease. Measurement of epinephrine offers the opportunity to identify short-term physiological changes which might otherwise go unnoticed, and which have important long-term sequelae. Cortisol is also thought to have an important influence on cardiovascular health, and to influence the immune system. Thus it may have profound effects on vulnerability to both non-infectious and infectious diseases (Sapolsky, Chapter 2; Flinn, Chapter 4).

In this chapter, we consider a growing body of data concerning the endocrine correlates of physical activity, sedentary work, and unemployment (Figure 5.1). We examine first energetic stress and second psychosocial stress, thought to influence levels of sex hormones and stress hormones, to link working behavior with health outcomes in adult men and women in various environments. We begin with a review of current research on physical work levels and related endocrine and health profiles across a range of populations from simple subsistence to modern industrial societies. We also summarize current understanding of how other metabolic loads, such as lactation, may compound energetic stress among reproductively active women. We then turn to studies linking sedentary work with hormonal variation and assess the evidence for models linking mental workload and perceptions of control at work with variation in epinephrine and cortisol levels. Such models have also been applied to predict hormonal changes following unemployment in industrial societies, and these in turn have been posited to underly some of the ill health known to be caused by unemployment. We examine the evidence that loss of work is associated with changes in stress hormone levels. Lastly, we consider the limited research available on children's hormonal profiles in their work environments. We argue that a more fine-grained integration of behavioral, endocrine, and health measures can provide a better understanding of many aspects of human biological variation.

5.2 Adult physical work

Biological anthropologists have long been interested in documenting working behavior and the relationships between health and energetic stress across a wide range of societies. Clarification of the ways in which hormones might mediate such relationships remains a challenge for current and future research. We summarize here a large literature on levels of physical work, socio-ecological contexts, and significance for health, together with the more limited, but expanding, research examining specifically the associations between energetic variables and hormonal variation.

5.2.1 Range of physical activity levels across populations

There now exist a good number of studies which provide quantitative data on the range of working activity between and within populations. Several sources of variability in work patterns have attracted specific interest: the type of economy (e.g., rural farmers versus urbanites), season (e.g., pre-harvest and post-harvest), age (e.g., adults versus children), gender, reproductive status (e.g., pregnant, lactating, non-pregnant and non-lactating), nutritional status, (e.g., low versus high body mass index), socio-economic status, and household characteristics (e.g., size and composition).

Table 5.1 details levels of physical activity for selected population samples, by type of economy, gender, and seasonality. The FAO/WHO/UNU (1985) has recommended classifying workloads in terms of light, moderate, or heavy physical activity levels (PAL) expressed as multiples of basal metabolic rate (Table 5.1). Basal metabolic rate (BMR), namely the energy expended under conditions of rest, can be predicted from an individual's age, sex, and body weight (FAO/WHO/UNU, 1985). Total energy expenditure (TEE in MJ/day) is usually estimated by one of three methods: the factorial method (where allocation of time is multiplied by the energy cost of specific tasks), the heart rate monitoring method (where heart beats reflect activity), or the measurement of doubly-labeled water (where tracing doses of stable isotopes indicates gross daily energy expenditure) (see Ulijaszek & Strickland, 1993; Ulijaszek, 1995 for methodological reviews). Thus physical activity levels (PAL = TEE/BMR, corrected for age, sex, and body weight) provide useful measures for comparing levels of work across populations.

For instance, studies of human energy expenditure determined by doubly-labeled water measurements have recently been used to document the upper and lower limits of physical activity (Black *et al.*, 1996). In modern industrial populations, minimum PAL (1.21) were recorded for chairbound subjects and very high PAL (above 2) were obtained for small samples (men and women) of Everest mountaineers and Nordic skiers over short periods of intense activity (Table 5.1). Habitual activity in sustainable lifestyles falls within a range of 1.2–2.5, representing minimal activity to very heavy work (Black *et al.*, 1996:74).

With respect to subsistence societies, a general interest in documenting habitual workloads has often begun with a "note of the fact that in many societies most adults, and particularly women, are working throughout the daylight hours and often into the night" (Harrison, 1996:1). Table 5.1 cites studies which have quantified workloads by gender, and where appropriate by season, for a number of subsistence economies. In rural Gambia, for example, men undertake very light PAL in the dry season and moderate PAL in the wet season, while women increase their PAL from moderate to very heavy (Lawrence and Whitehead, 1988). Farming women also sustain higher workloads than men in Upper Volta and Senegal (Bleiberg *et al.*, 1980; Brun *et al.*, 1981; Simondon *et al.*, 1993), whilst the converse is observed in highland Ecuador (Leonard *et al.*, 1995). Among the egalitarian Tamang agro-pastoralists in Nepal, both men and women sustain moderately heavy workloads in the dry winter and very heavy workloads in the wet monsoon. This entails spending between 6 and 7 hours/day in the winter, and between 8 and 9 hours/day in the monsoon, in outdoor subsistence activities, let alone domestic work (Panter-Brick, 1996a).

Specific attention to within population variation in workloads has proved particularly instructive. For example, examination of gender-specific consequences of seasonal changes in working behavior has helped researchers move beyond simple, population-level correlations of behavioral variables and health outcomes. Biological anthropologists are particularly concerned with intra-population variation in order to achieve rigorous analyses of socio-ecological contexts (Harrison, 1996), and avoid the dangers of "ecological fallacy", whereby variables are largely empirically correlated without paying much attention to specific causality (Wood, 1994:529–36).

5.2.2 Energetic variables and health outcomes

The impact of changes in working behavior on health-related outcomes, such as levels of food availability, dietary intake and composition, anthropometric status, physical capacity, development, aging, and reproductive

Table 5.1. *Total energy expenditure (TEE in MJ/d) and physical activity levels (PAL, multiples of basal metabolic rate) for males and females in various populations*

	Males			Females			
Population	TEE	PAL	level	TEE	PAL	level	Reference
Farmers							
Gambia							Ulijaszek & Strickland, 1993:77
Dry season	7.6	1.17	vL	9.6	1.68	M	Lawrence & Whitehead, 1988
Wet season	12.5	1.87	M	11.3	1.97	vH	Lawrence & Whitehead, 1988
Wet season				10.4	1.95	vH	Singh *et al.*, 1989
Upper Volta							Ulijaszek & Strickland, 1993:77
Dry season	10.1	1.55	L	9.7	1.88	H	Brun *et al.*, 1981; Bleiberg *et al.*, 1980
Wet season	14.4	2.15	H	12.1	2.33	vH	Brun *et al.*, 1981; Bleiberg *et al.*, 1980
Senegal							
Dry season	11.1	1.6	L	11.5	2.0	vH	Simondon *et al.*, 1993
Wet season	10.1	1.5	L	10.0	1.8	H	Simondon *et al.*, 1993
Papua New-Guinea							
Coast	9.8	1.5	L	7.7	1.5	L	Norgan *et al.*, 1974; Katzmarzyk *et al.*, 1994:728
Highland	10.8	1.6	L	9.4	1.8	H	Norgan *et al.*, 1974; Katzmarzyk *et al.*, 1994:728
Samoa	12.2	1.6	L	9.5	1.6	L–M	Pearson, 1990; Katzmarzyk *et al.*, 1994:728
Ecuador							
Coast	10.1	1.58	L	8.3	1.62	M	Leonard *et al.*, 1995
Highland	15.9	2.39	vH	10.3	1.94	H	Leonard *et al.*, 1995
Peru							
Machiguenga	13.4	2.2	vH	8.0	1.7	M	Montgomery & Johnson, 1977; Katzmarzyk *et al.*, 1994:728
India Tamil Nadu							
Overall (4 seasons)	12.0	1.96	MH	8.3	1.69	M	Gillepsie & McNeill 1992
Main harvest	—	1.98	MH	—	1.80	H	McNeill *et al.*, 1988
Hot season	—	1.84	M	—	1.54	L	McNeill *et al.*, 1988
Nepal							
Pre-monsoon	12.9	2.16	H	9.9	1.82	H	Strickland & Tuffrey, 1997
Post-monsoon	12.2	2.01	H	8.6	1.67	M	Strickland & Tuffrey, 1997
Monsoon	13.3	1.98	H	7.3	1.52	L	Strickland & Tuffrey, 1997

Agro-pastoralists							
Nepal							
Winter	11.8	1.88	M	9.1	1.77	M	Panter-Brick, 1996*a*
Monsoon	13.9	2.22	vH	10.5	2.01	vH	Panter-Brick, 1996*a*
Bolivia							
Highland Quechua	8.4	1.3	vL	6.7	1.3	L	Leonard, 1988; Katzmarzyk *et al.*, 1994:728
Highland Amarya	11.1	1.96	MH	9.8	2.04	vH	Kashiwazaki *et al.*, 1995
Pastoralists							
Kenya Turkana	9.0	1.3	L	7.3	1.3	L	Galvin, 1985; Katzmarzyk *et al.*, 1994:728
Foragers							
Paraguay Ache	13.9	2.2	vH	11.0	1.9	H	Leonard & Robertson, 1992; Katzmarzyk *et al.*, 1994:728
Bostwana !Kung	9.1	1.7	M	7.2	1.5	L	Leonard & Robertson, 1992; Katzmarzyk *et al.*, 1994:728
Herders/fishermen/villagers							
Siberia							
Evenki herders	11.9	1.8	M	8.8	1.6	M	Katzmarzyk *et al.*, 1994
Keto fishermen	11.4	1.7	M	7.8	1.5	L	Katzmarzyk *et al.*, 1994
Evenki villagers	10.0	1.48	vL	8.5	1.59	L	Leonard *et al.*, 1997
Russian villagers	10.9	1.39	vL	8.3	1.53	L	Leonard *et al.*, 1997
Canada							
Eskimo sedentary	10.5	1.5	L	9.8	1.8	H	Godin & Shepard, 1973; Katzmarzyk *et al.*, 1994:728
Modernized populations							
Handicapped	6.1	1.22	vL	6.1	1.22	vL	Bandini *et al.*, 1991; Black *et al.*, 1996
Scottish students	12.3	1.7	M	9.6	1.8	H	Durnin & Passmore, 1967; Katzmarzyk *et al.*, 1994:728
Belfast urbanites		1.88	M		1.77	MH	Livingstone *et al.*, 1991
Swimmers	16.7	2.08	H	10.0	1.75	H	Jones & Leitch, 1993; Black *et al.*, 1996
Everest mountaineers	14.7	2.44	vH	12.0	2.00	vH	Westerterp *et al.*, 1992; Black *et al.*, 1996
Nordic skiers	30.3	3.47	vH	18.3	2.81	vH	Sodin *et al.*, 1994; Black *et al.*, 1996

Note: This table includes only populations for which data existed for both men and women, and expressed as both TEE and PAL. Where available, a contrast of seasons is presented.
Methodology includes time allocation with indirect calorimetry, heart-rate monitoring, and doubly-labeled water techniques.
The FAO/WHO/UNU (1985) grades PAL as: light, 1.55; moderate, 1.78; heavy, 2.1 for men; light, 1.56; moderate, 1.64; heavy, 1.82 for women.
The grades vL (very light), MH (moderate to heavy) and vH (very heavy) are Panter-Brick's own.

Table 5.2. *Differences in salivary progesterone levels (pmol/l) for healthy, non-contracepting women (all ages and mid-reproductive career) in five contrasting populations*

Populations	Age range	*N*	Luteal Mean	Luteal SD	Mid-luteal Mean	Mid-luteal SD	References
Boston, USA	18–46	135	229	84	327	135	Lipson & Ellison 1992; and pers.comm.
	20–34	62	252	83	362	132	
	23–35	47	256	76	366	120	
Rural Poland	23–39	22	198	122	256	126	Jasienska 1996; and pers.comm.
	23–35	17	204	129	264	144	
Quechua of Bolivia	19–42	20	167	89	238	78	Vitzthum pers.comm.
	23–35	8	182	90	263	54	
Lese of Zaire	17–46	34	131	57	186	96	Ellison *et al.*, 1989; and pers.comm.
	20–34	15	127	67	186	109	
	23–35	13	146	66	218	113	
Tamang of Nepal	17–46	21	91	39	114	63	Panter-Brick *et al.*, 1993 and this volume
	20–34	10	104	49	131	78	
	23–35	6	130	46	188	77	

Note: All data were analyzed by the Reproductive Ecology Laboratory, Harvard University.
Progesterone profiles were aligned relative to day 0 of menstrual onset and averaged for the luteal (days − 1 to − 16 for USA, Bolivia, Zaire, and Nepal; − 1 to − 14 for Poland) and mid-luteal (days − 5 to − 9 for USA, Bolivia, Zaire, and Nepal; − 7 to − 11 for Poland) phases of the menstrual cycle.

performance, have been examined extensively (Norgan, 1992; Ulijaszek & Strickland, 1993). Of particular interest are the relationships between energetic variables and reproductive function.

Energetics, or energy availability, comprise at least three different variables, which may show quite distinct relationships with health outcomes and should thereby be explicitly differentiated. First, one must consider the level of *energy expenditure per se* (e.g., low/high physical activity, or overall workload). Second, one must appraise *energy store* (current nutritional status, e.g., body mass index). Third, one must examine *energy balance*, reflected in anthropometric changes (e.g., losses or gains of body weight, fat, and fat-free mass) over time, occasioned mostly from increased physical activity or reduced dietary intake. Furthermore, the type of activities or work schedule, which may demand long-term endurance from day-to-day, or sporadic bouts of intense effort, may well have different consequences for energetic availability, and ultimately for reproductive function (Panter-Brick, 1996*b*).

5.2.3 Energetics, sex hormones, and reproduction

The energetic stress of physical activity has been shown to significantly influence levels of *female* sex hormones in ways likely to affect both reproductive function (fecundity) and actual fertility. Research in the field of reproductive ecology has rapidly expanded (Ellison, 1990), and good evidence now exists for both interpopulation and intrapopulation differences in levels of reproductive steroid hormones, particularly progesterone and estradiol.

The work of Peter Ellison and colleagues (see Ellison, Chapter 6) deserves special attention. Comparisons of data sets from several Western and Third World populations were based on analyses of salivary hormones undertaken in the same laboratory, thus minimizing sources of error in comparing absolute levels across populations (Table 5.2). Figure 5.2 illustrates the significant inter-population variation observed in progesterone levels, serving to characterize female ovarian function, which can be related to energetic variables in different ecological contexts (Ellison *et al.*, 1993; Bentley, 1996). Women working in simple farming subsistence economies, such as the Lese in the Ituri forest of the Democratic Republic of Congo (formerly Zaire) and the Tamang of Nepal, have significantly lower progesterone profiles than women in urban USA. The Quechua of highland Bolivia and rural Polish farm women show intermediary profiles.

The Bolivian Quechua women have light workloads (Table 5.1), in both

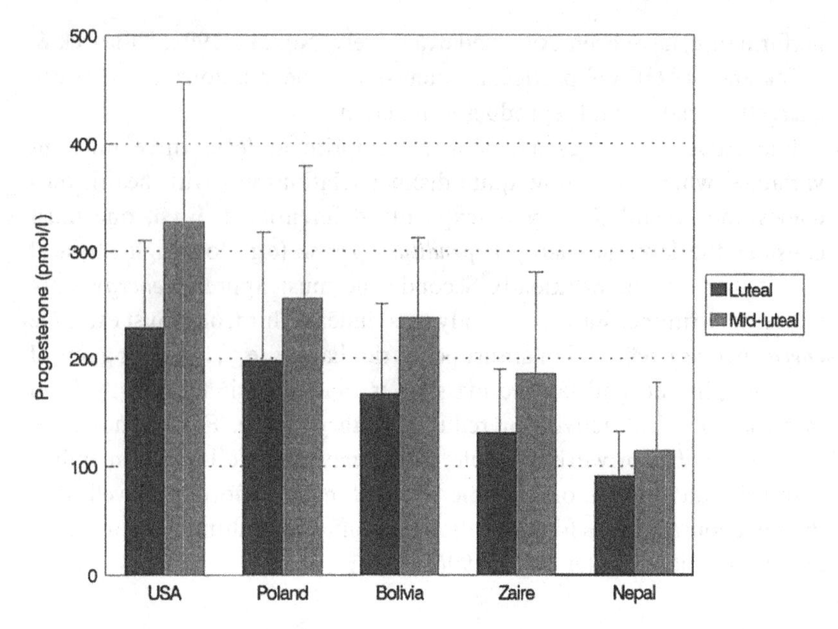

Figure 5.2. Interpopulation differences in progesterone levels (Mean and SD). Two variables serve to characterize progesterone values postovulation: the luteal and mid-luteal phases, relative to menstrual onset (see Table 5.2). The profiles for all women are unaffected by sample age composition.

the wet and dry seasons, as well as limited food intake (these data are for Nuñoa: Leonard, 1988, unpublished results). The Polish women average heavy workloads (PAL = 1.88 overall), which range from light (PAL = 1.57) in winter to very heavy (PAL = 2.02) in summer, and are unconstrained by food availability (Jasienska, 1996). The Tamang in Nepal sustain moderate to very heavy PAL (Table 5.1), and do not incur significant food shortages. No comparable data on energy expenditure have been published for the Boston and Lese women: Lese activity patterns have not been analysed in terms of TEE or PAL (Jenike, 1996). The Lese experience recurrent "hungry seasons" when weight loss is nearly universal (Ellison *et al.*, 1989).

While variation in ovarian function and steroid metabolism reflects a whole cluster of variables (such as age, dietary intake and composition, energy expenditure, energy stores and energy balance), levels of physical activity are clearly implicated (Ellison *et al.*, 1993). Furthermore, explanation for *inter*-population differences in progesterone levels involves consideration of sporadic versus habitual exposure to energetic stress, especially during the period of growth and development, which may reflect human

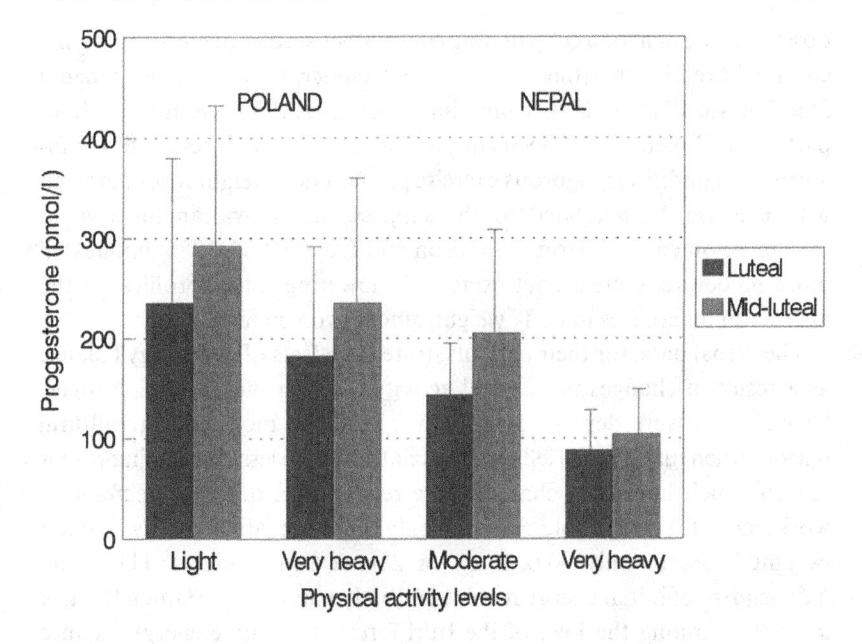

Figure 5.3. Intrapopulation differences in progesterone levels (Mean and SD) for rural women in Poland ($N = 22$) and Nepal ($N = 9$, 23–35 year-olds). For Polish women, physical activity levels are light (1.57) in winter and very heavy (2.02) in summer (Jasienska, 1996). For Tamang women of Nepal, PAL are moderate (1.77) in winter and very heavy (2.01) in the monsoon (Panter-Brick, 1996*a,b*).

adaptability to specific ecological settings (Bentley, 1996:49; Ellison, 1996 and Chapter 6 this volume; Vitzthum, 1997).

Particularly good evidence linking workloads with changes in female fecundity comes from *intra*-population studies of hormonal variation (Ellison *et al.*, 1993). Ovarian function has been shown to vary in relation to both the level of physical activity (overall effort) and net energy balance (weight loss) experienced in Western and non-Western populations. Figure 5.3 illustrates data for the Polish and Nepali populations, which show significant seasonal variation at times of relative leisure versus intense agricultural workloads.

The Polish data demonstrate the impact of levels of physical activity *per se*. Thus women with very demanding summer work schedules (PAL = 2.11) on the farm averaged lower luteal progesterone levels (P = 161 pmol/l) than those with moderately demanding physical activity (PAL = 1.67 and P = 234 pmol/l; Jasienska, 1996). Levels of physical activity drove this association as the two groups were similar in weight, fat and body mass index and did not lose body weight over the period of observation. Such data echo findings from other studies demonstrating the effect of exercise, whether strenuous or moderate, on female gonadal function (see Warren, 1980; Cumming & Wheeler, 1990; Rosetta, 1993). In particular, Bullen *et al.* (1985) adopted an experimental design to distinguish two conditions, rigorous exercise *per se* versus weight loss combined with exercise, both related to the suppression of ovarian function in Western women. Furthermore, Ellison and Lager (1985, 1986) implicated more moderate, recreational exercise in lowering luteal profiles, in the absence of differences in body weight, among women in Boston.

The Nepal data, for their part, illustrate the effects of net energy balance as a result of changes in physical activity. Among the Tamang, progesterone levels were depressed at the height of the monsoon agricultural season (when luteal levels fell by 27% relative to the less demanding winter season). Such hormonal changes were related to a dramatic increase in workloads (PAL in Table 5.1) and only slight negative energy balance (weight losses of only −0.99 kg, or 2% of body weight). They were independent of initial energy reserves, or body mass index (Panter-Brick *et al.*, 1993). Among the Lese of the Ituri forest, a negative energy balance (weight losses of 2 kg), which was associated with food shortages rather than physical activity, also led to the suppression of progesterone values (Ellison *et al.*, 1989).

In all, current understanding of female reproductive function has gathered complexity. A threshold model of menstrual function (menses versus non-menses) in relation to nutritional status was first considered (Frisch & Revelle, 1970). A more dynamic model is now proposed, whereby several grades of ovarian function (frank amenorrhea, ovulatory failure, luteal or follicular hormonal supression, and fully competent cycles) respond to changes in energy balance and/or levels of physical activity (Ellison *et al.*, 1993). The picture outlined above for progesterone is corroborated by data on other hormones such as estradiol. Most importantly, hormonal changes also relate to variation in birth rates or actual fertility (Panter-Brick, 1996*c*) and conception probabilities per menstrual cycle (Lipson & Ellison, 1996).

In men, the evidence that variation in testosterone levels is directly related to acute or chronic energetic stress is weak or equivocal, in marked contrast to the situation observed for female sex hormones. The acute response to physical effort is an increase in testosterone, but sustained exercise suppresses testosterone levels, as has been demonstrated in research on male athletes (Hackney, 1989). Research conducted by Kujala *et al.* (1990) suggests that gonadotrophin releasing hormone (GnRH) secretion is suppressed during exercise, and that the testes have a reduced capacity for testosterone secretion. The small cluster of field studies available give little support to the hypothesis that male gonadal function varies with physical activity, energy store, or energy balance. Salivary testosterone levels recorded for a number of non-Western populations are within the broad normal range (150–500 pmol/l; Read, 1993) reported for Western males, although towards the low end of the spectrum (Figure 5.4). In Nepal, Tamang agro-pastoralists and Kami blacksmiths, two castes cohabiting the same villages but contrasting in economic activities and anthropometric characteristics, show no seasonal variation in testosterone values or strong association with energetic variables (Ellison & Panter-Brick, 1996). Similarly, data for Lese horticulturalists in the Ituri (Bentley *et al.*, 1993), the Aymara in high-altitude Bolivia (Beall *et al.*, 1992) and !Kung foragers (Christiansen, 1991) show no relationship between testosterone levels and indices of acute or chronic energetic stress.

Most likely, variation in testosterone levels represents the regulation of somatic maintenance costs (e.g., promoting muscle anabolism) (Bribiescas, 1996), rather than as in females the modulation of reproductive outcome, which in males is less sensitive to energy balance or status (Ellison & Panter-Brick, 1996). Another possibility is that hormonal modulation may simply be a pleiotropic correlate of the responsivity of the female reproductive system to environmental cues (Campbell & Leslie, 1995): in other words, such effects evolved because they were adaptive in women, and appear in men because they share the same biochemical pathways. We do not yet understand testosterone variation sufficiently to be able to distinguish between these hypotheses.

5.2.4 Physical activity and catecholamines

We also have good evidence that physical activity causes an increase in the secretion of catecholamines (Ward & Mefford, 1985), which are responsible for the release of metabolites such as glucose and free fatty acids (Sapolsky, Chapter 2). Studies of manual workers in industrial societies have demonstrated a positive correlation between workload and both epinephrine and norepinephrine levels in men and women (Cullen *et al.*, 1979; Timio *et al.*, 1979; Cox *et al.*, 1982; Mulders *et al.*, 1982; Lundberg *et al.*, 1989). Physical activity appears to be the strongest determinant of norepinephrine levels; however, mental work seems to play a more important role than physical work in determining epinephrine levels. No good evidence on the effect of physical activity on catecholamine variation with regard to work in

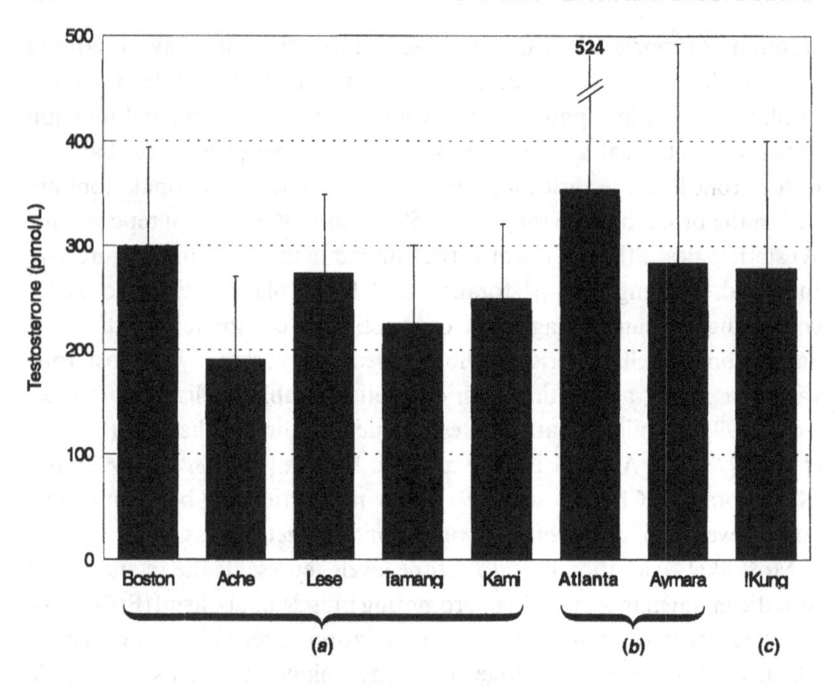

Figure 5.4. Interpopulation differences in morning testosterone levels (Mean and SD; analyses from three laboratories for: (*a*) Boston USA, Ache in Paraguay, Lese in Zaire, Tamang and Kami in Nepal, (*b*) Cleveland USA, Aymara in Bolivia, (*c*) !Kung of Namibia.) (From Ellison and Panter-Brick 1996, and references in text.)

subsistence-based societies is available, perhaps because, unlike steroids such as sex hormones, catecholamine levels cannot be readily assessed in saliva. It is therefore necessary to collect urine, which brings a number of methodological difficulties with regard to collection and assessment of hormones levels (Pollard, 1997).

Within the literature there is an implicit assumption that elevation of catecholamines brought about by physical exercise does not carry the same risks to health as elevation in the absence of physical activity. Thus, for example, free fatty acids which are metabolized during exercise are unlikely to alter serum lipid profiles in a deleterious manner and to constitute a risk of cardiovascular disease.

5.3 Women's work in reproduction

We now return briefly to female reproductive function to consider additional energetic costs known to affect fecundity, fertility, and reproductive health. Women's work entails not only an expenditure of energy in food production and household tasks, but also in childbearing. As detailed above, in many parts of the world women assume a physically demanding workload: moreover, when pregnant or lactating, they usually cannot afford to discontinue physical activity. Thus numerous ethnographies report that women continue to work throughout pregnancy, often giving birth at the worksite. Shostak (1990:178) described the situation of childbearing !Kung women of Botswana as follows: "Many women maintain their normal work routines until the day they give birth. Pregnancy is thought as a given; it is 'women's work'".

The view that working women cope with tight energy budgets for physical activity in conjunction with reproduction has been argued most forcefully for The Gambia (Prentice and Prentice, 1988). In rural Gambia, workloads are reduced in the last term of pregnancy (through curtailing time for non-essential housework and leisure, not time for actually farming) (Roberts *et al.*, 1982) and in the first 3 months of lactation (Lawrence and Whitehead, 1988), but then return to the heavy levels habitually sustained (Table 5.1). It has been shown that seasonal energetic constraints affect maternal nutritional status and reproductive performance, as well as subsequent childhood morbidity and mortality (for reviews see Ulijaszek, 1995; Lunn, 1996).

How is this situation reflected and accommodated by endocrine function? What are the hormonal profiles of women who are energy constrained through assuming both subsistence and childbearing responsibilities? As Lunn (1996:195) recently emphasized, the relationship between nutritional stress and reproduction has been the focus of immense concern, but the mechanisms of this association, via hormonal regulation, remain subjects of controversy.

We discuss below how the demands of physical activity and breastfeeding may combine to modulate variation in female fecundity. In the case of lactating women, workloads affect hormonal profiles and fecundity in at least two ways. First, they influence the actual time available for breastfeeding and supplementary feeding of infants during the working day. In Bangladesh for instance, seasonal variation in nursing times was related to the demands upon women's time for crop processing at the harvest season (Huffman *et al.*, 1980). In Nepal, nursing schedules have been described as governed by "opportunity" rather than "demand": Tamang women breastfeed their infants at the work-place during the time other workers pause for rest (Panter-Brick, 1991, 1992). Moreover, the range of women's activities away from home affects the timing and type of supplementary feeding, and thereby lactation. Variation in nursing behavior has been shown to influence, on the one hand, maternal levels of the hormone prolactin and the drive for milk production, and on the other, the hypothalamic release of GnRH which inhibits levels of luteinizing hormone and prolongs post-partum amenorrhea (McNeilly, 1993). Both the intensity of breastfeeding (namely the frequency, interval, or duration of daily feeds, not simply the overall duration of lactation in months post-partum) and supplementary feeding are thought to influence prolactin levels and variation in ovarian function. The hormonal mechanisms which link nursing behavior, prolactin, and ovarian function remain unclear, however (Vitzthum, 1994; Tay *et al.*, 1996).

Second, levels of physical work govern the degree of energetic stress which is experienced; for nursing women, lactation is an additional metabolic load. Indeed, Lunn (1996: 200) has argued convincingly that the nutritional stress of Third World lactating women occurs "primarily as a result of their excessive work-loads whilst lactating rather than ... an inadequate supply of food". In examining the links between energetic constraints and female ovarian function, both Rosetta (1992:304, 1993) and Lunn (1996:200) consider that a comparison of poor Third World women with Western athletes is particularly insightful. These women share many characteristics with respect to endurance physical activity, diet, and body composition (Rosetta, 1993). Bentley (1985) also argued that the pattern of energy expenditure among !Kung hunter–gatherers, best characterized as endurance exercise, would contribute, in addition to lactational practices, to lower female fecundity.

Lunn (1996) has reviewed the complex relationships between energy expenditure, nutrition and lactation driving fecundity variation among women (see also Rosetta, 1993; Wood, 1994; Ulijaszek, 1995). He outlines two main theories linking ovarian activity with, on the one hand, maternal nutritional status (Frisch, 1987), and on the other, the suckling stimulus from lactation (Lunn, 1985). Frisch (1987) argued that adipose tissue influenced estrogen metabolism and maintained the hypothesis that post-partum infecundity was a direct consequence of low body weight and/or fat. The relationship between weight/fat loss and post-partum amenorrhea, however, has not been confirmed. In particular, for lactating Gambian women, a large dietary supplementation (3000 kJ/d) had minimal effect on body weight and composition but considerably shortened the duration of post-partum amenorrhea (by 8 months) and increased fertility (Lunn *et al.*, 1984). Lunn (1985) proposed instead the concept of the "baby in the driving seat" whereby infant suckling behavior regulated maternal metabolism for milk production (indexed by prolactin) and indirectly modulated ovarian function. Lunn (1996) argues that these two facets of hormonal regulation might actually be related, such that nutritional stress (referred to in this chapter as energetic stress) may interact with the suckling stimulus to influence responses of the hypothalamus and ovarian activity. As Ellison *et al.* (1993:2254) noted, it is unfortunate that the two models regarding "the lactational and energetic modulation of female fecundity were initially perceived and presented as mutually exclusive alternatives".

Thus a simple, direct relationship between nursing behavior, hypothalamic or gonadal hormonal responses, and infecundity is under question. Prolactin is now considered only as a proxy measure of lactational infecundity, namely an index of nipple stimulation rather than direct ovarian regulation (reviewed in Wood, 1994:346–9). Other intervening variables, particularly energetic constraints (such as physical workloads and nutritional status) but also maternal age and psycho-social stress (see reviews in Vitzthum, 1994:314–15; Ulijaszek, 1995:107–15), are increasingly considered in the overall framework relating ecological variables to endocrine measures of ovarian function. Recent research has also highlighted the role of metabolic hormones (insulin and insulin-like growth factors) in regulating ovarian response to gonadotropin stimulation and androgen production (Nahum *et al.*, 1995); insulin levels, in turn, are sensitive to energetic constraints (McGarvey, Chapter 8). Such complexity might explain "anomalies" in what was previously conceived as a fairly straightforward relationship between maternal–infant behavior, hormones, and physiology or health outcomes. For example, intensive monitoring of UK women over 24 hours (during early lactation, after the introduction of baby foods, after first menses, and final weaning) showed that the duration of amenorrhea was unrelated to prolactin levels in response to nursing frequency, but was correlated to the timing of introduction of supplementary foods (Tay *et al.*, 1996). In Papua New Guinea, the Amele breastfeed 1- to 3-year-olds intensively (one to three times per hour for 1–3 minutes) yet have a surprisingly short period of lactational amenorrhea (11 months) and high fertility, possibly because they are not energetically constrained by food intake or energy expenditure (Worthman *et al.*, 1993). The evidence from Amele salivary progesterone and estradiol measures and serum prolactin levels indicates that among women with good nutritional status, energetic balance and intensive nursing may have little effect on reproductive function. In rural Nepal, hard-working agro-pastoralist Tamang women and housewives of low-caste Kami blacksmiths, co-resident in the same villages, experience different prolactin profiles despite adopting similarly intensive nursing schedules (once every 1–2 hours for 6–7 minutes) for 1- to 3-year-olds. Relative to the blacksmith housewives, the Tamang show lower fertility, longer birth intervals (by 9 months), longer post-partum amenorrhea (by 8 months), more sustained prolactin responses after a given breastfeed, and elevated mean prolactin levels for as long as 22 months post-partum (Stallings *et al.*, 1996, 1998). The sensitivity of neuro-endocrine responses to suckling is possibly heightened among the Tamang, as they combine prolonged breastfeeding with heavy energy expenditure in subsistence.

In sum, the regulation of ovarian function is modulated by a number of interrelated variables, such as the infant's suckling stimulus and the degree of maternal energetic stress, which in many communities, is directly or indirectly affected by physical workload. Functionally, the production of ovarian hormones is suppressed when women experience a heavy energetic burden compounded by lactation, poor nutritional status, and demanding physical activity. The effects of breastfeeding on ovarian function may be mediated via neuroendocrine mechanisms, involving GnRH, or metabolic-endocrine mechanisms, notably insulin (Figure 5.1).

5.4 Sedentary work

Investigation of the consequences of non-manual and sedentary economic behavior for health and well-being has traditionally been the domain of psychologists rather than anthropologists. It has become increasingly clear that sedentary work has potentially important consequences for physical as well as mental processes, and that many of these effects are mediated by endocrine changes. Here we review research (in industrial contexts) which has attempted to identify the key characteristics of sedentary work with respect to health outcomes, and then consider studies which have related these characteristics to hormone variation, principally in the "stress hormones" epinephrine and cortisol.

5.4.1 Occupational specialization and sedentary work

Early sociologists such as Marx, Durkheim, and Weber all identified occupational specialization with industrial society and contrasted it with the agricultural systems of earlier societies (Cameron, 1995). In recent years further economic changes have led to employment of many more people in service sectors, rather than in manufacturing, giving rise to an increase in non-manual professional and semi-professional occupations and a decline in manual employment. Thus people living in industrial contexts, in both rich and poor countries, are increasingly likely to conduct specialized non-manual work.

Non-manual jobs offer protection from the hazards and demands of physical work, but it is becoming increasingly clear that mental work and a sedentary lifestyle may also exact costs to health and well-being. For example, in the British Whitehall Study, Marmot and colleagues investigated mortality among men working in central government offices in London (Marmot & Madge, 1987). They found clear gradients of age-adjusted mortality by occupational status, with men in the lower grades exhibiting the highest mortality. Their results indicate that both socio-economic status and experiences at work are likely to play a role in generating these differences. The authors probed the the role of work-associated psychosocial stress by administering detailed questionnaires to another sample of government workers. Men in lower status positions reported their jobs to be more monotonous, and to provide less opportunity for control and for use of skills, all job characteristics thought to generate feelings of stress (Marmot & Madge, 1987). Furthermore, in the prospective Whitehall II Study, low job control at baseline was associated with increased risk of coronary heart disease at follow-up an average of 5 years later (Bosma *et al.*, 1997).

Karasek's model of job strain is now widely used to explain such effects. Karasek (1979) identified two primary job characteristics, demand and decision latitude, as the main influences on the psychosocial consequences of work in the industrialized workplace. From Figure 5.5, it can be seen that job strain is considered to result when demand is high and decision latitude is low. Karasek and Theorell (1990) have expanded on the meaning of the two central constructs of the model. They use demand to describe psychological, rather than physical, load. They describe it as a measure of "how hard you work". Thus they see task requirements or mental workload as the central component of demand in most jobs. They define decision latitude as a combination of decision authority or autonomy (control) and skill discretion or task variety (skill use).

Karasek (1979) proposed that job strain resulting from high demand and low decision latitude is likely to lead to impaired mental well-being and physical ill health. This prediction has been tested for cardiovascular disease, amongst other outcomes. In large-scale epidemiological analyses of data from random national samples in Sweden and the USA, Karasek and his colleagues have shown that men in high strain jobs had a greater prevalence of cardiovascular disease (Karasek *et al.*, 1981, 1988). This link

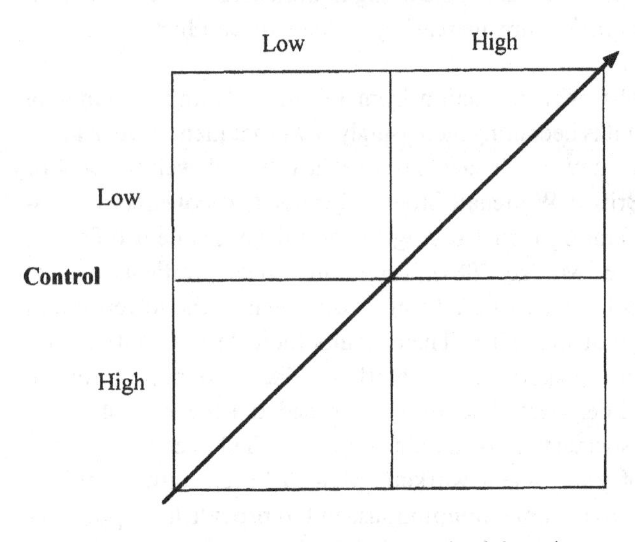

Figure 5.5. Illustration of Karasek's model (from Karasek and Theorell, 1990). Karasek suggests that job strain results when workload or demand is high and control is low. He predicts that "stress hormone" levels will be raised in people working in high strain jobs.

between job characteristics and cardiovascular disease was explained by considering the hormonal correlates of job strain. Karasek *et al.* (1982) considered chronic catecholamine and cortisol elevation to be part of a general catabolic response brought about by job strain. Transiently raised levels of epinephrine were not conceptualized as part of this response, but instead as part of a positive anabolic response associated with job activity or involvement, for people whose jobs involve high demand and high decision latitude. Thus, in essence, epinephrine is expected to rise with demand, whether control is high or low.

With respect to its predictions regarding hormonal variation at work, this model bears many similarities to that developed by Frankenhaeuser, on the basis of laboratory studies conducted in Sweden. She sees everyday work experiences as partial determinants of the subjective states she considers relevant to epinephrine and cortisol variation (Frankenhaeuser, 1989). Her model correlates effort at work with mental arousal and epinephrine secretion, and lack of control with negative feelings and cortisol secretion. Thus, as for Karasek, jobs which make high demands and offer little personal control are considered to lead to high levels of both epinephrine and cortisol. Unlike Karasek, Frankenhaeuser views demand and control as independent determinants of the activity of the sympathetic adrenal-medullary axis and the pituitary adrenal–cortical axis, respectively, via their effects on subjective state.

A number of studies have been carried out in the industrialized workplace to test these models. Latterly, anthropologists have made use of the techniques developed for such studies to explore hormonal variation associated with economic activity in a wider range of societies. Anthropological studies provide an important contrast to data from industrial societies, but they have not approached the sophistication of the work in Sweden and other European countries because of methodological difficulties involved with collecting hormonal data in the field.

5.4.2 Epinephrine in modernizing and industrial communities

Hormone assays conducted amongst men in modernizing Pacific islands have consistently revealed higher epinephrine levels in those integrating into developing cash economies and moving from traditional subsistence activities to forms of waged work (Hanna *et al.*, 1986). Jenner *et al.* (1987) compared 24-hour catecholamine excretion rates collected during normal working days in people from the Tokelau Islands, urban Nigeria, Britain, the USA, and Japan. In men, epinephrine levels were lowest in the Tokelau Islands, intermediate in urban Nigeria, Japan and Britain, and highest in the USA. Samples were collected from women only in the Tokelau Islands and in Britain; again levels were lower in the Tokelau Islanders, although the differences were less marked in women than in men (Figure 5.6).

Thus there is some evidence from comparative analyses that greater levels of epinephrine are secreted in those undertaking work in urban industrial societies compared to those living more traditional subsistence lifestyles, despite the fact that physical activity levels are generally lower. More detailed work conducted within industrial societies, often using longitudinal within subject designs, throws more light on the work situations in which secretion of epinephrine is high.

Two main kinds of comparisons have been made in investigations of work-related epinephrine variation in industrialized contexts. The first contrasts hormone levels in people at work and at leisure, usually using a within subject, longitudinal design. The other compares the effects of different types of work on hormone variation, usually using a between subjects, cross-sectional design. The comparison of work and leisure addresses the question of whether sedentary work *per se* causes an increase

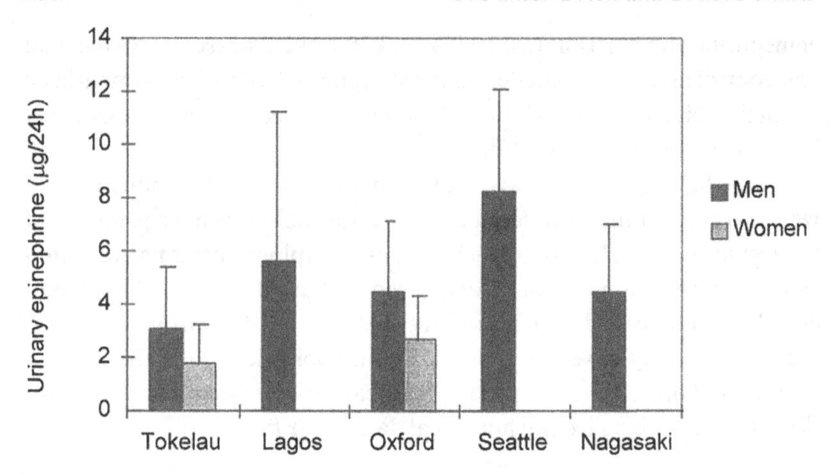

Figure 5.6. 24-hour urinary epinephrine excretion rate in selected populations (mean and SD from Jenner *et al.*, 1987).

in epinephrine secretion. If it does, the result will be chronically (given that most people in industrialized societies are at work for a large proportion of their waking hours) elevated hormone levels, which are likely to be deleterious to health. It is then relevant to ask what particular aspects of work are associated with elevated epinephrine levels.

Studies have led to the conclusion that urinary epinephrine excretion levels are higher in British men living in Oxfordshire when they are doing a variety of forms of mainly non-manual work relative to levels for the same men during their normal weekend days off work (Jenner *et al.*, 1980; Pollard *et al.*, 1996), and in Swedish men at work compared to a weekday spent at home (Frankenhaeuser *et al.*, 1989). These studies all included women as well, but findings were less consistent. Frankenhaeuser *et al.*'s study of a relatively small number of non-manual Volvo workers did not reveal any gender difference in the work day/rest day difference in epinephrine levels. Harrison *et al.* (1981) report on the results from women involved in the same project as the men of Jenner *et al.*'s (1980) study, comparing a normal working day with a Sunday in a much larger sample. They also found that women showed significantly higher urinary epinephrine levels on work compared to rest days. Pollard *et al.* (1996) report no significant difference between urinary epinephrine levels on work and rest days in British women in a range of mostly sedentary occupations, in contrast to their results for men.

Taken together, the results of a series of studies, mostly conducted in Sweden, can be interpreted to show that mental workload and time pressure are the key features leading to elevated epinephrine levels at work (Frankenhaeuser and Gardell, 1976; Rissler, 1977; Cox *et al.*, 1982). Further corroboration comes from Johansson and Aronsson (1984), who studied Swedish women working for an insurance firm. They found that women who mainly performed data entry using visual display units had higher urinary epinephrine excretion rates (compared to baseline) in the morning than in the afternoon. Self-reported effort followed the same pattern, which apparently resulted from the necessity to enter a day's quota of data with the knowledge that computer breakdown could disrupt work at any time. On the days chosen for the study the computer did not break down and mental workload in the afternoon was comparatively light, because the bulk of the data entry had been completed in the morning. Another group of women, who were typists and secretaries, reported an increase in effort, rush, and fatigue in the afternoon because of the need to type material left late in the day for posting the same evening. This group's epinephrine excretion increased in the afternoon.

Evans and Carrère (1991) found that self-reported control in male bus drivers in Los Angeles had a negative relationship with urinary epinephrine excretion rate change from a morning baseline level. There was also a significant positive relationship between exposure to traffic congestion and epinephrine level. The authors suggested that the traffic congestion decreased perceived control and therefore led to hormone elevation, although statistical evidence for this pathway was lacking. In contrast, Gardell (1987) attributed increased urinary epinephrine levels in Swedish bus drivers to greater perceived time pressure created by the demands of remaining on schedule and dealing with traffic congestion. Bus driving may be an occupation in which demand and control are strongly inversely correlated. It is certainly an occupation carrying high risks for ill health, particularly cardiovascular disease (Belkic *et al.*, 1994).

In a group of people from Oxford in a variety of occupations, Pollard *et al.* (1996) found that over a Sunday, Monday, and Tuesday, urinary epinephrine excretion co-varied with perceived demand in men, but not in women. Workload was much higher on Monday and Tuesday in both men and women, and in men urinary epinephrine was also higher on Monday and Tuesday. Tests showed that the increase in demand on workdays was probably largely responsible for the elevation in epinephrine levels seen on workdays in men. This study provides some evidence for the positive association between work demand and epinephrine secretion predicted by both Karasek and Frankenhaeuser's models.

Thus people, particularly men, whose employment requires them to undertake mentally demanding work in the context of an industrial society, are likely to have chronically elevated epinephrine levels. One of the main roles of epinephrine is to free metabolites, such as free fatty acids, in order to provide the body with an immediate source of energy. Those undertaking manual work are much more likely to make use of these metabolites than are those sitting at a desk 5 days a week. It has been suggested that unused free fatty acids may contribute to an elevated cholesterol level, and thus to the risk of cardiovascular disease (Niaura *et al.*, 1992). It could well be that the shift to sedentary work, in combination with a generally inactive lifestyle and calorific diet, is an important cause of high rates of cardiovascular disease in the wealthier, more developed, regions of the world.

5.4.3 Cortisol and "stress" at work

As a consequence of laboratory studies with people and other animals, cortisol, like epinephrine, is known as a stress hormone. The few studies which have examined the responses of epinephrine and cortisol simultaneously have shown that they do not respond to everyday work in the same way. As we have seen, Frankenhaeuser's model (1989), based on laboratory research, suggests that epinephrine levels tend to rise in response to increased demand but that cortisol increases when mental distress is experienced. Many studies have shown that feelings of lack of control lead to distress (Spector, 1986), and it is for this reason that Frankenhaueser suggested that perceptions of lack of control at work are likely to lead to elevated cortisol levels.

It is certainly true that people in developed nations generally report higher levels of distress when they are at work than during their leisure hours (Neale *et al.*, 1987; Pollard *et al.*, 1996). Studies have not shown, however, that cortisol levels are higher when people are at work (Cullen *et al.*, 1979; Frankenhaueser *et al.*, 1989; Lundberg *et al.*, 1989; Pollard *et al.*, 1992, 1996; van Eck & Nicolson, 1994), nor has the expectation of Frankenhaeuser's model that lack of control at work should result in elevated cortisol levels been fulfilled (Pollard *et al.*, 1996).

Several Swedish studies have examined the relationship between Karasek's job strain dimension and cortisol levels. Theorell *et al.* (1988, 1990) took morning plasma cortisol measures four times over 1 year from 50 men in six different occupations and measured job strain at the same time, using a Swedish modification of a questionnaire designed by Karasek. The highest levels of job strain were reported by waiters, who had relatively low cortisol values, the opposite finding to that predicted under Karasek's model.

Härenstam and co-workers (Härenstam *et al.*, 1988; Härenstam & Theorell, 1990) also took morning plasma cortisol measurements, but only once, in just over 2000 Swedish prison employees of both sexes. They obtained self-reports on many aspects of the working environment. Noting the failure of Theorell *et al.* (1988) to find any association between job strain and cortisol level, Härenstam and Theorell (1990) suggested that alcohol abuse may complicate such investigations because of its potential effect on liver function and hence the metabolism of cortisol. They therefore used measures of serum levels of gamma-glutamyl transpeptidase, as an indicator of liver function. They corrected for the positive correlation between this enzyme and plasma cortisol level by examining people with high levels of the enzyme and people with low levels in separate groups. In this way they found that men reporting the least and the most "psychic strain" had higher levels of cortisol than other men. The exact statistical test used is unclear and it was one of many. No such relationship was reported for women. Doncevic *et al.* (1988) collected data from three groups of female district nurses and also failed to find an association between job strain and morning plasma cortisol.

Thus there is very little evidence to support the predictions of Frankenhaueser with respect to job control and cortisol, or those of Karasek with respect to job strain and cortisol, and some of the studies designed to test these predictions may not have been the most appropriate. For example, many of the investigators took early morning salivary cortisol measurements, yet it seems more likely that work experience will affect hormone levels during the working day. Furthermore, variation in secretion of cortisol is so marked in the early morning that precise information on sleep and saliva collection schedules is needed, but not always provided in these studies.

Some studies have examined the relationship between mental workload and cortisol level. For example, Doncevic *et al.*'s results show a slight indication that nurses with the highest objectively measured mental workload (assessed using the number of people living in their district and number of patient consultations) had the highest morning plasma cortisol levels, but this finding was not statistically significant. Zeier *et al.* (1996) collected salivary cortisols from male air-traffic controllers in Switzerland. Here there was a positive correlation between mental workload and change in cortisol level from baseline over a working session. Interestingly, however, these men did not subjectively experience their job as distressing and they described their working sessions using words associated with positively connotated arousal.

In their study of air-traffic controllers, Rose *et al.* (1982) did not find any correlations between mood and plasma cortisol at work. Lundberg *et al.* (1989) found significant correlations between self-reported irritation, tenseness, and tiredness and urinary cortisol levels, in Swedish manual factor workers, but none of these correlations were significant over both occasions of measurement. Thus, while self-reported emotions have been correlated with cortisol levels in some studies of everyday working life, such findings have not been consistent and have not been confined solely to negative feelings (Pollard, 1995). Distress during everyday working life has not been shown to affect cortisol variation in the same way as distress caused by acute experimental stressors.

5.5 Adult unemployment

Unemployment is an important economic phenomenon in most industrialized societies and in many industrializing contexts, whereas in traditional societies it is not part of the economic structure. Within industrialized economies, levels of long-term unemployment appear to be increasing, and are considered by some to be the inevitable price we pay for low inflation and a "flexible" labour market (Bartley, 1994). Cameron (1995) notes that in Britain, at least, there is little prospect now of a return to so-called full employment.

5.5.1 Unemployment and ill health

Unemployment is generally thought to have a detrimental impact on health and has been linked to increased mortality, for example by Brenner (1977), who interpreted longitudinal data sets from the USA to indicate that peaks in mortality from a variety of causes follow peaks in unemployment. The interpretation of such correlational trends is difficult, and more detailed studies of the effects of unemployment on health are also plagued by methodological problems, such as the possibility that selection of unhealthy people into unemployment explains the higher morbidity and mortality consistently seen in this group (Cameron, 1995). Furthermore, people who experience spells of unemployment also tend, when they are in work, to hold jobs which carry the greatest risks for health (for example, jobs with high demand and offering little personal control, such as assembly-line work; Bartley, 1994). Only prospective studies can distinguish between such selection effects and causal effects of unemployment on health. The few that are available do appear to confirm that there is a causal relationship. For example, Morris *et al.* (1994) found that amongst a group of British men who took part in the British Regional Heart Study, and who were in stable employment at the start of the study, those who lost employment were twice as likely to die as continuously employed men at a 5.5-year follow-up.

Apart from selection effects, Bartley (1994) notes a number of mechanisms by which the relative general ill health of the unemployed might be explained. Unemployment may have effects via its impact on finances, i.e., increased relative poverty is likely to have harmful consequences for health. Second, being unemployed may affect people's health-related behavior, such that they smoke, drink or eat more, or take less exercise, when unemployed. Morris *et al.* (1992) showed that men in the British Regional Heart Study who left work gained more weight than men who remained in employment, presumably because of such changes. Importantly for the consideration of selection effects, these men were also more likely to smoke or drink heavily before they became unemployed, indicating that those with more unhealthy behavior may have been selected out of employment. Thus the evidence with respect to the causality of associations between unemployment and unhealthy behavior may not always be clear-cut.

The final mechanism identified by Bartley (1994) concerns the effects that job loss may have as a stressful life event. Unemployment is generally an unwelcome experience for emotional as well as financial reasons. Jahoda (1942) pointed to the loss of benefits of employment, such as having a time structure to the day, maintenance of self-esteem, social support, and respect from others, as some of the major costs of unemployment. Warr (1987) has demonstrated the negative consequences of unemployment for mental health, particularly depression, and has suggested that they often reach a peak about 3 months after leaving work (Warr and Jackson, 1985). Worries can begin with job insecurity or anticipation of job loss (Ferrie *et al.*, 1995) and can thus affect an even greater number than those unemployed at any one time. We might expect endocrine changes to be associated with the stressful experiences of job insecurity and unemployment, and that such changes could mediate the effects of anxiety on physical health. In a now classic study, Kasl *et al.* (1968) demonstrated, using a prospective design, a rise in total cholesterol levels in men who became unemployed, and, more recently, Mattiasson *et al.* (1990) showed a rise in total cholesterol levels relative to controls in male Swedish shipyard workers threatened with unemployment. It is possible that unmeasured changes in "stress hormone" levels were at least partially responsible for these rises in cholesterol, and thus probably elevated risk of cardiovascular disease.

5.5.2 Endocrine variation associated with unemployment

Unfortunately there have been, as yet, very few studies of the consequences of either unemployment or job insecurity for hormone levels. Of those that have been published, most have addressed the possible impact of distress generated by these situations on cortisol levels. Arnetz *et al.* (1988) examined a range of physiological measures in a group of mainly female Swedish manual workers during anticipation of job loss, actual job loss, and short and long-term unemployment, following the group over a period of more than 2 years. Levels of well-being, as assessed by the General Health Questionnaire and other measures, were lowest just before job loss and highest immediately after (presumably due to the resolution of uncertainty and, perhaps, enjoyment of not having to work). Well-being fell again after 3 months and then gradually rose until the end of the study. Levels of serum cortisol were assessed in blood samples taken in the early morning, as were levels of prolactin and growth hormone, as there is also some evidence that these hormones may rise in response to psychosocial stress (Delahunt & Mellsop, 1987). Levels of all three hormones changed significantly over the course of the study, showing peaks at the anticipatory stage and 1 year after leaving work. Serum prolactin and growth hormone levels were raised both before and immediately after job loss, declining after 6 months of unemployment. Here there appears to be some evidence of endocrine changes associated with the process of becoming unemployed, although these do not appear to correspond closely to changes in well-being.

Claussen (1994) reports the results of a study of similar basic design, this time conducted in Norway with a group of both women and men who had already been unemployed for at least 12 weeks at the start of the study. Again, serum samples were taken for the assessment of cortisol and prolactin, as well as testosterone, but the timing of sample collection was not standardized between subjects or over the two assessments, making the results difficult to interpret because of the potentially overwhelming effects of normal circadian variation. Although those who were re-employed at the 2-year follow-up showed an increase in mental well-being, there were no differences in the changes in endocrine values between the re-employed and continuously unemployed groups. Hall and Johnson (1987) found no differences in serum cortisol levels between unemployed Swedish women

2 years after losing their blue-collar jobs and employed women in similar occupations, although the unemployed women reported higher levels of depression.

Toivanen *et al.* (1996) also measured plasma hormone levels, in this case epinephrine, norepinephrine, and cortisol. The assessment of catecholamines in blood obtained by venepuncture is, unfortunately, particularly suspect, as these hormones have been shown to rise very quickly with anticipation of this procedure (Engel *et al.*, 1980). Nevertheless, the comparison between changes in levels over the 6-month long prospective study in six women who happened to be laid off during this time, with a control group of permanently employed subjects, is worth mentioning. The unemployed women showed an increase in cortisol levels and a decrease in epinephrine levels compared to the controls.

Ockenfels *et al.* (1995) took a different approach to the investigation of the effects of unemployment on cortisol levels. They were interested in effects on diurnal variation in hormone levels and examined salivary cortisol levels in employed and unemployed (for an average of 12.6 months) men and women in the USA over 2 week days. The groups did not differ in their overall cortisol excretion or in cortisol reactivity to daily stressors.

Thus it seems that the acute experience of anticipation of job loss may well be associated with an elevation in cortisol levels, but there is some evidence that those who have been unemployed for some time show normal levels. It could be that the impact of unemployment on cortisol levels, as opposed to the largely null findings with respect to the impact of variation in working experience, reflects the fact that the consequences for the mental well-being of an individual are more severe than any kind of variation in experience within employment. It is also likely that hormonal changes with job insecurity and unemployment will vary markedly according to the economic and personal consequences of job loss. Some of the variation in the results of the studies cited may arise due to the different experiences of unemployment in these countries. For example, state intervention to protect the financial well-being of those without work has traditionally been much greater in Sweden than in the USA. At present the data are insufficient to allow any conclusions to be drawn with respect to the impact of unemployment on hormones other than cortisol.

5.6 Children's stress in work environments

In the final section of this chapter, we briefly consider research on the hormonal correlates of children's behavior, health, and well-being in diverse work-related environments. Broadly speaking, children's main "work" in modern industrial societies takes place at school, while in non-industrial contexts, children's work includes informal or formal participation in subsistence activities and wage labor, with or without periods at school. We discuss here some innovative endocrine research, first for young children in school settings, and second for homeless street-boys who face serious hazards in their working life.

Such research is concerned not so much with the physical aspects of children's activities (energetic stress) as with the psychological dimension of individual performance and social interactions (psychosocial stress) in school or other environments. There is of course an extensive literature focusing attention on child health in relation to physical activity, particularly the allegedly low levels of physical exertion amongst Western children, and extensive work responsibilities outside school in the developing world (for review see Panter-Brick, 1998). This literature incorporates little or no hormonal data. Cortisol variation, for example, while related to physical exertion (Kirschbaum & Hellhammer, 1994), is rarely analyzed with respect to physical activity in studies of child health (see Flinn, Chapter 4).

Interest in children's everyday behaviors and hormonal profiles has thus focused on their significance for psychosocial well-being. Comparatively few endocrine studies have been conducted on children, in either Western or non-Western contexts. Kirschbaum and Hellhammer's (1989, 1994) often-cited reviews of the literature on salivary cortisol, for instance, mention only one study of 9- to 14-year-olds and one of infants in the 1989 paper, and four studies (of infants) in the 1994 article, all from Western populations. Few studies of "childhood stress", e.g., at school (Sears & Milburn, 1990), have incorporated measures of hormonal variation. Yet endocrine output in relation to achievement and social adjustment at school had been investigated early on (Johansson *et al.*, 1973; Bergman & Magnusson, 1979).

In Western industrial settings, most studies of children's endocrine profiles (post-infancy) have examined responses to activities on school versus home days or holidays. For example, Long *et al.* (1993) examined differences in home and school levels of both urinary catecholamines and corticosteroids in British 4- to 9-year-olds attending primary school. School levels of cortisol (as well as cortisone and norepinephrine, the latter taken to indicate physical activity), but not epinephrine, were significantly lower relative to the home day. The authors thought their data might reflect a structured school environment with clearly defined routines, and emphasized that results might not be generalizable to other institutional settings. What they did demonstrate was that cortisol and epinephrine, the two stress hormones, showed distinct profiles in home and school environments; as with adults, these hormones may respond to everyday situations in different ways.

Lundberg and colleagues have conducted interesting research relating children's neuroendocrine profiles to cardiovascular measures (blood pressure and heart rates) and behavioral characteristics (antecedents of type A behavior, namely aggressive, competitive, impatient dispositions), to arrive at a composite picture of stressful interactions in preschool and home contexts (Lundberg *et al.*, 1991/2; 1993). Lower levels of urinary cortisol, but higher levels of urinary epinephrine, were reported for 3- to 6-year-olds at the day-care center as compared to home (Lundberg, 1983; Lundberg *et al.*, 1993). Decreased cortisol levels were interpreted as indicating feelings of control, whilst increased heart rates and epinephrine were taken to indicate challenging activities at day-care. Children were given physical and emotional challenges, in order to observe physiological arousal and type A behavior in stressful situations. There was little evidence for an association between physiological reactivity and type A behavior in the day-care environment (Lundberg *et al.*, 1991/2).

In turn, Spangler (1995) examined behavioral measures and salivary cortisol among 6-year-old German children, on days of primary school and holidays. He found no cortisol differences related to school attendance *per se*, but did observe lower values in children with type A disposition and children with poor school performance (measured by school achievement, motivation during homework, and school-related problems). Tennes and Kreye (1985), conversely, reported higher urinary cortisol levels for underachieving 6- to 9-year-olds at two North American schools, noting however that hostility towards teachers and low peer popularity correlated with lower cortisol. In their study, cortisol was positively associated both with school work preoccupation and positive social involvement, and negatively associated with aggressive behaviors. In younger children entering a North American preschool, Gunnar *et al.* (1997) noted high median cortisol values among 3- to 5-year-olds characterized by poor self-control, and high cortisol reactivity among children who were less outgoing and socially competent. The relationships between cortisol regulation, reactivity, and behavioral stress were thought to reflect both individual and contextual differences (see also Granger *et al.*, 1994). Such research suggest that relationships between adrenocorticol profiles and childhood responses to contrasting environments, including reactions to potentially stressful demands of work and social interactions, are quite complex.

Far fewer studies have been conducted on children in Third World settings. Flinn's ethnographic study (Chapter 4) is one outstanding example, examining endocrine responses to changes in social and demographic environments and individual personality. Recently, Durbrow *et al.* (in press) have examined variation in adrenocorticol reactivity in Caribbean (St Vincent) village children, 5 to 12-years-old, in relation to their behavior problems, temperament, peer relations, and performance in academic work. Levels of salivary cortisol, collected on school days and on weekends, were higher immediately prior to coming to school, and upon school arrival (relative to home), and were associated with poor schoolwork grades (especially for children with problems of inattention/internalizing behaviors). The differences between home and school probably reflected anxiety about school-tests, reprimands from teachers, or interactions with peers.

In brief, cortisol profiles, as well as catecholamines, appear to be associated both with school-work performance and social interactions in school environments. Moreover, chronically elevated cortisol levels have been associated with indicators of ill health. Elevation of cortisol, upon school entry, has been associated with significant changes in immune responses, namely increased B lymphocyte cells and decreased antibody response (Boyce *et al.*, 1995). Fernald and Grantham-McGregor (in press) have also reported links between salivary cortisol, emotional withdrawal, and mild stunting of growth among 8- to 10-year-old urban Jamaican children.

A wider range of environments and health indicators was considered by Panter-Brick and colleagues, in a study of 10- to 14-year-old Nepali children in four diverse socio-economic and urban/rural contexts: homeless street-boys as well as children in a poor urban slum, a privileged school, and a remote village. The homeless boys subsist principally from begging or picking out scrap metals and plastics from household and municipal dumps, to sell for junkyards for recycling. They live independently from their families, spending nights on the street or taking refuge with a non-governmental organization. Poor slum-dwelling children also work on the streets, picking junk for resale, but live at home in crowded settlements. The urban middle-class boys are at a private fee-paying school, while the village children are routinely involved in an agro-pastoral subsistence economy in addition to attending the local school.

This study integrated observations on children's activities and social networks with a range of biological indicators measuring variation in hormonal, cardiovascular, immune, growth, and morbidity status (Baker *et al.*, 1996, Baker, 1997; Panter-Brick *et al.*, 1996*a b*). Measures of lifestyles and well-being were examined across the range of environments, and detailed knowledge of individual children was used to interpret biological variables, particularly cortisol variation. The salivary cortisol profiles of these Nepali children were intriguing. Average mid-morning values were low for all four populations (0.18 – 0.27 μg/dl), in comparison to the range of published values for US children (0.23 – 0.66 μg/dl), although similar to mid-morning values obtained in the same laboratory at Emory University for Atlanta school-boys (0.10 μg/dl) (C. Worthman, personal communication). They are very close to the values reported for Caribbean children, awake for a comparable period of time (0.22 μg/dl, 2 hours after waking) (Flinn and England, 1997:39). In Nepal, mean levels were higher for urban homeless and school boys relative to slum and village children (Table 5.3). The values for school-boys are perhaps explained by the demands of school-work in preparation for examinations. The values for homeless street-boys are far from high, although elevated by comparison to slum-dwelling boys, who share a similar day-time environment but return to their families at night. These data have been extensively validated, including comparison of village children with their parents (the Tamang described earlier on pp. 143 and pp. 148–51), who show similarly low cortisol values.

The picture given by day-to-day cortisol variability, as opposed to mean values, is also revealing (Table 5.3). The homeless boys show low cortisol variance relative to boys living at home. This would suggest an increased arousal threshold to day-to-day circumstances, or put more simply, that homeless boys have a blunted response to "stress" in their environments. Conversely, less reactive boys may be those who become or remain homeless on the street.

Detailed knowledge of children's lives is obviously needed to shed light on cortisol variation. One particularly striking incident during fieldwork illustrates the potential value of a more fine-grained approach. It concerns two homeless children, threatened by a gang while picking junk for resale on the street (Figure 5.7). One boy (LLR, age 13 years, homeless for the past year) was stopped by the gang of street-boys, who tried to extort money and threatened to "cut his arm" if he failed to give them anything of value when they next met. Two days later, he was apprehended again and, with no money to hand over, had his arm cut with a razor blade. The boy explained this matter-of-factly once researchers noticed his injury, but had not reported these incidents when asked to detail his work and other

Table 5.3. *Salivary cortisol values for four groups of Nepali 10–14-year-old boys habituated to saliva collection*

	Boys	Mean (μg/dl)		Variance (%)	
	N	Mean	SD	Mean	SD
Urban homeless	27	0.27	0.08	46.4	16.4
Urban school	30	0.24	0.08	52.9	19.6
Urban slum	20	0.18	0.06	52.1	16.9
Rural village	30	0.18	0.07	69.5	20.6

Note: Data are based on 753 am samples (pm values showed no detectable variation). Time since waking (2 h on average) and since consuming food (1–1.5 hour) did not affect group comparisons.
Differences are significant for homeless ($p < 0.001$) and school ($p < 0.01$) boys compared with either slum and village boys, in mean values; for village ($p < 0.01$) relative to urban groups, in variance (coefficient of variation, SD/mean*100, for 2–15 sample days).
From Panter-Brick *et al.* (1996*c*).

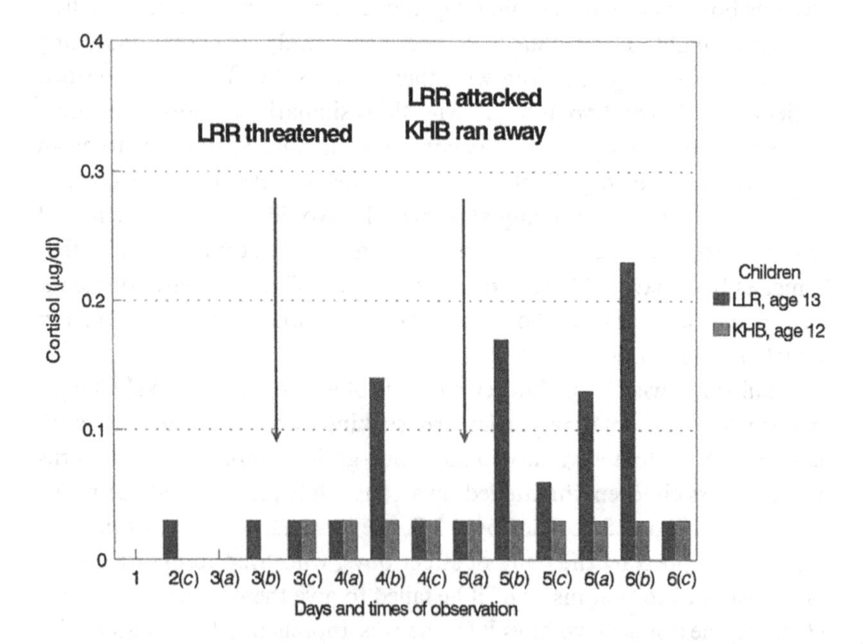

Figure 5.7. Cortisol variation in two homeless Nepali boys threatened while working on the streets. Morning (*a*), lunchtime (*b*), and evening (*c*) samples from the 2nd to the 6th of August.

activities, or rate his day, overall, as a good or bad day. In his case, cortisol values were elevated in response to the threats, and remained high over the subsequent days of observation (Figure 5.7). The other boy (KHB, age 12 years, homeless for the previous 6 months), who was with LRR on the day he sustained his injury, ran away instead of facing the gang. Remarkably, his cortisol levels remain at a very low level, showing no diurnal variation throughout the period of observation. This profile might indicate a child coping with stress by downregulation of the hypothalamo–pituitary–adrenal axis. In brief, one child showed marked physiological responses to a particularly stressful event, but did not report the incident as such, while his companion fled the threat and maintained very low cortisol levels.

These data raise many issues regarding cortisol regulation and variation. Perhaps homeless boys show low baseline values because they habituate to the hazards of working life on the streets. Perhaps they represent a select population of "survivors" in this regard, characterized by a blunted hormonal response to stressful events. Such issues are yet to be answered. Relationships between cortisol and personality (measured through a simple rank exercise completed by the children), perceptions of life circumstances (ascertained in ethnographic enquiries), illnesses (from self-reports), and physical activity (from 24-h heart rate monitoring) have yet to be explored. A great deal more understanding on endocrine variation may be gained from the fine-grained study of individual children, although interpretation of hormonal data may prove more challenging than actual collection of samples!

As a final word of caution, measures of hormonal variation often relate to indicators of lifestyle, health, and well-being in rather complex ways. In the Nepal study, measures included serum hemoglobin, albumin, and alpha-1-antichymotrypsin (ACT, a measure of inflammatory responses to high pathogen exposure) in addition to those above mentioned. While mean cortisol values showed little relationship with other health indicators, such as measures of growth status, higher cortisol variances were significantly correlated with low ACT responses, indicative of a low pathogen load (Panter-Brick *et al.*, 1996*c*). Such relationships are not simple to interpret, but do serve as markers of health for children in their natural environments.

5.7 Conclusion

Examination of ways in which physiological variables mediate the relationships between lifestyle and health in contrasting social and ecological contexts, as well as during the entire human lifespan, is a task of considerable importance. Working behavior, incorporating both aspects of physical exertion and psychosocial stress, is a hugely important component of lifestyle, and it is certainly related to variation in health outcomes. We are just beginning to learn how endocrine changes may mediate many of these effects. So far, rather different investigations, and different hormones, have been undertaken in industrial and non-industrial contexts. Clearly, comparative research across diverse environments can enrich our understanding of hormonal variation and the range of biological outcomes expected. Studies in day-to-day environments of children, not just adults, are also important in this respect. We expect future research will exploit the potential offered by recent methodological advances, such as analyses of salivary samples, to explore the issues raised by current inquiries.

5.8 References

Arnetz, B., Brenner, S., Hjelm, R., Levi, L., & Petterson, I. (1988). *Stress Reactions in Relation to Threat of Job Loss and Actual Unemployment: Physiological, Psychological and Economic Effects of Job Loss and Unemployment.* Department of Stress Research, National Institute for Psychosocial Factors and Health, Sweden.

Baker, R. (1997). Runaway street-children in Nepal: Social competence away from home. In *Children and Social Competence: Arenas of Action*, ed. I. Hutchby & J. Moran-Ellis, pp. 46–63. London: Falmer Press.

Baker, R., Panter-Brick, C., & Todd, A. (1996). Methods used in research with children in Nepal. *Childhood*, **3**, 171–93.

Bandini, L. G., Schoeller, D. A., Fukagawa, N. K., Wykes, L. J. & Dietz, W. H. (1991). Body composition and energy expenditure in adolescents with cerebral palsy or myelodysplasia. *Pediatric Research*, **29**, 70–7.

Bartley, M. (1994). Unemployment and ill health: Understanding the relationship. *Journal of Epidemiology and Community Health*, **48**, 333–7.

Beall, C. M., Worthman, C. M., Stallings, J., Strohl, K. P., Brittenham, G. M. & Barragan. M. (1992). Salivary testosterone concentration of Aymara men native to 3600 m. *Annals of Human Biology*, **19**, 67–78.

Belkic, K., Savic, C., Theorell, T., Rakic, L., Ercegovac, D., & Djordjevic, M. (1994). Mechanisms of cardiac risk among professional drivers. *Scandinavian Journal of Work Environment and Health*, **20**, 73–86.

Bentley, G. R. (1985). Hunter–gatherer energetics and fertility: A reassessment of the !Kung San. *Human Ecology* **13**, 79–109.

Bentley, G. R. (1996). Evidence for interpopulation variation in normal ovarian function and consequences for hormonal contraception. In *Variability in Human Fertility*, ed. L. Rosetta & C. G. N. Mascie-Taylor, pp. 46–65. Cambridge: Cambridge University Press.

Bentley, G. R., Harrigan, A. M., Campbell, B., & Ellison, P. T. (1993). Seasonal effects on salivary testosterone levels among Lese males of the Ituri forest, Zaire. *American Journal of Human Biology*, **5**, 711–17.

Bergman, L., & Magnusson, D. (1979). Overachievement and catecholamine excretion in an achievment-demanding situation. *Psychosomatic Medicine*, **41**, 181–8.

Black, A. E., Coward, W. A., Cole, T. J., & Prentice, A. M. (1996). Human energy expenditure in affluent societies: An analysis of 574 doubly-labelled water measurements. *European Journal of Clinical Nutrition* **50**, 72–92.

Bleiberg, F. M., Brun, T. A., & Goihman, S. (1980). Duration of activities and energy expenditure of female farmers in dry and rainy seasons in Upper-Volta. *British Journal of Nutrition*, **43**, 71–82.

Bosma, H., Marmot, M., Hemingway, H., Nicholson, A., Brunner, E., & Stansfeld, S. (1997). Low job control and risk of coronary heart disease in Whitehall II (prospective) study. *British Medical Journal*, **314**, 558–65.

Boyce, W. T., Adams, S., Tschann, J. M., Cohen, F., Wara, D., & Gunnar, M. R. (1995). Adrenocorticol and behavioral predictors of immune responses to starting school. *Pediatric Research*, **38**, 1009–17.

Brenner, M. (1977). Health costs and benefits of economic policy. *International Journal of Health Services*, **7**, 581–95.

Bribiescas, R. G. (1996). Testosterone levels among Ache hunter–gatherer men: A functional interpretation of population variation among adult males. *Human Nature*, 163–88.

Brun, T., Bleiberg, F., & Goihman, S. (1981). Energy expenditure of male farmers in dry and rainy seasons in Upper-Volta. *British Journal of Nutrition*, **45**, 67–75.

Bullen, B. A., Skrinar, G. S., Beitins, I. Z., Von Mering, G., Turnbull, B. A. & McArthur, J. W. (1985). Induction of menstrual disorders by strenuous exercise in untrained women. *New England Journal of Medicine*, **312**, 1349–53.

Cameron, A. (1995). Work and change in industrial society: a sociological perspective. In *Work and Health: An introduction to occupational health care*, ed. M. Bamford, pp. 61–100. London: Chapman and Hall.

Campbell, B., & Leslie, P. (1995). Reproductive ecology of human males. *Yearbook of Physical Anthropology*, **38**, 1–26.

Christiansen, K. H. (1991). Serum and saliva sex hormone levels in !Kung San men. *Ammerican Journal of Physical Anthropology*, **86**, 37–44.

Claussen, B. (1994). Psychologically and biochemically assessed stress in a follow-up study of long-term unemployed. *Work and Stress*, **8**, 4–18.

Cox, S., Cox, T., Thirlaway, M., & Mackay, C. (1982). Effects of simulated repetitive work on urinary catecholamine excretion. *Ergonomics*, **25**, 1129–41.

Cullen, J., Fuller, R., & Dolphin, C. (1979). Endocrine stress responses of drivers in a 'real-life' heavy goods vehicle driving task. *Psychoneuroendocrinology*, **4**, 107–15.

Cumming, D. C., & Wheeler, G. D. (1990). Exercise-associated changes in reproduction: A problem common to women and men. *Progress in Reproduction and Biological Medicine*, **14**, 125–35.

Delahunt, J., & Mellsop, G. (1987). Hormone changes in stress. *Stress Medicine*, **3**, 123–34.

Doncevic, S., Theorell, T., & Scalia-Tomba, G. (1988). The psychosocial work environment of district nurses in Sweden. *Work and Stress*, **2**, 341–51.

Durbrow, E., Gunnar, M. R., Bozoky, I., Adam, E., Jimerson, S., & Chen, C. (in press). The stress of coming to school: Variation in adrenocortical reactivity in Caribbean village children. *International Journal of Behavioral Development*.

Durnin, J. V. G. A., & Passmore, R. (1967). *Energy, Work and Leisure*. London: Heinemann.
Ellison, P. T. (1990). Human ovarian function and reproductive ecology: New hypotheses. *American Anthropologist*, **92**, 933–52.
Ellison, P. T. (1996). Developmental influences on adult ovarian hormonal function. *American Journal of Human Biology*, **8**, 725–34.
Ellison, P. T., & Lager, C. (1985). Exercise-induced menstrual disorders. *New England Journal of Medicine*, **313**, 825–6.
Ellison, P. T., & Lager, C. (1986). Moderate recreational running is associated with lowered salivary progesterone profiles in women. *American Journal of Obstetrics and Gynecology*, **154**, 1000–3.
Ellison, P. T., & Panter-Brick, C. (1996). Salivary testosterone levels among Tamang and Kami males of central Nepal. *Human Biology*, **68**, 955–65.
Ellison, P. T., Panter-Brick, C., Lipson, S. F., & O'Rourke, M. T. (1993). The ecological context of human ovarian function. *Human Reproduction*, **8**, 2248–58.
Ellison, P. T., Peacock, N., & Lager, C. (1989). Ecology and ovarian function among Lese women of the Ituri forest, Zaire. *American Journal of Physical Anthropology*, **78**, 519–26.
Engel, R., Mueller, F., Muench, U., & Ackenheil, M. (1980). Plasma catecholamine response and autonomic functions during short-time psychological stress. In *Catecholamines and Stress: Recent Advances*, ed. E. Usdin, R. Kvetnansky, & I. Kopin, pp. 461–6. Amsterdam: Elsevier.
Evans, G., & Carrère, S. (1991). Traffic congestion, perceived control, and psychophysiological stress among urban bus drivers. *Journal of Applied Psychology*, **76**, 658–63.
FAO/WHO/UNU Expert Consultation. (1985). Energy and protein requirements. Technical Report Series 724, World Health Organization, Geneva.
Fernald, L. C., & Grantham-McGregor, S. (in press). Stress response in children who have experienced growth retardation. *American Journal of Clinical Nutrition*.
Ferrie, J., Shipley, M., Marmot, M., Stansfeld, S., & Smith, G. (1995). Health effects of anticipation of job change and non-employment: Longitudinal data from the Whitehall II study. *British Medical Journal*, **311**, 1264–8.
Flinn, M. V., & England, B. G. (1997). Social economics of childhood glucocorticoid stress response and health. *American Journal of Physical Anthropology*, **102**, 33–53.
Frankenhaeuser, M. (1989). A biopsychosocial approach to work life issues. *International Journal of Health Services*, **19**, 747–58.
Frankenhaeuser, M., & Gardell, B. (1976). Underload and overload in working life: Outline of a multidisciplinary approach. *Journal of Human Stress*, **2**, 35–46.
Frankenhaeuser, M., Lundberg, U., Fredrikson, M., Melin, B., Tuomisto, M., Myrsten, A., Hedman, M., Bergman-Losman, B., & Wallin, L. (1989). Stress on and off the job as related to sex and occupational status in white-collar workers. *Journal of Organizational Behavior*, **10**, 321–46.
Frisch, R. E. (1987). Body fat, menarche, fitness and fertility. *Human Reproduction*, **2**, 521–33.
Frisch, R. E., & Revelle, R. (1970). Height and weight at menarche and a hypothesis of critical body weight and adolescent events. *Science*, **169**, 397–9.

Galvin, K. A. (1985). Food procurement, diet, activities and nutrition of Ngisonyoka, Turkana pastoralists in an ecological and social context. Doctoral dissertation. State University of New York, Binghamton.
Gardell, B. (1987). Efficiency and health hazards in mechanized work. In *Work Stress*, ed. J. Quick, R. Bhagat, J. Dalton, & J. Quick, pp. 50–71. New York: Praeger.
Gillepsie, S., & McNeill, G. (1992). *Food, Health and Survival in India and Developing Countries*. Delhi: Oxford University Press.
Godin, G., & Shepard, R. J. (1973). Activity patterns of the Canadian Eskimo. In *Polar Human Biology*, ed. O. G. Edolhm & E. K. E. Gunderson, pp. 193–215. London: Heinemann Medical.
Granger, D., Stansbury, K., & Henker, B. (1994). Preschoolers' behavioral and neuroendocrine responses to social challenge. *Merill-Palmer Quarterly*, **40**, 190–211.
Gunnar, M. R., Tout, K., de Haan, M., & Pierce, S. (1997). Temperament, Social Competence, and Adrenocortical Activity in Preschoolers. *Developmental Psychobiology*, **31**, 65–86.
Hackney, A. C. (1989). Endurance training and testosterone levels. *Sports Medicine*, **8**, 117–27.
Häerenstam, A., & Theorell, T. (1990). Cortisol elevation and serum gamma-glutamyl transpeptidase in response to adverse job conditions: How are they interrelated. *Biological Psychology*, **31**, 157–71.
Häerenstam, A., Palm, U., & Theorell, T. (1988). Stress, health and the working environment of Swedish prison staff. *Work and Stress*, **2**, 281–90.
Hall, E., & Johnson, J. (1987). Depression in unemployed Swedish women. *Social Science and Medicine*, **27**, 1349–55.
Hanna, J. M., James, G. D., & Martz, J. M. (1986). Hormonal measures of stress. In *The changing Samoans: Behavior and Health in Transition*, ed. P. T. Baker, J. M. Hanna & T. S. Baker, pp. 203–21. New York: Oxford University Press.
Harrison, G. A. (1996). Introduction: the biological anthropological approach. In *Variability in Human Fertility*, ed. L. Rosetta & G. G. N. Mascie-Taylor, pp. 1–3. Cambridge: Cambridge University Press.
Harrison, G. A., Palmer, C., Jenner, D., & Reynolds, V. (1981). Associations between rates of urinary catecholamine excretion and aspects of lifestyle among adult women in some Oxfordshire villages. *Human Biology*, **53**, 617–33.
Huffman, S. L., Chowdhury, A. K. M. A., Chakraborty, J., & Simpson, N. K. (1980). Breastfeeding patterns in rural Bangladesh. *American Journal of Clinical Nutrition*, **33**, 144–54.
Jahoda, M. (1942). Incentives to work – a study of unemployed adults in a special situation. *Occupational Psychology*, **16**, 20–30.
Jasienska, G. (1996). Energy expenditure and ovarian function in rural women from Poland. PhD thesis, Harvard University, Cambridge, Massachusetts.
Jenike, M. R. (1996). Activity reduction as an adaptive response to seasonal hunger. *American Journal of Human Biology*, **8**, 517–34.
Jenner, D., Harrison, G. A., Prior, I. A. M., Leonetti, D. C., Fujimoto, W. J., Kabuto, M. (1987). Inter-population comparisons of catecholamine excretion. *Annals of Human Biology*, **14**, 1–9.
Jenner, D., Reynolds, V. & Harrison, G. (1980). Catecholamine excretion rates and occupation. *Ergonomics*, **23**, 237–46.
Johansson, G., & Aronsson, G. (1984). Stress reactions in computerized administrative work. *Journal of Occupational Behavior*, **5**, 159–81.

Johansson, G., Frankenhauser, M., & Magnusson, D. (1973). Catecholamine output in school children as related to performance and adjustment. *Scandinavian Journal of Psychology*, **14**, 20–8.

Jones, P. J., & Leitch, C. A. (1993). Validation of doubly labelled water for measurement of caloric expenditure in collegiate swimmers. *Journal of Applied Physiology*, **74**, 2909–14.

Karasek, R. (1979). Job demands, job decision latitude, and mental strain: Implications for job redesign. *Administrative Science Quarterly*, **24**, 285–308.

Karasek, R., & Theorell, T. (1990). *Healthy work: Stress, productivity, and the reconstruction of working life*. New York: Basic Books.

Karasek, R., Baker, D., Marxer, F., Ahlbom, A. & Theorell, T. (1981). Job decision latitude, job demands, and cardiovascular disease: A prospective study of Swedish men. *American Journal of Public Health*, **71**, 694–705.

Karasek, R., Russell, R., & Theorell, T. (1982). Physiology of stress and regeneration in job related cardiovascular illness. *Journal of Human Stress*, **8**, 29–42.

Karasek, R., Theorell, T., Schwartz, J., Schnall, P., Pieper, C., & Michela, J. (1988). Job characteristics in relation to the prevalence of myocardial infarction in the US Health Examination Survey (HES) and the Health and Nutrition Examination Survey (HANES). *American Journal of Public Health*, **78**, 910–18.

Kashiwazaki, H., Dejima, Y., Orias-Rivera, J., & Coward, W. A. (1995). Energy expenditure determined by the doubly-labeled water method in Bolivian Aymara living in a high altitude agropastoral community. *American Journal of Clinical Nutrition*, **62**, 901–10.

Kasl, S., Cobb, S., & Brooks, G. (1968). Changes in serum uric acid and cholesterol levels in men undergoing job loss. *Journal of American Medical Association*, **206**, 1500–07.

Katzmarzyk, P. T., Leonard, W. R., Crawford, M. H., & Sukernik, R. I. (1994). Resting metabolic rate and daily energy expenditure among two indigeneous Siberian populations. *American Journal of Human Biology*, **6**, 719–30.

Kirschbaum, C., & Hellhammer, D. H. (1989). Salivary cortisol in psychobiological research: An overview. *Neuropsychobiology*, **22**, 150–69.

Kirschbaum, C., & Hellhammer, D. H. (1994). Salivary cortisol in psychoendocrine research: Recent developments and applications. *Psychoneuroendocrinology*, **19**, 313–33.

Kujala, U. M., Alen, M., & Huhtaniemi, I. T. (1990). Gonadotropin-releasing hormone and human chorionic gonadotropin tests reveal that both hypothalamic and testicular endocrine functions are suppressed during acute prolonged physical exercise. *Clinical Endocrinology*, **33**, 219–25.

Lawrence, M., & Whitehead, R. G. (1988). Physical activity and total energy expenditure of child-bearing Gambian village women. *European Journal of Clinical Nutrition*, **42**, 145–60.

Leonard, W. R. (1988). The impact of seasonality on caloric requirements of human populations. *Human Ecology*, **16**, 343–6.

Leonard, W. R., & Robertson, M. L. (1992). Nutritional requirements and human evolution: A bioenergetics model. *American Journal of Human Biology*, **4**, 179–95.

Leonard, W. R., Galloway, V. A., & Ivakine, E. (1997). Underestimation of daily energy expenditure with the factorial method: Implications for anthropological research. *American Journal of Physical Anthropology*, **103**,

443–54.

Leonard, W. R., Katzmarzyk, P. T., Stephen, M. A., & Ross, A. G. P. (1995). Comparison of the heart-rate monitoring and factorial method: Assessment of energy expenditure in highland and coastal Ecuadoreans. *American Journal of Clinical Nutrition*, **61**, 1146–52.

Lipson, S. F., & Ellison, P. T. (1992). Normative study of age variation in salivary progesterone profiles. *Journal of Biosocial Science*, **24**, 233–44.

Lipson, S. F., & Ellison, P. T. (1996). Comparison of salivary steroid profiles in naturally ocurring conception and non-conception cycles. *Human Reproduction*, **11**, 2090–6.

Livingstone, M. B. E., Strain, J. J., Prentice, A. M., *et al.* (1991). Potential contribution of leisure activity to the energy expenditure patterns of sedentary populations. *British Journal of Nutrition*, **65**, 145–55.

Long, B. L., Ungpakorn, G., & Harrison, G. A. (1993). Home–school differences in stress hormone levels in a group of Oxford primary school children. *Journal of Biosocial Science*, **25**, 73–8.

Lundberg, U. (1983). Sex differences in behaviour pattern and catecholamine and cortisol excretion in 3–6 year old day-care children. *Biological Psychology*, **16**, 109–17.

Lundberg, U., Granqvist, M., Hansson, T., Magnusson, M., & Wallin, L. (1989). Psychological and physiological stress responses during repetitive work at an assembly line. *Work and Stress*, **3**, 143–53.

Lundberg, U., Rasch, B., & Westermark, O. (1991/1992). Physiological reactivity and Type A behavior in preschool children: A longitudinal study. *Behavioral Medicine*, **17**, 149–57.

Lundberg, U., Westermark, O., & Rasch, B. (1993). Cardiovascular and neuroendocrine activity in preschool children: Comparison between day-care and home levels. *Scandinavian Journal of Psychology*, **34**, 371–8.

Lunn, P. G. (1985). Maternal nutrition and lactational infertility: the baby in the driving seat. In *Maternal Nutrition and Lactational Infertility*. Nestlé Nutrition Workshop Series, ed. J. Dobbing, vol. 9, pp. 1–16, Raven Press.

Lunn, P. G. (1996). Breast-feeding practices and other metabolic loads affecting human reproduction. In *Variability in Human Fertility*, ed. L. Rosetta & G. G. N. Mascie-Taylor, pp. 195–216. Cambridge: Cambridge University Press.

Lunn, P. G., Austin, S., Prentice, A. M., & Whitehead, R. G. (1984). The effect of improved nutrition on plasma prolactin concentrations and postpartum infertility in lactating Gambian women. *American Journal of Clinical Nutrition*, **39**, 227–35.

Marmot, M., & Madge, N. (1987). An epidemiological perspective on stress and health. In *Stress and Health: Issues in Research Methodology*, ed. S. Kasl & C. Cooper, pp. 3–26. London: John Wiley.

Mattiasson, I., Lindgarde, F., Nilsson, J., & Theorell, T. (1990). Threat of unemployment and cardiovascular risk factors: Longitudinal study of quality of sleep and serum cholesterol concentrations in men threatened with redundancy. *British Medical Journal*, **301**, 461–6.

McNeilly, A. S. (1993). Lactational amenorrhoea. *Endocrinology and Metabolism Clinics of North America*, **22**, 59–73.

McNeill, G., Payne, P. R., Rivers, J. P. W., Enos, A. M.T., de Britto, J., & Mukarji, D. S. (1988). Socio-economic and seasonal patterns of adult energy nutrition in a South Indian village. *Ecology of Food and Nutrition*, **22**, 85–95.

Montgomery, E., & Johnson, A. (1977). Machiguenga energy expenditure.

Ecology of Food and Nutrition, **6**, 97–105.
Morris, J., Cook, D., & Shaper, A. (1992). Non-employment and changes in smoking, drinking, and body weight. *British Medical Journal*, **304**, 536–41.
Morris, J., Cook, D., & Shaper, A. (1994). Loss of employment and mortality. *British medical journal*, **308**, 1135–9.
Mulders, H., Meijman, T., O'Hanlon, J., & Mulder, G. (1982). Differential psychophysiological reactivity of city bus drivers. *Ergonomics*, **25**, 1003–11.
Nahum, R., Thong, K. J., & Hillier, S. G. (1995). Metabolic regulation of androgen production by human thecal cells *in vitro*. *Human Reproduction*, **10**, 75–81.
Neale, J., Hooley, J., Jandorf, L., & Stone, A. (1987). Daily life events and mood. In *Coping with Negative Life Events: Clinical and Social Psychological Perspectives*, ed. C. Snyder & C. Ford, pp. 161–89. New York: Plenum Press.
Niaura, R., Stoney, C., & Herbert, P. (1992). Lipids in psychological research: The last decade. *Biological Psychology*, **34**, 1–43.
Norgan, N. ed. (1992). *Physical Activity and Health*. Cambridge: Cambridge University Press.
Norgan, N. G., Ferro-Luzzi, A., & Durnin, J. V. G. A. (1974). The energy and nutrient intake and the energy expenditure of 204 New Guinean adults. *Philosophical Transactions of the Royal Society London B*, **268**, 309–48.
Ockenfels, M., Porter, L., Smyth, J., Kirschbaum, C., Hellhammer, D., & Stone, A. (1995). Effect of chronic stress associated with unemployment on salivary cortisol: Overall cortisol levels, diurnal rhythm, and acute stress reactivity. *Psychosomatic medicine*, **57**, 460–7.
Panter-Brick, C. (1991). Lactation, birth spacing and maternal work-loads among two castes in rural Nepal. *Journal of Biosocial Science*, **23**, 137–54.
Panter-Brick, C. (1992). Women's working behavior and maternal–child health in rural Nepal. In *Physical Activity and Health*, ed. N. Norgan, pp. 190–206. Cambridge: Cambridge University Press.
Panter-Brick, C. (1996*a*). Seasonal and sex variation in physical activity levels of agro-pastoralists in Nepal. *American Journal of Physical Anthropology*, **100**, 7–21.
Panter-Brick, C. (1996*b*). Physical activity, energy stores, and seasonal energy balance among men and women in Nepali households. *American Journal of Human Biology*, **8**, 263–74.
Panter-Brick, C. (1996*c*). Proximate determinants of birth seasonality and conception failure in Nepal. *Population Studies*, **50**, 203–20.
Panter-Brick, C. (1998). Biological anthropology and child health: context, process and outcome. In *Biosocial Perspectives on Children*, ed. C. Panter-Brick, pp. 66–101. Cambridge: Cambridge University Press.
Panter-Brick, C., Lotstein, D. S., & Ellison, P. T. (1993). Seasonality of reproductive function and weight loss in rural Nepali women. *Human Reproduction*, **8**, 684–90.
Panter-Brick, C., Todd, A., & Baker, R. (1996*a*). Growth status of homeless Nepali boys. *Social Science and Medicine*, **43**, 441–51.
Panter-Brick, C., Todd, A., Baker, R., & Worthman, C. M. (1996*b*). Heart rate monitoring of physical activity among village, school and homeless Nepali boys. *American Journal of Human Biology*, **8**, 661–72.
Panter-Brick, C., Worthman, C. M., Lunn, P., Baker, R., & Todd, A. (1996*c*). Urban-rural and class differences in biological markers of stress among Nepali children [Abstract]. *American Journal of Human Biology*, **8**, 126.

Pearson, J. D. (1990). Estimation of energy expenditure in Western Samoa, American Samoa, and Honolulu by recall interviews and direct observation. *American Journal of Human Biology*, **2**, 313–26.
Pollard, T. M. (1995). Use of cortisol as a stress marker: Practical and theoretical problems. *American Journal of Human Biology*, **7**, 265–74.
Pollard, T. M. (1997). Physiological consequences of everyday psychosocial stress. *Collegium Antropologicum*, **21**, 17–28.
Pollard, T., Ungpakorn, G., & Harrison, G. A. (1992). Some determinants of population variation in cortisol levels in a British urban community. *Journal of Biosocial Science*, **24**, 477–85.
Pollard, T. M., Ungpakorn, G., Harrison, G. A., & Parkes, K. R. (1996). Epinephrine and cortisol responses to work: A test of the models of Frankenhaeuser and Karasek. *Annals of Behavioral Medicine*, **18**, 229–37.
Prentice, A., & Prentice, A. (1988). Reproduction against the odds. *New Scientist*, **118**, 42–6.
Read, G. F. (1993). Status report on measurement of salivary estrogens and androgens. *Annals of the New York Academy of Sciences*, **694**, 146–60.
Rissler, A. (1977). Stress reactions at work and after work during a period of quantitative overload. *Ergonomics*, **20**, 577–80.
Roberts, S. B., Paul, A. A., Cole, T. J., & Whitehead, R. G. (1982). Seasonal changes in activity, birth weight and lactational performance in rural Gambian women. *Transactions of the Royal Society for Tropical Medicine and Hygiene*, **76**, 668–78.
Rose, R., Jenkins, C., Hurst, M., Herd, J. & Hall, R. (1982). Endocrine activity in air traffic controllers at work. II. Biological, psychological and work correlates. *Psychoneuroendocrinology*, **7**, 113–23.
Rosetta, L. (1992). Aetiological approach of female reproductive physiology in lactational amenorrhoea. *Journal of Biosocial Science*, **24**, 301–15.
Rosetta, L. (1993). Female reproductive dysfunction and intense physical training. *Oxford Reviews of Reproductive Biology*, **15**, 114–41.
Sears, S. J., & Milburn, J. (1990). School-age stress. In *Childhood Stress*, ed. I. E. Arnold, pp. 223–46. New York: John Wiley.
Shostak, M. (1990). *Nisa: The Life and Words of a !Kung Woman*. London: Earthscan Publications. (new edition).
Simondon, K. B., Bénéfice, E., Simondon, F., Delaunay, V. & Chahnazarian, A. (1993). Seasonal variation in nutritional status of adults and children in rural Senegal. In *Seasonality and Human Ecology*, ed. S. J. Ulijaszek & S. S. Strickland, pp. 166–83. Cambridge: Cambridge University Press.
Singh, J., Prentice, A. M., Diaz, E., Coward, W. A., Ashford. J., Sawyer, M., & Whitehead, R. G. (1989). Energy expenditure of Gambian women during peak agricultural activity measured by the doubly-labelled water method. *British Journal of Nutrition*, **62**, 315–29.
Södin, A. M., Andersson, A. B., Högberg, J. M., & Westerterp, K. R. (1994). Energy balance in cross country skiers. A study using doubly labelled water. *Medicine and Science in Sports and Exercise*, **26**, 720–4.
Spangler, G. (1995). School performance, type A behavior and adrenocortical activity in primary school children. *Anxiety, Stress and Coping*, **8**, 299–310.
Spector, P. (1986). Perceived control by employees: A meta-analysis of studies concerning autonomy and participation at work. *Human Relations*, **39**, 1005–16.
Stallings, J. F., Worthman, C. M., & Panter-Brick, C. (1998). Biological and

behavioral factors influence group difference in prolactin levels among breastfeeding Nepali women. *American Journal of Human Biology* **10**: 191–210.

Stallings, J. F., Worthman, C. M., Panter-Brick, C., & Coates, R. J. (1996). Prolactin response to suckling and maintenance of postpartum amenorrhea among intensively breastfeeding Nepali women. *Endocrine Research*, **22**, 1–28.

Strickland, S. S., & Tuffrey, V. R. (1997). *Form and Function – A study of nutrition, adaptation and social inequality in three Gurung villages of the Nepal Himalaya.* London: Smith-Gordon.

Tay, C. C. K., Glasier, A. F., & McNeilly, A. S. (1996). Twenty-four hour patterns of prolactin secretion during lactation and the relationship to suckling and the resumption of fertility in breast-feeding women. *Human Reproduction*, **11**, 950–5.

Theorell, T., Karasek. R., & Eneroth, P. (1990). Job strain variations in relation to plasma testosterone fluctuations in working men – a longitudinal study. *Journal of Internal Medicine*, **227**, 31–6.

Theorell, T., Perski, A., Akerstedt, T., Sigala, F., Ahlberg-Hulten, G., Svensson, J., & Eneroth, P. (1988). Changes in job strain in relation to changes in physiological state. *Scandinavian Journal of Work Environment and Health*, **14**, 189–96.

Timeo, M., Gentili, S., & Pede, S. (1979). Free adrenaline and noradrenaline excretion related to occupational stress. *British Heart Journal*, **42**, 471–4.

Tennes, K., & Kreye, M. (1985). Children's adrenocortical responses to classroom activities and tests in elementary school. *Psychosomatic Medicine*, **47**, 451–60.

Toivanen, H., Lansimies, E., Jokela, V., Helin, P., Penttila, I., & Hanninen, O. (1996). Plasma levels of adrenal hormones in working women during an economic recession and the threat of unemployment: Impact of regular relaxation training. *Journal of Psychophysiology*, **10**, 36–48.

Ulijaszek, S. J. (1995). *Human Energetics in Biological Anthropology.* Cambridge: Cambridge University Press.

Ulijaszek, S. J., & Strickland, S. S. (1993). Seasonality and nutrition. *Nutritional Anthropology: Prospects and Perspectives*, London: Smith-Gordon and Company Ltd.

van Eck, M., & Nicolson, N. (1994). Perceived stress and salivary cortisol in daily life. *Annals of Behavioral Medicine*, **16**, 221–7.

Vitzthum, V. J. (1994). Comparative study of breastfeeding structure and its relation to human reproductive ecology. *Yearbook of Physical Anthropology*, **37**, 307–49.

Vitzthum, V. J. (1997). Flexibility and paradox: The nature of adaptation in human reproduction. In *The Evolving Female: A Life History Perspective*, ed. M. E. Morbeck, A. Galloway & A. L. Zihlman, pp. 242–58, Princeton, New Jersey: Princeton University Press.

Ward, M., & Mefford, I. (1985). Methodology of studying catecholamine response to stress. In *Clinical and Methodological Issues in Cardiovascular Psychophysiology*, ed. A. Steptoe, H. Rueddel & H. Neus, pp. 131–43. Berlin: Springer-Verlag.

Warr, P. (1987). *Work, Unemployment and Mental Health.* Oxford, Clarendon Press.

Warr, P., & Jackson, P. (1985). Factors influencing the psychological impact of prolonged unemployment and of re-employment. *Psychological Medicine*, **15**, 795–807.

Warren, W. P. (1980). The effects of exercise on pubertal progression and reproductive function in girls. *Journal of Clinical Endocrinological Metabolism*, **51**, 1150–7.
Westerterp, K. R., Kayser, B., Brouns, F., Herry, J. P., & Saris, W. H. M. (1992). Energy expenditure climbing Mt. Everest. *Journal of Applied Physiology*, **73**, 1815–19.
Wood, J. W. (1994). *Dynamics of Human Reproduction – Biology, Biometry, Demography*. New York: Aldine de Gruyter.
Worthman, C. M., Jenkins, C. L., Stallings, J. F., & Lai, D. (1993). Attenuation of nursing-related ovarian suppression and high fertility in well-nourished, intensively breast-feeding Amele women of lowland Papua New Guinea. *Journal of Biosocial Science*, **25**, 425–43.
Zeier, H., Brauchli, P., & Joller-Jemelka, H. (1996). Effects of work demands on immunoglobulin A and cortisol in air traffic controllers. *Biological Psychology*, **42**, 413–23.

6

Reproductive ecology and reproductive cancers

PETER T. ELLISON

6.1 Introduction

The study of human reproductive ecology has in recent years greatly expanded our appreciation of the natural variability of human reproductive physiology. In particular, gonadal function in both males and females has been shown to occur along a broad continuum of variation, both within and between individuals and between populations. Furthermore, such variation has been shown to be interpretable as a healthy, adaptive response to ecological conditions (Ellison *et al.*, 1993*a*; Ellison 1994, 1995). In keeping with the broader perspective of Darwinian medicine (Ewald, 1980; Nesse & Williams, 1994), the perspective of reproductive ecology has entailed making a distinction between adaptive responses to stressful situations and the etiology of those situations themselves. Under energetic stress, for example, it may be advantageous for ovarian function to be suppressed. That does not mean that energetic stress itself is advantageous. Seeking to treat suppressed ovarian function under these conditions as if it were a pathology, rather than seeking to relieve the energetic stress, would imply a basic misunderstanding of the problem, its etiology, and its appropriate treatment. It is only recently, for example, that clinicians have begun to realize that the amenorrhea associated with athletic activity in many women is not itself a pathology to be rectified by "driving" cycles of endometrial maturation with exogenous steroids. (For a fuller treatment of this area, see Panter-Brick and Pollard, Chapter 5.)

In addition to expanding our understanding of the healthy functioning of the human reproductive system, the new perspective of reproductive ecology can help us improve our understanding of true reproductive pathologies, including reproductive cancers. The goal of this chapter is to demonstrate how the results and insights of reproductive ecology can serve to link an understanding of the pathogenesis of reproductive cancers to an understanding of cancer epidemiology and even to suggest practical strategies for the reduction of risk. While the discussion applies to reproductive cancers generally, breast cancer will often be used as a specific example. It is now estimated that one in every eight North American women will be diagnosed with breast cancer in their lives, an incidence that could rise to one in five by the end of the decade. Breast cancer has become the leading cause of cancer mortality among women in the USA, and the single leading cause of death among women in their forties (McPherson *et al.*, 1994; Kramer & Wells, 1996). As this chapter will argue, these alarming statistics may soon be visited upon the developing world as well, making the breast cancer "epidemic" a serious, if somewhat more slowly developing, rival to the AIDS epidemic in its projected impact on human life and suffering.

6.2 Carcinogenesis and tumor growth

Malignant tumors, including those of the reproductive system, are caused by the unregulated clonal expansion of somatic cells often accompanied by inappropriate gene expression. Through its own physiological activity and its interference with the activity of other physiological symptoms, the tumor can severely undermine the health of the affected individual and eventually cause death. Unlike most physiological responses to infection or trauma, there is nothing about the pathophysiology of malignant tumors that can be understood as functional or adaptive from the perspective of the affected individual.

Current understanding of cancer pathogenesis identifies two conceptually distinct processes: the initial transformation of a normal cell to a cancerous state, and the subsequent growth and pathological expression of the resulting tumor. Tumors are most likely to develop in mitotically active tissues where each generation of cell division represents an opportunity for a carcinogenic mutation to occur (Albanes & Winick, 1988; Cohen & Ellwein, 1990). The risk of cellular transformation in a daughter cell thus increases linearly with the number of mitotic divisions that a cell undergoes. Epithelial tissues that undergo constant or periodic renewal, such as the skin and linings of the respiratory and digestive systems, secretory epithelia of glands, lymphocytes that undergo clonal expansion as part of the immune response, glial cells in the brain that serve analogous functions to lymphocytes in the circulation, and hemopoetic stems cells in bone marrow, are among the tissues most susceptible to carcinogenesis by virtue of their mitotic activity. Genetic factors contributing to specific cancer risks, such as the breast cancer genes, BRCA-1 and BRCA-2, often increase the likelihood of a transforming event, as do exogenous carcinogens. The original event of cellular transformation can occur years before the clinical diagnosis of a malignancy, however. The intervening period, and the prognosis once the cancer is identified, are both largely functions of the rate of tumor growth, a process separate from the probability of its initial formation. Tumor growth can be influenced by endogenous growth factors, especially those that are effective in promoting normal proliferation of the affected tissue. Positive energy balance and high caloric intake have been associated with generally elevated cancer risks in many tissues (Hocman, 1988; Simopoulos, 1990), very likely because they are associated with higher levels of generalized growth factors such as insulin, growth hormone, and insulin-like growth factors.

The pathogenesis of reproductive cancers conforms to this general outline. Reproductive cancers, like all cancers, are most likely to form in mitotically active tissues, such as the epithelia of the cervix, uterine endometrium, lobular–alveolar ducts of the breast, and the prostate, as well as the germinal epithelium of the testis and the granulosa cells of ovarian follicles. These are all tissues which undergo mitotic renewal in reproductively mature individuals, chronically in the case of the prostate and testis, periodically in the case of the cervix and endometrium, episodically in the case of the follicle, or both periodically and episodically in the case of lobular–alveolar ducts.

The lobular-alveolar duct system of the breast has a two-stage pattern of developmental maturation that results in a particular period of enhanced susceptibility to a transforming event. The duct system undergoes initial, but incomplete, proliferation during pubertal breast development stimulated by ovarian estrogens. The final proliferation of the secretory milk ducts occurs under the peculiar hormonal regimen of late pregnancy, involving in addition to estrogen stimulation high levels of progesterone, prolactin, and cortisol, in preparation for postpartum milk production. At that time the cells of the secretory epithelium undergo terminal functional differentiation (Russo *et al.*, 1982). Waves of mitotic activity and tissue regression will accompany subsequent pregnancies, and to a lesser extent normal menstrual cycles, but never to the same extent as during the phase of primary tissue differentiation.

Given this pattern of development, the period between puberty and first pregnancy appears to be a period of heightened susceptibility to the development of breast cancer (MacMahon *et al.*, 1970; Kvåle *et al.*, 1987;

Kvåle & Heuch, 1987; Layde *et al.*, 1989; Pike *et al.*, 1993; Kelsey *et al.*, 1993; Hulka & Stark, 1995). The length of this interval is consistently found to be an important positive risk factor in epidemiological studies. It remains unclear, however, whether this heightened risk represents a greater probability of a transforming event in tissue that is incompletely differentiated, or whether it is the intense mitotic activity of the first pregnancy itself that "amplifies" the cumulative risk of the preceding period by promoting the dramatic growth of any cancerous cell lines that might exist at that time.

The elevated risk associated with the interval between puberty and first pregnancy plays a significant role in the recent increase of breast cancer incidence among women in the developed Western countries where, among higher socio-economic groups especially, the first pregnancy is often delayed until the late twenties or early thirties. In many developing countries, and in the lower socio-economic groups of developed countries, average ages at first pregnancy are considerably earlier with a resultant lower lifetime breast cancer risk (Henderson & Bernstein, 1991; Kelsey & Horn-Ross, 1993).

Other reproductive cancers do not show evidence of particular periods of heightened risk in the life cycle, other than the period of mature reproductive function itself. In these cases cancer incidence increases relatively linearly with time since puberty, reflecting the cumulative risk of a transforming event as well as increasing opportunity for even slowly developing tumors to be diagnosed. The end of regular menstruation at menopause represents a disjunction in the risk of many reproductive cancers in women, such as cervical, ovarian, and endometrial cancer (Henderson *et al.*, 1988). Age-specific incidence of these cancers continues to rise after menopause, but at a much slower rate than before. In part, this represents a dramatic reduction in the probability of a transforming event in tissues that are no longer subject to monthly cycles of mitotic activity. In part, it represents a reduction in the rate of growth of any tumors present at that time.

6.3 Cancer growth and endogenous steroids

Tumor growth is often responsive to levels of endogenous growth factors, especially those that regulate the normal mitotic activity of the untransformed tissue (Henderson *et al.*, 1988; Cohen & Ellwein, 1990). Among the most potent of such stimuli to the growth of reproductive cancers are gonadal steroids (Siiteri *et al.*, 1976; Soto & Sonnenschein, 1987; Clarke & Sutherland, 1990). These hormones, such as testosterone in males and estradiol and progesterone in females, stimulate mitotic activity in target tissues of the reproductive tract, particularly those that undergo periodic regrowth or high secretory activity. Estradiol, for example, produced by the developing follicles of the ovary, stimulates the proliferation of the endometrium in the uterus and of the follicular granulosa cells during the follicular phase of a normal menstrual cycle. Progesterone, secreted by the corpus luteum after ovulation, halts the proliferation of the endometrium and supports its secretory activity during the luteal phase of the cycle (Clark & Markaverich, 1988). Testosterone promotes the secretory activity of the prostate gland in the male and the mitotic activity of the germinal epithelium of the seminiferous tubules that is necessary for sperm production (Coffey, 1988). Both estradiol and progesterone stimulate mitotic activity of the lobular–alveolar epithelium of the breast, with progesterone also promoting preparation for the intense secretory activity of milk production (Tucker, 1988).

The role of endogenous gonadal steroids in promoting the growth of reproductive cancers was first made apparent by the dramatic effect of gonadectomy in slowing tumor growth and improving prognosis (Crawford, 1992; Beatson, 1896). This surgical intervention is still used under certain clinical circumstances (Early Breast Cancer Trialists' Collaborative Group, 1996; Barley, 1996; Labrie *et al.*, 1983; van Tinteren and Dalesio, 1993; Prostate Cancer Trialists' Collaborative Group, 1995). The decline in incidence of breast, ovarian, and uterine cancers post-menopausally also implicates endogenous ovarian steroids as promoters of reproductive cancers (Henderson *et al.*, 1988; Pike *et al.*, 1993). Obesity, both post-menopausal and pre-menopausal, has been implicated as a cause of elevated risk of endometrial cancer due to the elevated levels of circulating estrogens resulting from the peripheral conversion of adrenal androgens in adipose tissue (MacDonald *et al.*, 1978; Siiteri, 1990). Exogenous hormone exposures have also been linked to increased risk of certain cancers, especially breast cancer (Brinton & Schairer, 1993; Malone *et al.*, 1993; Stanford & Thomas, 1993; Malone, 1993). High dose estrogen contraceptives and regimes of post-menopausal estrogen replacement have both been causally linked to increased risks of endometrial cancer (Key & Pike, 1988*a*; Henderson *et al.*, 1988). Lowering the estrogen dose in these treatments and combining the estradiol dose with progesterone, which counters the effect of estradiol in promoting endometrial proliferation, has largely attenuated this risk.

As mitotic activity of the lobular-alveolar epithelium of the breast is stimulated by both estrogens and progesterone, combining these steroids does not necessarily neutralize their effect on this tissue (Key & Pike, 1988*b*; Pike *et al.*, 1993; Colditz *et al.*, 1995), nor do all breast cancers display equivalent densities of steroid receptors or respond equally to steroid stimulation. Nevertheless, many of the most effective pharmacological treatments for breast cancer rely on neutralizing the growth promoting effects of endogenous steroids (Santen *et al.*, 1990; Early Breast Cancer Trialists' Collaborative Group, 1992; Henderson, 1993; Henderson *et al.*, 1993). Tamoxifen, for example, one of the most widely used pharmacological agents, acts as a specific estradiol antagonist at the receptor level (Bush & Hezloyer, 1993; Jordan & Allen, 1980).

In part because it has the greatest epidemiological impact of all reproductive cancers, we know a good deal about risk factors associated with breast cancer incidence and prognosis. Many of them, especially those with the greatest quantitative importance, can be interpreted as reflecting the impact of acute and chronic exposure to endogenous gonadal steroids (Bernstein & Ross, 1993). Prognosis, for example, is strongly related to age, with cancers in younger women having much higher rates of mortality and recurrence (Figure 6.1; de la Rochfordière *et al.*, 1993). This is true even after correcting for tumor type and stage at time of diagnosis. Younger women also have higher average levels of estradiol and progesterone than older women, even controlling for factors such as weight, fatness, and exercise (Figure 6.2; O'Rourke & Ellison, 1993*b*; Lipson & Ellison, 1992, 1994). The higher steroid milieu of younger women is thus implicated in the more vigorous growth of breast cancers in this group. So powerful is this effect that chemical or surgical castration is often considered an advisable treatment (Early Breast Cancer Trialists' Collaborative Group, 1996).

Pregnancy and lactation also have effects on breast cancers that can be understood as mediated by gonadal steroid exposure (Kelsey *et al.*, 1993; McPherson *et al.*, 1994, Hulka & Stark, 1995). Recent pregnancy is a risk factor both for breast cancer incidence and recurrence with the risk diminishing as time since last pregnancy increases (Figure 6.3; Bruzzi *et al.*, 1988; Guinee *et al.*, 1994). Late pregnancy is a period of particularly high acute steroid exposure (although the steroids in this case are predominantly of placental origin), one function of which is to promote mitotic activity in the lobular–alveolar duct system of the breast in preparation for lactation. Hence, the effect of this reproductive state in stimulating tumor growth and increasing the risk of a transforming event is readily understood. Lactation itself, on the other hand, although a state of high secretory activity, is associated with the suppression of gonadal steroid production for greater or lesser periods of time. Hence the cumulative amount of time

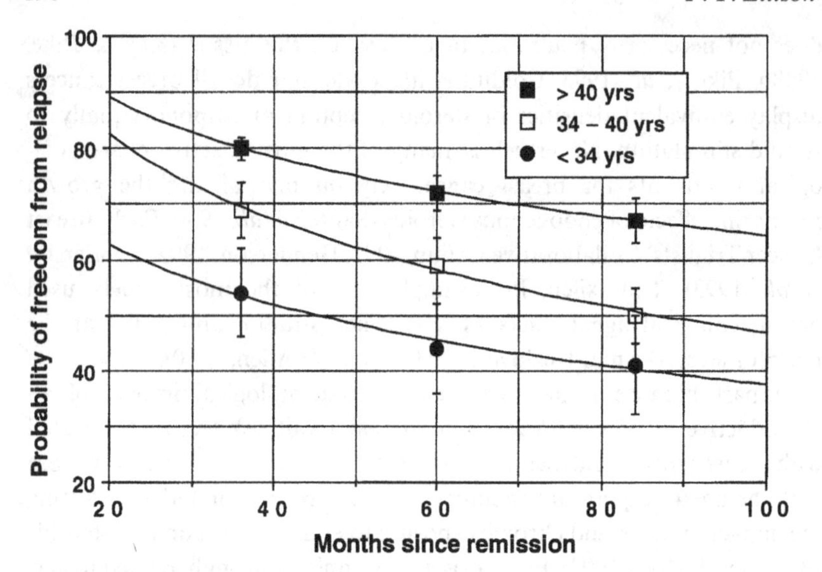

Figure 6.1. Probability of freedom from relapse of breast cancer by months since initial remission, stratified by subject age (data from de la Rochfordière *et al.*, 1993). Younger women have a higher probability of relapse than older women at all stages after initial cancer remission.

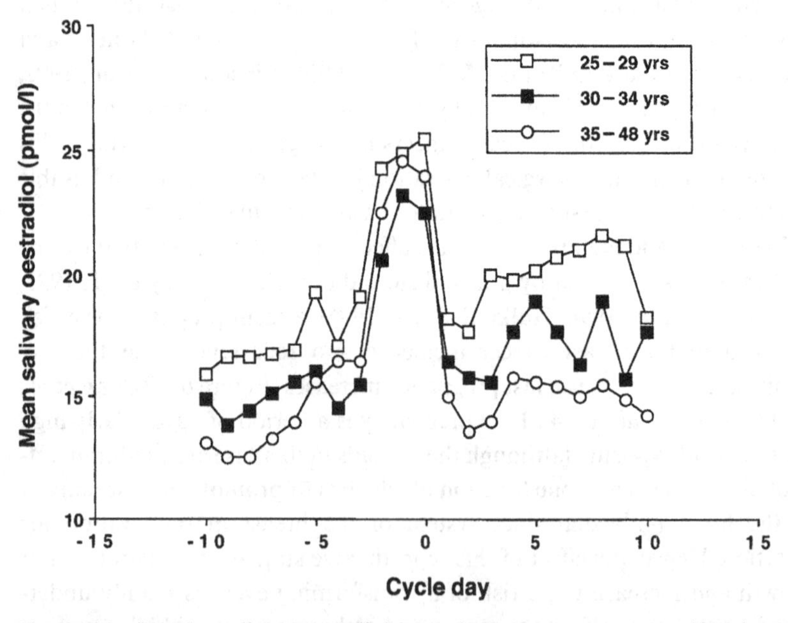

Figure 6.2. Average salivary estradiol levels in Boston women stratified by age (O'Rourke & Ellison, 1993*a*, 1993*b*). Younger women have higher levels of salivary estradiol (an index of free estradiol in circulation) than older women in both follicular and luteal phases of the menstrual cycle.

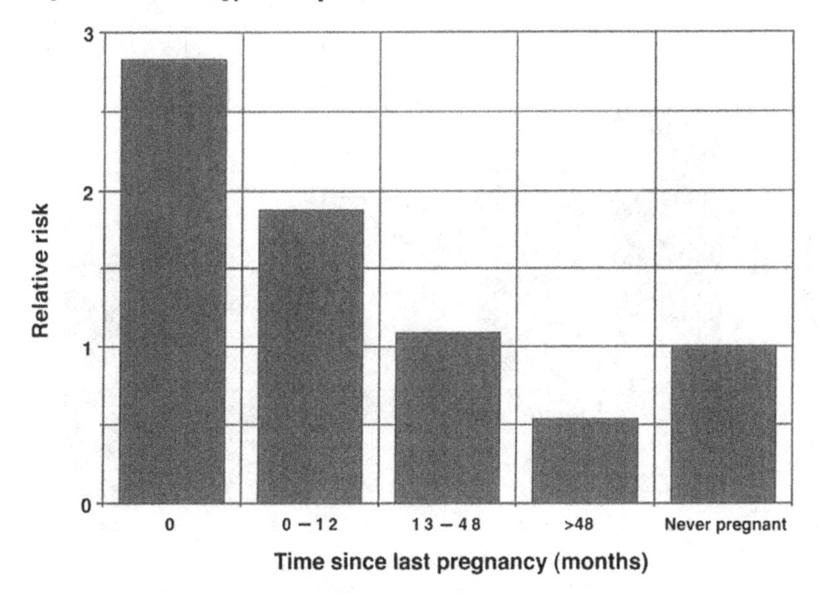

Figure 6.3. Relative risk of breast cancer incidence by time since last pregnancy controlled for age (data from Guinee *et al.*, 1994). Compared with women who have never been pregnant, women who have been pregnant in the last 12 months have an elevated risk of breast cancer incidence. Notably, women who have been pregnant, but not in the last 4 years, have a lower relative risk than women who have never been pregnant. This may be due to the extended period of heightened risk prior to the first pregnancy for the latter group. See the discussion of this effect in the text.

spent lactating is negatively associated with lifetime risk of breast cancer (Figure 6.4; Byers *et al.*, 1985; McTiernan & Thomas, 1986; Siskind *et al.*, 1989; Newcomb *et al.*, 1994).

Early age at menarche and late age at menopause are also significantly associated with elevated breast cancer risk (Figure 6.5; Apter & Vihko, 1983; Kelsey *et al.*, 1993; Stoll *et al.*, 1994). Early age at menarche is, of course, confounded with a long interval between menarche and first pregnancy, but remains a significant risk factor even after correction for age at first pregnancy. Late age at natural menopause extends the cumulative period of lifetime exposure to gonadal steroids even as early menopause, whether natural or surgical, shortens it. Thus the effect of these two aspects of gynecological history on breast cancer risk can also be understood as mediated by steroid exposure (Kelsey & Gammon, 1991; Kelsey *et al.*, 1993).

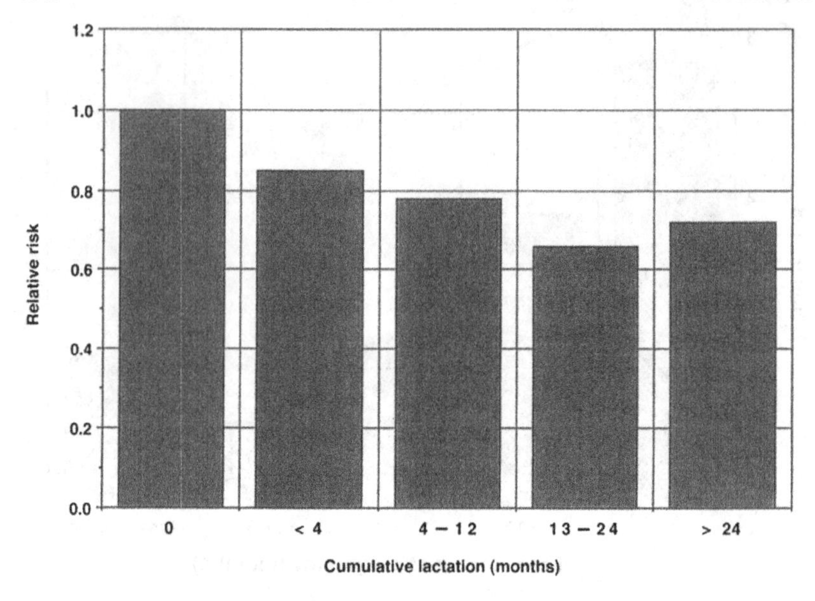

Figure 6.4. Relative risk of breast cancer incidence by cumulative time spent lactating (data from Newcomb *et al.*, 1994). Relative to women who have borne children but who did not breastfeed them, the relative risk of breast cancer declines with cumulative time spent lactating.

6.4 Reproductive ecology and endogenous steroids

The study of human reproductive ecology has revealed that endogenous steroid production in both males and females is sensitive to environmental, constitutional, and behavioral factors (Ellison, 1993, 1995). Rather than being fixed or narrowly channelled features of our biology, levels of gonadal steroid activity vary hugely, in the absence of pathology, within individuals over time, between individuals within populations, and between populations. Many of the patterns that this variation presents can now be understood in terms of adaptive responses of the human reproductive system to environmental and constitutional conditions. The same patterns also help to illuminate many issues in the epidemiology of reproductive cancers.

Aspects of chronic and acute energy availability have been shown to exert a strong influence on both female and male gonadal function (see Panter-Brick and Pollard, Chapter 5). In females the responsiveness of ovarian function to energy balance and energy expenditure is particularly finely tuned. Weight loss of as little as 1 kg over a month, or a sustained loss of 2 kg over a season, has been associated with lower progesterone levels

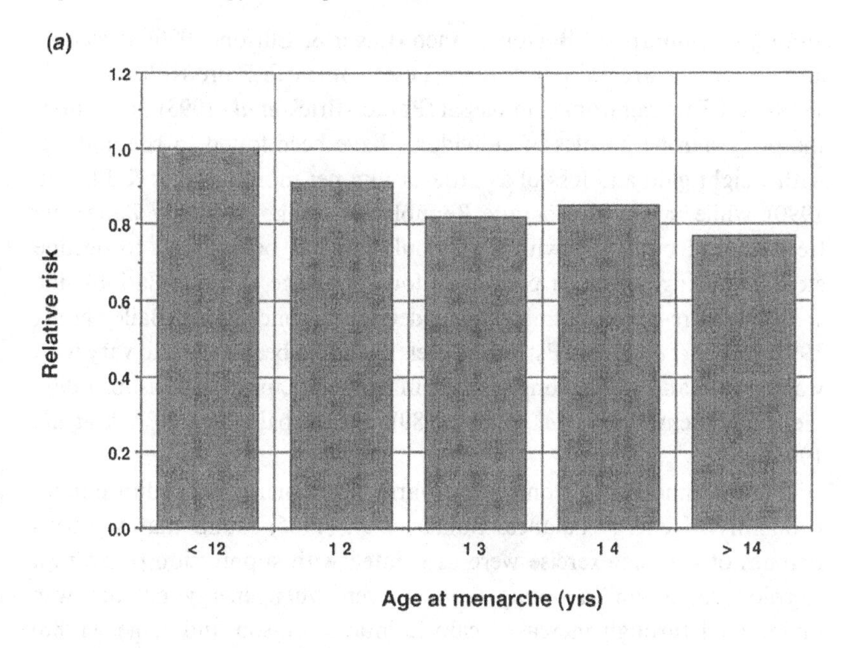

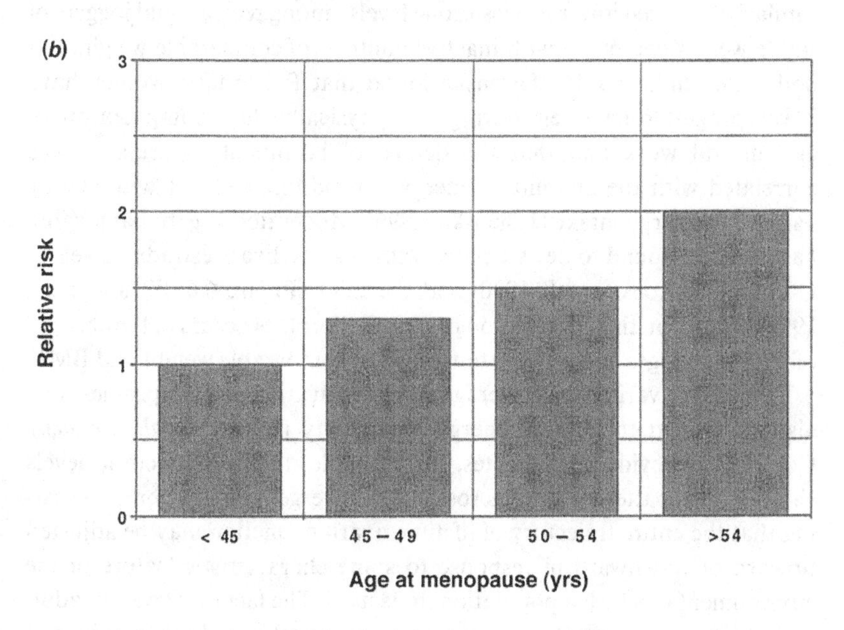

Figure 6.5. Relative risk of breast cancer incidence by (*a*) age at menarche and (*b*) age at menopause (data from Kelsey *et al.*, 1993). Compared with women who reach menarche before age 12, those with later menarcheal ages have a lower risk of breast cancer. Compared with women who experience menopause before age 45, women with later menopausal ages have a higher risk of breast cancer.

among well-nourished Boston women (Lager & Ellison, 1990) as well as among more marginally nourished Lese women in Zaire (Ellison *et al.*, 1989) and Tamang women in Nepal (Panter-Brick *et al.*, 1993). In Boston, the progesterone profiles of individuals have been found to rise and fall with weight gain and loss of as little as 1 kg per month (Lager & Ellison, 1990), while in the Democratic Republic of Congo (formerly Zaire) the frequency of ovulation within a population has been found to decline steadily during sustained periods of food shortage and population-wide weight loss, resulting ultimately in a decline in conceptions (Bailey *et al.*, 1992; Jenike *et al.*, 1996). Estradiol levels have also been shown to vary with weight gain and loss among women in Boston (Lipson & Ellison, 1996), Germany (Schweiger *et al.*, 1988, 1989), and Nepal (Panter-Brick *et al.*, 1996).

Ovarian function in women also varies with energy expenditure independently of energy balance. Bullen *et al.* (1985) found that vigorous regimes of aerobic exercise were associated with suppression of ovarian steroid production in young women even when energy balance was maintained through increased caloric intake. Ellison and Lager (1986) similarly observed lower progesterone levels among recreational joggers of stable weight compared with inactive controls of comparable weight and body mass index (BMI). Jasienska found that Polish farm-women have lower progesterone levels during the physically demanding season of agricultural work and that the degree of hormonal suppression was correlated with the amount of energy expenditure but not with energy balance or energy intake (Jasienska, 1996). Moderate recreational jogging has also been found to be associated with lower salivary estradiol levels in Germany (Broocks *et al.*, 1990) and the USA (Figure 6.6; Ellison *et al.*, 1996). In Boston this suppression of estradiol profiles occurs in the absence of weight changes and relative to women of comparable weight and BMI.

Population variation in average levels of ovarian steroid profiles may also be related to chronic energy availability (Ellison *et al.*, 1993*a*,*b*). Comparative evidence indicates, for example, that progesterone levels show parallel patterns by age across widely divergent populations, suggesting that the entire trajectory of lifetime ovarian function may be adjusted upward or downward in response to some characteristic feature of the environment in which a population finds itself. The fact that levels of adult ovarian function are also related to menarcheal age, both within and between populations, suggests that chronic energy availability may effect such modulation of adult ovarian function as it effects similar modulation of other aspects of growth and development (Ellison, 1996*a*,*b*).

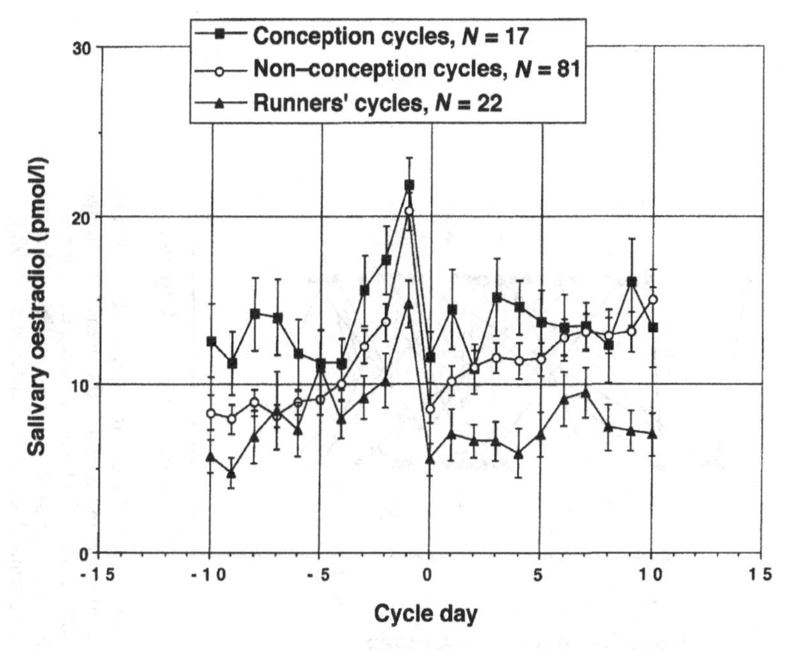

Figure 6.6. Average estradiol profiles in recreational runners, compared with conception and non-conception cycle profiles from age-matched non-exercising controls (Ellison *et al.*, 1996). Runners have significantly lower levels of salivary estradiol than either conception or non-conception cycles in inactive women matched for age, height, and weight.

In men the relationship between gonadal steroid production and energy balance and energy expenditure is somewhat different. The response to acute changes in energy balance is not as sensitive as it is in women. Dramatic or extreme weight loss, as during fasting or famine, is associated with significant suppression of testosterone levels. The effects of moderate variation in energy balance are less clear. Lese men in Zaire, for example, who undergo periodic weight loss due to food shortages of similar magnitude and duration to that observed in Lese women, do not show evidence of parallel changes in testosterone levels (Figure 6.7; Bentley *et al.*, 1993). Similarly, Nepali men do not show seasonal patterns of change in testosterone levels associated with changes in energy balance comparable to the progesterone and estradiol changes observed among women in the same population (Ellison and Panter-Brick, 1996). Population variation in testosterone levels is suggestive of a chronic and/or developmental effect of energy shortage on male testosterone production in a pattern similar to that observed among women (Ellison *et al.*, 1989; Bribiescas, 1996, Panter-Brick and Pollard, Chapter 5).

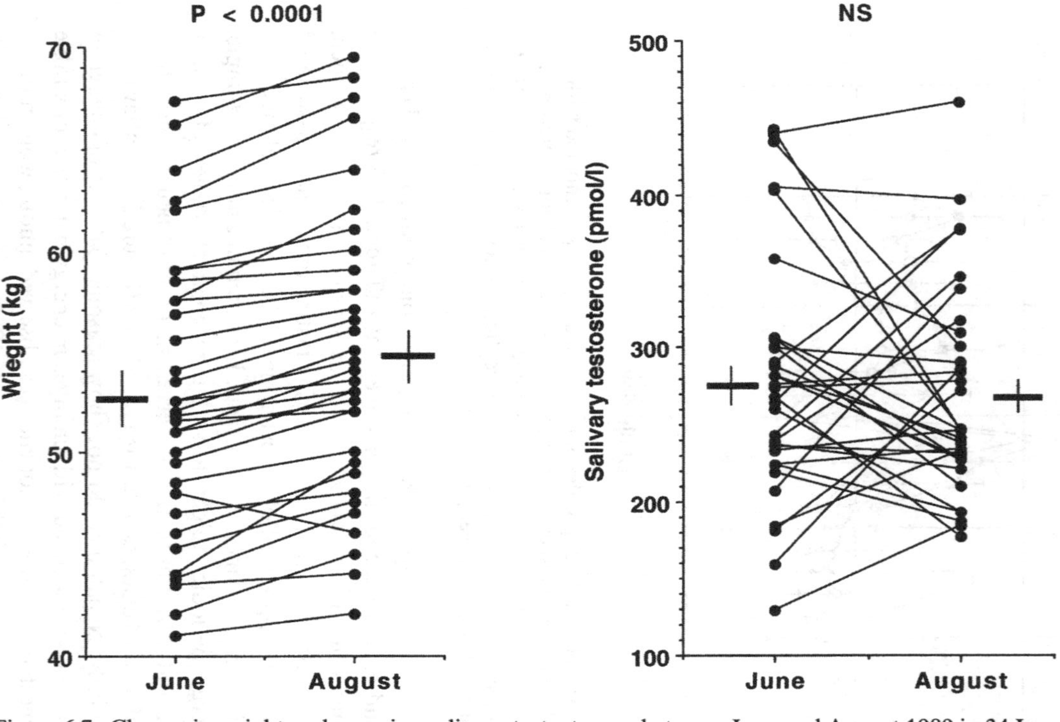

Figure 6.7. Change in weight and morning salivary testosterone between June and August 1989 in 34 Lese men from the Ituri Forest, Zaire. Weight gain is highly significant between the end of the preharvest hunger season in June and the post-harvest season in August ($P < 0.0001$, two-tailed paired t-test), but there is no significant pattern of change in testosterone levels. Mean values and standard errors are indicated for each month by the horizontal and vertical lines adjacent to the data points. (Details in Bentley *et al.*, 1993.)

Other aspects of nutritional ecology have also been associated with differences in steroid exposure. Vegetarian and high fiber diets, for example, have been associated with low levels of estradiol compared with omnivorous, low fiber diets (Armstrong *et al.*, 1981; Goldin *et al.*, 1982, 1986; Persky *et al.*, 1992; Barr *et al.*, 1994). This association is ascribed to the capacity of fiber in the gut to absorb free estradiol, increasing the rate of fecal excretion (Adlercreutz *et al.*, 1976; Adlercreutz & Järvenpää, 1982; Goldin *et al.*, 1986). It should be noted that a reduction in circulating estradiol levels mediated by excretion may have a paradoxical effect of increasing ovarian production by decreasing negative feedback effects at the level of the hypothalamus and pituitary. This can lead to greater FSH and LH stimulation of follicular development and higher intrafollicular steroid levels. This effect is analogous to that produced by exogenous estrogen antagonists such as clomiphene when used to stimulate follicular development and ovulation. In a recent study in our laboratory of 11 women whose customary diets include levels of dietary fiber at least one standard deviation above the national average (17.42 ± 1.25 g/day) and eight age, height, and weight-matched women who consumed average levels of dietary fiber (8.92 ± 0.60 g/day), it was found that the high fiber consumers had shorter overall cycle lengths (26.4 ± 0.9 versus 30.8 ± 1.5 days, $p < 0.01$) and shorter follicular phase lengths (13.9 ± 8 versus 18.0 ± 1.9 days, $p < 0.05$) than the controls (Halperin, 1997).

Recently Schweiger *et al.* (1992) have demonstrated an association between suppressed ovarian steroid levels and "restrained eating" (eating significantly less than appetite would dictate) even when weight is stable. The mechanisms that might be responsible for this effect are unclear; however, restrained eating may be a confounding factor in vegetarian diets (Barr *et al.*, 1994). High fat diets, conversely, have been associated with elevated gonadal steroid levels (Lustig *et al.*, 1990).

The patterns of responsiveness of the reproductive system in females and males to energy availability and expenditure can be understood as adaptive. In females, reduced levels of gonadal function in association with negative energy balance and elevated energy expenditure probably represent modulation of fecundity in response to energy availability. Successful female reproduction depends on partitioning significant amounts of energy towards the requirements of gestation and lactation from the beginning of pregnancy, and a state of negative energy balance or high energy expenditure makes this task more difficult. It is similarly adaptive for an individual who develops under conditions of chronic energy scarcity to lower lifetime fecundity and increase the average interval between successive pregnancies in anticipation of the challenge of maintaining long-term energy balance (Ellison *et al.*, 1993*a*; Ellison, 1996*a,b*).

In males, modulation of gonadal steroid production in association with energy availability probably represents an adaptive modulation of metabolic mass, rather than fecundity (Bribiescas, 1996). Muscle represents the largest component of active metabolic mass that can be adjusted without deleterious effects on survival. (The other major components are brain and vital organs.) Adjustment of muscle mass is not necessarily an effective strategy to meet short-term reductions of energy availability. These can better be met through mobilization of stored fat reserves or reductions in physical activity. Chronic reductions in energy availability, however, may exceed the capacity of such short-term adjustment strategies and require longer-term adjustment of energy requirements. As in the case of females, chronic energy shortage during growth and development may elicit permanent adjustments in adult physiology, such as reductions in baseline levels of gonadal function, resulting in baseline adjustments in metabolic mass.

6.5 Reproductive ecology and reproductive cancer risk

The relationship between reproductive ecology and endogenous steroid levels provides a single framework for understanding several aspects of reproductive cancer epidemiology. Many of the constitutional variables that have been linked to elevated risks of reproductive cancers, such as tall stature, high BMI, high caloric intake, high fat/low fiber diet, and early menarcheal age in women are also associated with elevated exposure to endogenous gonadal steroids (Hunter & Willet, 1993; Stoll *et al.*, 1994; Thorling, 1996). Developmental effects on adult gonadal function may be particularly important in certain of these risk factors, such as stature and early maturity, as these risk factors are interpretable as markers of high set-points for adult gonadal function (Ellison, 1996*a*, *b*).

Most importantly, however, the perspective of reproductive ecology provides a framework for understanding global patterns of reproductive cancer incidence. The highest incidence of such cancers is found among the privileged populations of developing countries (Henderson & Bernstein, 1991). These same populations generally mature at the earliest ages, postpone reproduction to the latest ages, and have the highest average caloric intakes and lowest average energy expenditures. As a consequence, their levels of gonadal steroid production and exposure are also the highest (Ellison *et al.*, 1993*b*). These high steroid levels and consequent risks of reproductive cancers can thus be interpreted as additional manifestations of the general syndrome of accelerated growth and maturation known as the secular trend. Like adult levels of gonadal function, the incidence of reproductive cancers shows a high correlation with average ages at maturation across populations. Other subpopulation contrasts that often correlate with different stages of the secular trend, such as rural/urban contrasts, are also associated with differential incidence of reproductive cancers (Rimpela & Pukkala, 1982). The accelerated transition to high cancer risk seen in many migrant populations that have come from settings of low cancer incidence to settings of high incidence is in the same way related to the accelerated secular trend in such groups (Trichopoulos *et al.*, 1984; Shimizu *et al.*, 1991; Thomas & Karagas, 1987; Ziegler *et al.*, 1993). Both transitions are the consequence of numerous changes in nutrition, health, and lifestyle that are integrated in general patterns of growth, maturation, and gonadal function. Seeking monocausal explanations of differential cancer rates in specific dietary components makes little sense in either case. One corollary of this perspective is the sobering prediction that rates of reproductive cancer will rise in the developing world along with living standards and general health even as we expect such populations to pass through a secular trend in growth and maturation.

Potentially the most important contribution that reproductive ecology can make to the study of reproductive cancers is by drawing attention to potential strategies for prevention. Many of the known risk factors for reproductive cancers are depressingly intractable. The recognition of specific genetic factors, for example, does not yet provide any effective opportunity for intervention, only ethical dilemmas related to testing. Similarly, many of the most potent risk factors associated with breast cancer cannot be easily modified and/or are unlikely to provide for intervention strategies that would be widely acceptable. Age at reproductive maturation, for example, is not easily postponed. Some have suggested that proscribed regimens of intensive exercise throughout childhood and adolescence could achieve this effect (Frisch *et al.*, 1985), but the practicality of such a suggestion for a broad spectrum of the population is doubtful. Drastic reduction of caloric intake might achieve the same maturational delay, but it is not clear that such an effect would be neutral in all other respects. Even in the absence of negative health effects, it is not clear that the emotional and psychosocial effects of late maturity would be neutral, or that late maturity as an elective option would appeal to a broad spectrum of the population.

Late age at first pregnancy is another powerful risk factor for breast cancer for which few practical interventions suggest themselves. The choice to postpone pregnancy is tightly bound to educational, career, and lifestyle choices of women and their partners, and cannot be easily changed without affecting all of these important aspects of modern life. Some have suggested that pregnancy might be pharmacologically simulated in adolescent women, thus achieving the effect of natural pregnancy in driving terminal differentiation of breast lobular–alveolar duct tissue and lowering breast cancer incidence (Eaton *et al.*, 1994). It remains uncertain, however, whether such a strategy would ever prove popular. At the very least it would involve prolonged exposure to exogenous steroids, chemicals with potent effects on both the central nervous system and systemic physiology. Even if the more conventional physiological side-effects of hormonal changes during pregnancy proved acceptable to many girls and their parents, a proposition that is itself doubtful, the uncertain effects of the regimen on other aspects of developmental biology would be a serious concern.

Advancing menopause through elective surgical or chemical castration after a desired number of births is another possible strategy. This approach would run counter to the trend currently represented by hormonal replacement therapy to counteract many of the undesirable physiological consequences of the termination of ovarian steroid production, such as increases in the risk of osteoporosis and decreases in libido and sexual function.

Dietary changes to incorporate high levels of fiber, soy products, or other specific nutrients believed to reduce steroid levels or steroid effects is another possibility. It remains unclear, however, whether the levels of specific nutrients necessary to obtain significant protective effects are compatible with diets that would be broadly acceptable; nor are the collateral effects of such diets well understood. As noted above, increased steroid clearance rates, by decreasing the level of negative feedback at the hypothalamus and pituitary, can lead to increases in gonadotropin stimulation of ovarian activity. Among the consequences of such increased gonadotropin stimulation could be increased rates of multiple folliculogenesis and dizygotic twinning. Similarly, although some nutrients might block estrogenic activity at estrogen receptors in cancerous tissue, their potential for simultaneous effect on activity at other oestrogen receptors, both systemically and in the central nervous system, is unknown. The acceptability, and indeed the advisability, of this strategy is thus also uncertain.

As so many of the interventions to reduce breast cancer risk through modulating identified risk factors are so intractable, the identification of a "modifiable risk factor" has become an objective of central importance in the development of preventive strategies (Bernstein *et al.*, 1991). Reproductive ecology suggests that moderate increases in exercise and moderate dietary restraint could represent important opportunities for intervention. The potential effect of exercise on breast cancer risk was suggested by early cohort studies of women who had engaged in high levels of sports activities in college (Frisch *et al.*, 1985). It was unclear whether exercise later in life could produce a similar beneficial effect. It was also unclear whether the exercise involved had to be of sufficient intensity to actually disrupt menstrual function in order to confer protection. One of the most compelling results of studies of human reproductive ecology has been the demonstration of a suppressive effect of even moderate energy expenditure on ovarian steroid production. Such a quantitative suppression of ovarian steroid levels is entirely compatible with regular menstrual function and not dependent on a high threshold of energy expenditure. Recent studies of physical activity and breast cancer incidence using case-control designs (Figure 6.8; Bernstein *et al.*, 1994; Friedenreich & Rohan, 1995; Mittendorf *et al.*, 1995; Thune *et al.*, 1997) have also shown that moderate levels of exercise or occupational physical activity in adulthood are associated with reduced cancer risk.

Moderate exercise thus emerges as a practical prescription for reducing cancer risk. The levels of exercise required are attainable by almost all women without major disruption of conventional lifestyles; exercise of this nature is already compatible with social goals and trends developed for other reasons in a broad spectrum of the population; moderate exercise is associated with collateral health benefits, such as reduced risk of heart disease and osteoporosis, and is not associated with significant negative health effects; the suppressive effect on ovarian function is reversible and so entails no permanent change in reproductive capacity.

Whether moderate exercise would have a similar effect in reducing reproductive cancer risk in men is less clear. Moderate exercise does not appear to have a noticeable effect on testosterone production in men and may even promote the tissue anabolic effects of testosterone in target muscle tissues (Bribiescas, 1996). The levels of exercise necessary to achieve a protective effect for prostate and testicular cancer may therefore be much higher than the levels that confer effective protection from breast and uterine cancer in women.

Moderate dietary restraint, not necessarily to the point of weight loss, may prove to be an alternative strategy for reduction of reproductive cancer risk in both sexes, although more research on this possibility is needed (Hocman, 1988). A better understanding of the effect of dietary restraint on endogenous steroid production is a good place to start. It may

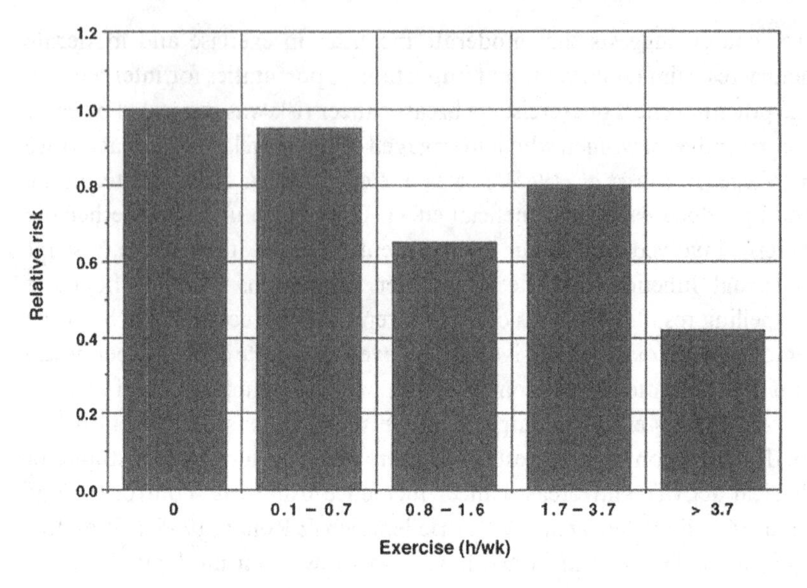

Figure 6.8. Relative risk of breast cancer incidence by average hours spent in recreational exercise. (Data from Bernstein *et al.*, 1994.)

well turn out that the effect of dietary restraint depends on other aspects of endogenous nutritional physiology, such as levels of adiposity, insulin, and/or leptin.

What does seem clear is that the study of reproductive ecology, by identifying patterns of healthy, non-pathological responsiveness of the human reproductive system to environmental, constitutional, and behavioral variables, can help in the important effort to develop acceptable strategies of reproductive cancer prevention (Figure 6.9). The high incidence of these cancers among people in developed countries is very likely a reflection of the profound transformation of human biology that has accompanied the ecological changes of industrial modernization. This realization stands in contrast to the idea that the high incidence of these and other cancers is a consequence of some ecological "pathology", such as high exposures to particular pollutants, environmental estrogenic compounds, or other discrete aberrations. The intervention strategy suggested by the pollution hypothesis involves increased investment in identifying and eliminating the offending cause. In some sense this strategy is loosely analogous to the standard medical approach to identifying and eliminating a pathogen in a sick organism. The strategy suggested by the perspective of reproductive ecology involves an adjustment of behavior and ecology at the level of the individual which reduces cumulative risk. In this sense it is

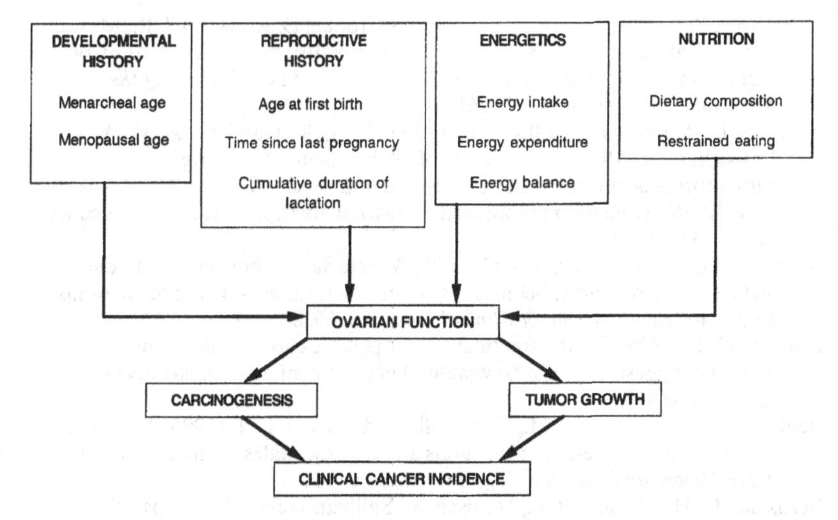

Figure 6.9. A schematic representation of the model discussed in this chapter, initially developed in the context of reproductive ecology. Female ovarian function serves as a common pathway by which many categories of risk factors for breast cancer affect both the initial appearance and the subsequent growth of cancerous tissue.

loosely analogous to the approach of behavioral medicine to reducing the risk of pathological states with complex etiology. Although we are culturally conditioned to expect the medical system to provide us with pills or inoculations that will confer effortless protection from the perils of disease, we are perhaps well advised to remember that the modern reduction of infectious disease mortality and consequent extension of life expectancy owes more to changes in lifestyle and ecology than to the cumulative contributions of medical technology. Rather than wait for a magic bullet, we may have to work, or at least exercise, to reduce the risk of reproductive cancers.

6.6 References

Albanes, D., & Winick, M. (1988). Are cell number and cell proliferation risk factors for cancer? *Journal of the National Cancer Institute*, **80**, 772–5.

Adlercreutz, H., & Järvenpää, P. (1982). Assay of estrogen in human feces. *Journal of Steroid Biochemistry*, **17**, 639–45.

Adlercreutz, H., Martin, F., Pulkkinen, Dencker, H., Rimer, U., Sjoberg, N. O., & Tikkanen, M. J. (1976). Intestinal metabolism of estrogens. *Journal of Clinical Endocrinology and Metabolism*, **43**, 497–505.

Apter, D., & Vihko, R. (1983). Early menarche, a risk factor for breast cancer, indicates early onset of ovulatory cycles. *Journal of Clinical Endocrinology and Metabolism*, **57**, 82–6.

Armstrong, B. K., Brown, J. B., Clarke, H. T., Crooke, D. K., Hahnel, R., Masarel, J. R., & Ratajczak, T. (1981). Diet and reproductive hormones: A study of vegetarian and nonvegetarian postmenopausal women. *Journal of the National Cancer Institute*, **67**, 761–7.

Bailey, R. C., Jenike, M. R., Ellison, P. T., Bentley, G.R., Harrigan, A. M., & Peacock, N. R. (1992). The ecology of birth seasonality among agriculturalists in Central Africa. *Journal of Biosocial Science*, **24**, 393–412.

Barley, V. (1996). Time for a reappraisal of ovarian ablation in early breast cancer. *Lancet*, **348**, 1184.

Barr, S. I., Janelle, K. C., Prior, J. C. (1994). Vegetarian vs. nonvegetarian diets, dietary restraint, and subclinical ovulatory disturbances: Prospective 6-mo. study. *American Journal of Clinical Nutrition*, **60**, 887–94.

Beatson, G. T. (1896). On the treatment of inoperable cases of carcinoma of the mamma: suggestions for a new method of treatment, with illustrative cases. *Lancet*, **2**, 104–7.

Bentley, G. R., Harrigan, A. M., Campbell, B., & Ellison, P. T. (1993). Seasonal effects on salivary testosterone levels among Lese males of the Ituri Forest, Zaïre. *American Journal of Human Biology*, **5**, 711–17.

Bernstein, L., Henderson, B. E., Hanisch, R., Sullivan-Halley, J., & Ross, R. K. (1994). Physical exercise and reduced risk of breast cancer in young women. *Journal of the National Cancer Institute*, **86**, 1403–8.

Bernstein, L., & Ross, R. K. (1993). Endogenous hormones and breast cancer risk. *Epidemiologic Review*, **15**, 48–65.

Bernstein, L., Ross, R. K., & Henderson, B. E. (1991). Prospects for the primary prevention of breast cancer. *American Journal of Epidemiology*, **135**, 142–52.

Bribiescas, R. G. (1996). Testosterone levels among Ach hunter-gatherer men: a functional interpretation of population variation among adult males. *Human Nature*, **7**, 163–88.

Brinton, L. A., & Schairer, C. (1993). Estrogen replacement therapy and breast cancer risk. *Epidemiologic Review*, **15**, 66–79.

Broocks, A., Pirke, K. M., Schweiger, U., Tuschl, R. J., Laessle, R. G., Strowitzki, T., Hö rl, T., Haas, W., & Jeschke, D. (1990). Cyclic ovarian function in recreational athletes. *Journal of Applied Physiology*, **68**, 2083–6.

Bruzzi, P., Negri, E., La Vecchia, C., Decarli, A., Palli, D., Parazzini, F., & Del Turco, M. R. (1988). Short term increase in risk of breast cancer after full term pregnancy. *British Medical Journal*, **297**, 1096–8.

Bullen, B. A., Skrinar, G. S., Beitins, I. Z., von Mering, G., Turnbull, B. A., & McArthur, J. W. (1985). Induction of menstrual disorders by strenuous exercise in untrained women. *New England Journal of Medicine*, **312**, 1349–53.

Bush, T. L., & Hezloyer, K. J. (1993). Tamoxifen for the primary prevention of breast cancer: a review and critique of the concept and trial. *Epidemiologic Review*, **15**, 233–4.

Byers, T., Graham, S., Rzepka, T., & Marshall, J. (1985). Lactation and breast cancer: Evidence for a negative association in premenopausal women. *American Journal of Epidemiology*, **121**, 664–74.

Clark, J. H., & Markaverich, B. M. (1988). Actions of ovarian steroid hormones. In *The Physiology of Reproduction*, ed. E. Knobil & J. Neill, pp. 675–726. New York: Raven.

Clarke, C. L., & Sutherland, R. L. (1990). Progestin regulation of cellular proliferation. *Endocrine Reviews*, **11**, 266–301.

Coffey, D. (1988). Androgen action and the accessory tissues. In *The Physiology of Reproduction*, ed. E. Knobil & J. Neill, pp. 1081–119. New York: Raven.
Cohen, S. M., & Ellwein, L. B. (1990). Cell proliferation in carcinogenesis. *Science*, **249**, 1007–11.
Colditz, G. A., Hankinson, S. E., Hunter, D. J., Willet, W. C., Manson, J. E., Stampfer, M. J., Hennekens, C., Rosner, B., & Speizer, F. E. (1995). The use of estrogens and progestins and the risk of breast cancer in postmenopausal women. *New England Journal of Medicine*, **332**, 1589–93.
Crawford, E. D. (1992). Challenges in the management of prostate cancer. *British Journal of Urology*, **70**, Suppl. 1, 33–8.
de la Rochfordière A., Asselain, B., Campana, F., Scholl, S. M., Fenton, J., Vilcoq, J. R., Durand, J. C., Pouillart, P., Magdelenat, H., & Fourquet, A. (1993). Age as a prognostic factor in premenopausal breast carcinoma. *Lancet*, **341**, 1039–43.
Early Breast Cancer Trialists' Collaborative Group. (1992). Systematic treatment of early breast cancer by hormonal, cytotoxic, or immune therapy. *Lancet*, **339**, 1–15, 71–85.
Early Breast Cancer Trialists' Collaborative Group. (1996). Ovarian ablation in early breast cancer: overview of the randomized trials. *Lancet*, **348**, 1189–96.
Eaton, S. B., Pike, M. C., Short, R. V., Lee, N. C., Trussell, J., Hatcher, R. A., Wood, J. W., Worthman, C. M., Blurton-Jones, N. G., Konner, M. J., Hill, K. R., Bailey, R., & Hurtado, A. M. (1994). Women's reproductive cancers in evolutionary context. *Quarterly Review of Biology*, **69**, 354–67.
Ellison, P. T. (1993). Salivary steroids and natural variation in human ovarian function. *Annals of the N Y Academy of Science*, **709**, 287–98.
Ellison, P. T. (1994). Advances in human reproductive ecology. *Annual Reviews in Anthropology*, **23**, 255–75.
Ellison, P. T. (1995). Understanding natural variation in human ovarian function. In *Human Reproductive Decisions: Biological and Social Perspectives*, ed. R. I. M. Dunbar, pp. 22–51. New York: St. Martin's Press.
Ellison, P. T. (1996*a*). Developmental influences on adult ovarian function. *American Journal of Human Biology*, **8**, 725–34.
Ellison, P. T. (1996*b*). Age and developmental effects on adult ovarian function. In *Variability in Human Fertility: A Biological Anthropological Approach*, ed. L. Rosetta, & C. N. G. Mascie-Taylor, pp. 69–90. Cambridge University Press, Cambridge.
Ellison, P. T., & Lager, C. (1986). Moderate recreational running is associated with lowered salivary progesterone profiles in women. *American Journal of Obstetrics and Gynecology*, **154**, 1000–3.
Ellison, P. T., Lipson, S., & Meredith, M. (1989). Salivary testosterone levels in males from the Ituri Forest, Zaire. *American Journal of Human Biology*, **1**, 21–4.
Ellison, P. T., Lipson, S. F., O'Rourke, M. T., Bentley, G.R., Harrigan, A. M., Panter-Brick, C., & Vitzthum, V. J. (1993*b*). Population variation in ovarian function. *Lancet*, **342**, 433–4.
Ellison, P. T., Lipson, S. F., & Sukalich, S. (1996). Recreational running is associated with low profiles of salivary estradiol. *American Journal of Physical Anthropology*, Suppl. 22, 102.
Ellison, P. T., & Panter-Brick, C. (1996). Salivary testosterone levels among Tamang and Kami males of central Nepal. *Human Biology*, **68**, 955–65.
Ellison, P. T., Panter-Brick, C., Lipson, S. F., & O'Rourke, M. T. (1993*a*). The

ecological context of human ovarian function. *Human Reproduction*, **8**, 2248–58.

Ellison, P. T., Peacock, N. R., & Lager, C. (1989). Ecology and ovarian function among Lese women of the Ituri Forest, Zaire. *American Journal of Physical Anthropology*, **78**, 519–26.

Ewald, P. W. (1980). Evolutionary biology and the treatment of signs and symptoms of infectious disease. *Journal of Theoretical Biology*, **86**, 169–76.

Friedenreich, C. M., & Rohan, T. E. (1995). Physical activity and risk of breast cancer. *European Journal of Cancer Prevention*, **4**, 145–51.

Frisch, R. E., Wyshak, G., Albright, N. L., albright, T. E., Schiff, I., Jones, K. P., Witschi, J., Shiang, E., Kogg, E., & Marguglio, M. (1985). Lower prevalence of breast cancer and cancers of the reproductive system among former college athletes compared to non-athletes. *British Journal of Cancer*, **52**, 885–91.

Goldin, B. R., Adlercreutz, H., Gorbach, A. L. *et al.* (1982). Estrogen excretion patterns and plasma levels in vegetarian and omnivorous women. *New England Journal of Medicine*, **307**, 1542–7.

Goldin, B. R., Aldercreutz, H., Gorbach, A. L., Woods, M. N., Dwyer, J. T., Conlon, T., Bohn, E., & Gershoff, S. N. (1986). The relationship between estrogen levels and diets of Caucasian American and Oriental immigrant women. *American Journal of Clinical Nutrition*, **44**, 945–53.

Guinee, V. F., Olsson, H., Mö ller, T., Hess, K. R., Taylor, S. H., Fahey, T., Gladikov, J. V., van den Blink, J., Bonichon, F., Dische, S., Yates, J. W., & Cleton, F. J. (1994). Effect of pregnancy on prognosis for young women with breast cancer. *Lancet*, **343**, 1587–9.

Halperin, F. (1997). The effect of dietary fiber on ovarian function. Honors Thesis, Department of Anthropology, Harvard University, Cambridge, MA.

Henderson, B. E., Bernstein, L. (1991). The international variation in breast cancer rates: an epidemiological assessment. *Breast Cancer Research and Treatment*, **18**, Suppl. 1, 11–17.

Henderson, B. E., Ross, R. K., & Berstein, L. (1988). Estrogens as a cause of human cancer: the Richard and Hilda Rosenthal Foundation award lecture. *Cancer Research*, **48**, 246–53.

Henderson, B. E., Ross, R. K., & Pike, M. C. (1993). Hormonal chemoprevention of cancer in women. *Science*, **259**, 633–8.

Henderson, M. (1993). Current approaches to breast cancer prevention. *Science*, **259**, 630–1.

Hocman, G. (1988). Prevention of cancer: restriction of nutritional energy intake (joules). *Comparative Biochemistry and Physiology*, **91A**, 209–20.

Hulka, B. S., & Stark, A. T. (1995). Breast cancer: cause and prevention. *Lancet*, **346**, 883–7.

Hunter, D. J., & Willet, W. C. (1993). Diet, body size, and breast cancer. *Epidemiologic Review*, **15**, 110–32.

Jasienska, G. (1996). Energy expenditure and ovarian function in rural women from Poland. PhD diss., Harvard University, Cambridge, MA, USA.

Jenike, M. R., Bailey, R., Ellison, P. T., Bentley, G. R., Harrigan, A. M., & Peacock, N. R. (1996). Variation saisonnière de la production alimentaire, statut nutritionnel, fonction ovarienne et fécondité en Afrique centrale. In *L'Alimentation en Foret Tropical: Interactions Bioculturelles et Perspectives de Developpement*, ed. C. M. Hladik, A. Hladik, H. Pagezy, O. F. Linares, A. Froment, pp. 605–23. Paris: UNESCO.

Jordan, V. C., & Allen, K. E. (1980). Evaluation of the antitumor activity of the nonsteroidal antioestrogen monohydroxytamoxifen in the DMBA-induced rat mammary carcinoma model. *European Journal of Cancer*, **16**, 239–51.

Kelsey, J. L., & Gammon, M. D. (1991). The epidemiology of breast cancer. *CA-A Cancer Journal for Clinicians*, **41**, 146–65.

Kelsey, J. L., Gammon, M. D., & John, E. M. (1993). Reproductive factors and breast cancer. *Epidemiologic Review*, **15**, 36–47.

Kelsey, J. L., & Horn-Ross, P. L. (1993). Breast cancer: magnitude of the problem and descriptive epidemiology. *Epidemiologic Review*, **15**, 7–16.

Key, T. J. A., & Pike, M. C. (1988*a*). The role of oestrogens and progestagens in the epidemiology and prevention of breast cancer. *European Journal of Clinical Oncology*, **24**, 29–43.

Key, T. J. A., Pike, M. C. (1988*b*). The dose-effect of "unopposed" oestrogens and endometrial mitotic rate: its central role in explaining and predicting endometrial cancer risk. *British Journal of Cancer* **57**, 205–12.

Kramer, M. M. & Wells, C. L. (1996). Does physical activity reduce risk of estrogen-dependent cancer in women? *Medicine and Science in Sports and Exercise*, **28**, 322–34.

Kvåle, G., Heuch, I., & Eide, G. E. (1987). A prospective study of reproductive factors and breast cancer: I. Parity. *American Journal of Epidemiology*, **126**, 831–41.

Kvåle, G., & Heuch, I. (1987). A prospective study of reproductive factors and breast cancer: II. Age at first and last birth. *American Journal of Epidemiology*, **126**, 842–50.

Labrie, F., Dupont, A., & Belanger, A. (1983). New approach in the treatment of prostatic cancer. Complete instead of partial withdrawal of androgens. *Prostate*, **4**, 579–94.

Lager, C., & Ellison, P. T. (1990). Effect of moderate weight loss on ovarian function assessed by salivary progesterone measurements. *American Journal of Human Biology*, **2**, 303–12.

Layde, P. M., Webster, L. A., Baughman, A. L., wingo, P. A., Rubin, G. L., & Ory, H. W. (1989). The independent associations of parity, age at first full term pregnancy, and duration of breastfeeding with the risk of breast cancer. *Journal of Clininical Epidemiology*, **42**, 963–73.

Lipson, S. F., & Ellison, P. T. (1992). Normative study of age variation in salivary progesterone profiles. *Journal of Biosocial Science*, **24**, 233–44.

Lipson, S. F., & Ellison, P. T. (1994). Reference values for luteal progesterone measured by salivary radioimmunoassay. *Fertility and Sterility*, **61**, 448–54.

Lipson, S. F., & Ellison, P. T. (1996). Comparison of salivary steroid profiles in naturally occurring conception and nonconception cycles. *Human Reproduction*, **11**, 2060–7.

Lustig, R. H., Hershcopf, R. J., & Bradlow, H. L. 1990. The effects of body weight and diet on estrogen metabolism and estrogen-dependent disease. In *Adipose Tissue and Reproduction, Progress in Reproductive Biological Medicine*, ed. R. E. Frisch, **14**, 107–24. Basel: Karger.

MacDonald, P. C., Edman, C. D., Hemsell, D. L., Porter, J. C., & Siiteri, P. K. (1978). Effect of obesity on conversion of plasma androstenedione to estrone in postmenopausal women with and without endometrial cancer. *American Journal of Obstetrics and Gynecology*, **130**, 448–55.

MacMahon, B., Cole, P., Lin, T. M., Lowe, C. R., Mirra, A. P., Ravnihar, B., Salber, E. J., Valaoras, V. G., & Yuasa, S. (1970). Age at first birth and breast

cancer risk. *Bulletin WHO*, **43**, 209–21.
Malone, K. E. (1993). Diethylstilbestrol (DES) and breast cancer. *Epidemiologic Review*, **15**, 108–9.
Malone, K. E., Daling, J. R., & Weiss, N. S. (1993). Oral contraceptives and breast cancer risk. *Epidemiologic Review*, **15**, 80–97.
McPherson, K., Steel, C. M., Dixon, J. M. (1994). Breast cancer – epidemiology, risk factors, and genetics. *British Medical Journal*, **309**, 1003–6.
McTiernan, A., & Thomas, D. B. (1986). Evidence for a protective effect of lactation on risk of breast cancer in young women: Results from a case-control study. *American Journal of Epidemiology*, **124**, 353–8.
Mittendorf, R., Langnecker, M. P., Newcomb, P. A., Dietz, A. T., Greenberg, E. R., Bogdan, G. F., Clapp, R. W., & Willet, W. C. (1995). Strenuous physical activity in young adulthood and risk of breast cancer (United States). *Cancer Causes and Control*, **6**, 347–53.
Nesse, R. M., & Williams, G. C. (1994). *Why we get sick*. New York: Random House.
Newcomb, P. A., Storer, B. E., Longnecker, M. P., Mittendorf, R., Greenberg, E. R., Clapp, R. W., Burke, K. P., Willet, W. C., & MacMahon, B. (1994). Lactation and a reduced risk of premenopausal breast cancer. *New England Journal of Medicine*, **330**, 81–7.
O'Rourke, M. T., & Ellison, P. T. (1993*a*). Age and prognosis in premenopausal breast cancer. *Lancet*, **342**, 60.
O'Rourke, M. T., Ellison, P. T. (1993*b*). Salivary estradiol levels decrease with age in healthy, regularly-cycling women. *Endocrine Journal*, **1**, 487–94.
Panter-Brick, C., Ellison, P. T., Lipson, S. F., & Sukalich, S. (1996). Energy balance and salivary estradiol among Tamang women in Nepal. *American Journal of Physical Anthropology*, suppl. 22, 182.
Panter-Brick, C., Lotstein, D. S., & Ellison, P. T. (1993). Seasonality of reproductive function and weight loss in rural Nepali women. *Human Reproduction*, **8**, 684–90.
Persky, V. W., Chatterton, R. T., Van Horn, L. V., Grant, M. D., Langenberg, P., & Marvin, J. (1992). Hormone levels in vegetarian and nonvegetarian teenage girls: Potential implications for breast canncer risk. *Cancer Research*, **52**, 578–83.
Pike, M. C., Spicer, D. V., Dahmoush, L., & Press, M. F. (1993). Estrogens, progestins, normal breast proliferation, and breast cancer risk. *Epidemiologic Review*, **15**, 17–35.
Prostate Cancer Trialists' Collaborative Group. (1995). Maximum androgen blockade in advanced prostate cancer: an overview of 22 randomized trials with 3283 deaths in 5710 patients. *Lancet*, **346**, 265–9.
Rimpella, A. H., & Pukkala, E. I. (1982). Cancers of affluence:positive social class gradient and rising incidence trend in some cancer forms. *Social Science and Medicine*, **24**, 601–6.
Russo, J., Tay, L. K., & Russo, I. H. (1982). Differentiation of the mammary gland and susceptibility to carcinogenesis. *Breast Cancer Research Treatment*, **2**, 5–73.
Santen, R. J., Manni, A., Harvey, H., & Redmond, C. (1990). Endocrine treatment of breast cancer in women. *Endocrine Reviews*, **11**, 221–65.
Schweiger, U., Laessle, R., Schweiger, M., Herrmann, F., Riedel, W., & Pirke, K. M. (1988). Caloric intake, stress, and menstrual function in athletes. *Fertility and Sterility*, **49**, 447–50.

Schweiger, U., Tuschl, R. J., Laessle, R. G., Broocks, A., & Pirke, K. M. (1989). Consequences of dieting and exercise on menstrual function in normal weight women. In *The Menstrual Cycle and Its Disorders*, ed. K. M. Pirke, W. Wuttke, & U. Schweiger, pp. 142–9. Berlin: Springer-Verlag.

Schweiger, U., Tuschl, R. J., Platte, P., Broocks, A., Laessle, R. G., & Pirke, K. M. (1992). Everyday eating behavior and menstrual function in young women. *Fertility and Sterility*, **57**, 771–5.

Shimizu, H., Ross, R. K., Bernstein, L., Yatani, R., Henderson, B. E., & Mack, T. M. (1991). Cancers of the prostate and breast among japanese and white immigrants in Los Angeles County. *British Journal of Cancer*, **63**, 963–6.

Siiteri, P. K. (1990). Obesity and peripheral estrogen synthesis. In *Adipose Tissue and Reproduction*, ed. R. E. Frisch, pp. 70–84. Basel: Karger.

Siiteri, P. K., Williams, J. E., & Takaki, N. K. (1976). Steroid abnormalities in endometrial and breast carcinoma: a unifying hypothesis. *Journal of Steroid Biochemistry*, **7**, 897–903.

Simopoulos, A. P. (1990). Energy imbalance and cancer of the breast, colon and prostate. *Medical Oncology and Tumor Pharmacotherapy*, **7**, 109–20.

Siskind, V., Schofield, F., Rice, D., & Bain, C. (1989). Breast cancer and breastfeeding: Results from an Australian case-control study. *American Journal of Epidemiology*, **130**, 229–36.

Soto, A. M., & Sonnenschein, C. (1987). Cell proliferation of estrogen sensitive cells: the case for negative control. *Endocrine Reviews*, **8**, 44–52.

Stanford, J. L. & Thomas, D. B. (1993). Exogenous proteins and breast cancer. *Epidemiologic Reviews*, **15**, 98–107.

Stoll, B. A., Vatten, L. J., & Kvinnsland, S. (1994). Does early physical maturity influence breast cancer risk? *Acta Oncologica*, **33**, 171–6.

Thomas, D. B., & Karagas, M. R. (1987). Cancer in first and second generation Americans. *Cancer Research*, **47**, 5771–6.

Thorling, E. B. (1996). Obesity, fat intake, energy balance, exercise and cancer risk: A review. *Nutritional Research*, **16**, 315–68.

Thune, I., Brenn, T., & Gaard, M. (1997). Physical activity and the risk of breast cancer. *New England Journal of Medicine*, **336**, 1269–75.

Tricholpoulos, D., Yen, S., Brown, J., Cole, P., & MacMahon, B. (1984). Effect of westernization on urine estrogens, frequency of ovulation, and breast cancer risk: a study of ethnic Chinese women in the Orient and the USA. *Cancer*, **53**, 187–92.

Tucker, H. A. (1988). Lactation and its hormonal control. In *The Physiology of Reproduction*, ed. E. Knobil, & J. Neill, pp. 2235–64. New York: Raven.

van Tinteren, H., & Dalesio, O. (1993). Systematic overview (meta-analysis) of all randomized trials of treatment of prostate cancer. *Cancer*, **72**, 3847–50.

Ziegler, R. G., Hoover, R. N., Pike, M. C., Hildesheim, A., Nomura, A. M. Y., West, D. W., Wu-Williams, A. H., Kolonel, L. N., Horn-Ross, P. L., Rosenthal, J. F., & Hyer, M. B. (1993). Migration patterns and breast cancer risk in Asian–American women. *Journal of the National Cancer Institute*, **85**, 1819–27.

7

Diet, hormones, and health: an evolutionary–ecological perspective

PATRICIA L. WHITTEN

7.1 Introduction

The old adage "you are what you eat" may be more apt than we ever could have imagined. Diet not only influences human health and development but also has shaped the evolution of our behavior and physiology. Although anthropologists traditionally have viewed culture as an intentional process obviating biological adaptation, there is good evidence that cultural practices can produce both physiological and evolutionary changes in human populations (Cohen & Armelagos, 1984; Durham, 1991). Biological responses, in turn, exert selective pressures on cultural traits (Rindos, 1989). Plant exploitation, cultivation, and consumption are good examples of these interactions.This chapter examines the influence of diet on cancer risk from an adaptive and phylogenetic perspective. It describes the growth of cancer and other chronic diseases in Western populations in relation to nutritional and other dietary constituents that directly and indirectly influence the development and function of the reproductive system. The links between diet and reproductive cancer are explained as an outcome of human reproductive strategies, adaptations of the reproductive system that coordinate reproduction with optimal nutritional conditions. Dietary practices that elevate cancer risk are related to human food preferences and underlying perceptual mechanisms that may reflect ancestral foraging strategies. These foraging and reproductive strategies are argued to be ancestral traits that humans share with other apes, reflecting our evolutionary history as frugivorous primates adapted to variable and unpredictable food resources, a heritage that molded the domestication of human crops in prehistory and that propels modern human populations toward chronic disease as our food preferences are increasingly realized.

7.2 Health, hormones, and diet

Western industrialized populations are plagued by a set of diseases associated with chronic hormone exposure. Cancers of the reproductive system like breast cancer, endometrial cancer, and prostate cancer along with colon cancer and coronary heart disease are much more common in Western Europe and North America than in Asia, Africa, or Eastern Europe (Trowell & Burkitt, 1983: Yu *et al.*, 1991). These differences in morbidity are thought to reflect differences in lifestyle since the incidence of these diseases rises over time in developing populations as they adopt a "Westernized lifestyle" (Adlercreutz, 1990). Moreover, immigrants to Western countries initially exhibit the incidence rate typical of their country of origin but assume the risk of their host country by the second generation (Shimizu *et al.*, 1991). The primary risk factor for reproductive cancers appears to be cumulative lifetime exposure to gonadal steroids, particularly estrogens (Marshall, 1993; Henderson *et al.*, 1993). Earlier onset of reproductive cycles or higher serum estrogen levels are associated with a higher risk of breast cancer (Ellison, Chapter 6) whereas lower circulating levels of estrogens and androgens in women in Japan and China and recent Asian emigrants to the USA are associated with a reduced risk of breast cancer (Goldin *et al.*, 1986; Key *et al.*, 1990; Shimizu *et al.*, 1990).

An additional, and closely linked, risk factor is diet (Adlercreutz, 1990). The eclectic nature of the human diet results in considerable differences in dietary composition both within and across human populations. Although it is clear that there are associations between cancer risk and diet, the dietary factors that are primarily responsible for differences in cancer incidence are still a matter of considerable debate. Much attention has been devoted to nutritional components like fat, calories, and energy balance that are believed to influence levels of endogenous steroids (Greenwald, 1989; Panter-Brick & Pollard, Chapter 5; Ellison, Chapter 6). Altering nutritional factors in clinical trials, however, has had variable success in lowering cancer incidence. For example, some studies have shown that the risk of breast cancer increases with increasing fat intake (Howe *et al.*, 1990), but other studies have found no evidence of increased risk (Willett *et al.*, 1987). These inconsistencies may be partly due to the fact that fat intake is so high in Western societies that within-population variation in diet produces relatively small changes in fat intake, with neglible impact on circulating estrogen (Adlercreutz, 1990). More recent efforts have focused on dietary factors that may be chemopreventative, rather than cancer-causing (Barnes *et al.*, 1995). Diets that are low in animal fat and protein also contain higher concentrations of "non-nutritive factors", fiber and other chemical products of plants that are not considered nutrients but nonetheless influence endocrine and cellular processes (Adlercreutz, 1990; Steinmetz & Potter, 1991).

Both nutrients, like lipids, and non-nutritive factors, like fiber, may influence cancer risk by altering the availability of endogenous estrogens (see Figure 7.1). Dietary fat and fiber exert mechanical effects on estrogen availability by altering the excretion of estrogens. Estrogens are eliminated from the body by conjugation with gluronic acid or sulfate in the liver, enhancing their solubility to facilitate excretion into urine or bile. The biliary route allows a recycling process known as enterohepatic recirculation in which conjugated estrogens are hydrolyzed by bacteria enzymes and resorbed into plasma, prolonging the residence of estrogens in the body (Gorbach & Goldin, 1987). Fat and protein slow digestion, allowing more time for resorption of estrogens whereas fiber speeds the passage of digesta through the intestinal tract, providing less time for reabsorption. Some fibrous components of plants appear to play a role in preventing cancer, but research to date has subsumed a variety of substances under the term "fiber". Fiber is composed of a heterogeneous group of structural and non-structural polysaccharides that are resistant to digestion (Slavin, 1987). Water-insoluble fibers like cellulose and lignin are more resistant to digestion and have physiological effects that differ from the more digestible, water-soluble, fibers like pectin, gum, and mucilage (Eastwood *et al.*, 1986). Relatively insoluble fibers like cellulose and lignin increase fecal bulk, effectively reducing the concentration of hydrolyzing enzymes and carcinogens, and hasten gut transit time, reducing the opportunity for absorption, which results in increased fecal excretion of estrogens (Adlercreutz, 1990). Soluble fibers like pectin do not affect fecal bulk because they are largely fermented (Eastwood *et al.*, 1986), but the short chain fatty acids that are the products of fermentation may themselves reduce serum cholesterol and alter cellular proliferation and differentiation (Jacobs, 1986).

Endocrine profiles of women with different diets provide evidence that dietary fiber can alter the bioavailability of endogenous steroids. Cross-sectional population studies suggest that fiber reduces serum estrogen levels (Feng *et al.*, 1993) but provide conflicting evidence for a stimulatory role of dietary fat (Feng *et al.*, 1993; Dorgan *et al.*, 1996; Kaneda *et al.*, 1997). More compelling evidence is provided by changes in hormonal profiles of individual women following experimental manipulations of dietary composition. The latter investigations show that elevations in dietary fiber suppress serum estrogen levels, as do reductions in dietary fat,

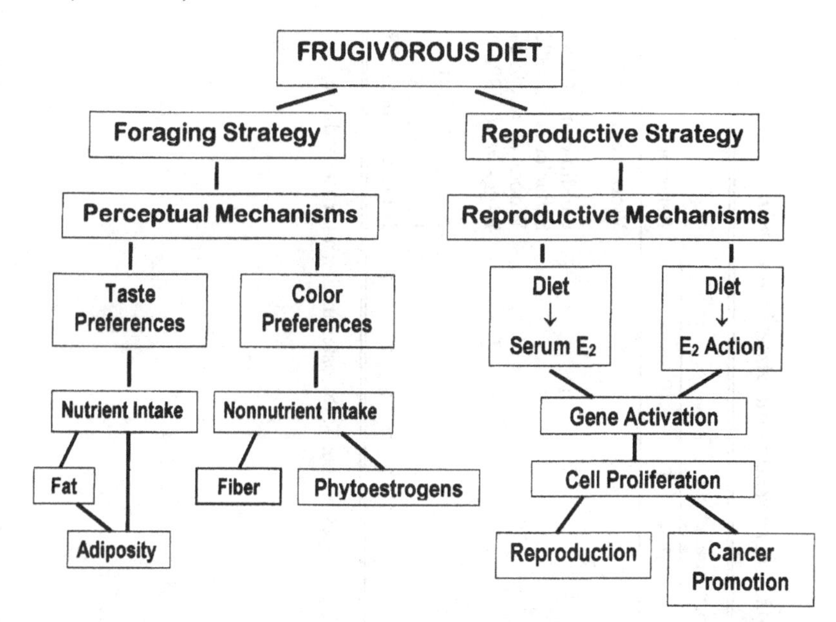

Figure 7.1. Hypothesized relationship between an ape ancestral heritage of reproductive and foraging strategies, based on a frugivorous diet, and observed links of contemporary diets with human reproduction and health. E_2, estradiol.

but do not affect ovulation, cycle length, or luteal phase progesterone (Table 7.1). These dietary effects on the hormonal profiles of Western women are in striking contrast to the profiles reported for women in traditional non-Western societies where the primary difference from cycles of Western women is low luteal phase progesterone levels (Ellison *et al.*, 1993), a difference that may be attributable to differences in energy balance (Panter-Brick and Pollard, Chapter 5; Ellison, Chapter 6) or to plant constituents like phytoestrogens that more directly affect estrogen action.

7.3 Phytoestrogen actions

"Phytoestrogens" are plant chemicals that resemble the vertebrate steroidal estrogens in structure and act as estrogens or estrogen antagonists in animals, providing a direct link between diet and reproductive function. Estrogen-like substances (such as lignans, indoles, and isoflavones) often accompany fiber in plant material and may be responsible for some of the reported chemopreventative actions of fiber or vegetarian diets (Steinmetz & Potter, 1991). Consumption of cereal bran, cruciferous vegetables like

Table 7.1. *Effects of dietary fat and fiber on serum hormone concentrations in premenopausal American women*

Dietary change	*N*	Fat levels % of calories	Fiber levels g/day	Hormonal effect						Reference
				E_2	E_1	E_1S	P_4	T	A	
High fat–low fiber to low fat–high fiber	17	40 → 25	12 → 40			↓				Woods *et al.*, 1989
	48	40 → 25	12 → 40	(↓)	↓	↓		↓	↓	Goldin *et al.*, 1994
	22	40 → 25	12 → 40			↓				Schaefer *et al.*, 1995
	21	40 → 25	12 → 40	↓		↓			↑	Woods *et al.*, 1996
	12	30 → 10	20 → 40	↓	↓					Bagga *et al.*, 1995
High to low fiber	62		15 → 30	↓	↓					Rose *et al.*, 1991
High to low fat	16	35 → 21		↓	↓					Rose *et al.*, 1987
	6	46 → 25								Hagerty *et al.*, 1988

Note: Hormone abbrevations: E_2: estradiol; E_1: estrone; E_1S: estrone sulfate; P_4: progesterone; T: testosterone; A: androstenedione.

cabbage, or soybeans, is associated with protection against carcinogenesis in both experimental and epidemiological research (Adlercreutz, 1990; Wattenberg, 1992; Messina *et al.*, 1994). The "mammalian lignans" are produced by gut bacteria from precursors common in oilseeds, vegetables, and cereals (Setchell *et al.*, 1984). Lignans suppress estrogen biosynthesis (Adlercreutz, 1990) and inhibit estrogen actions such as cellular proliferation (Mousavi & Adlercreutz, 1992). The "indoles" and "isothiocyanates", acrid metabolites of glucosinolates that are common in cruciferous vegetables, inhibit steroid metabolizing hydroxylases that change estrogens to a more potent and lasting form (Bradlow *et al.*, 1991), an action opposite to that of fatty acids, which enhance the production of the more potent form (Telang *et al.*, 1992). The "isoflavonoids", common in soybeans and other legumes (Franke *et al.*, 1995), display a mixture of estrogen agonism and antagonism, augmenting estrogen action at low doses and inhibiting estrogen action at higher concentrations (Whitten *et al.*, 1992, 1994, 1997).

Dietary isoflavones and lignans could provide cancer protection by altering estrogen profiles through negative feedback effects on estrogen secretion; however, investigations of ovarian hormone responses of British and American women to dietary manipulations have produced inconsistent results (Table 7.2). Dietary supplementation with flax seed, which appears to be exceptionally high in lignan content (Thompson, 1994), lengthened the luteal phase but did not alter serum hormone concentrations (Phipps *et al.*, 1993). Soy isoflavones and phytoestrogens extracted from alcoholic beverages reduced LH (luteinizing hormone) secretion in post-menopausal women (van Thiel *et al.*, 1991; Nicholls *et al.*, 1995). Dietary supplements of soy have both increased (Cassidy *et al.*, 1994; Lu *et al.*, 1996) and reduced (Lu *et al.*, 1996) cycle length with variable effects on serum estradiol and progesterone. These inconsistent results reflect variation in dose and duration of treatment, which vary considerably across studies (Table 7.2), and the ability of phytoestrogens, like other estrogens, to produce biphasic actions. Chronic estrogen treatment is contraceptive, enhancing the normal negative feedback between estrogen and gonadotropin secretion, but low or intermittent estrogen can initiate ovarian cycles in acyclic females by briefly suppressing LH which then rebounds at levels sufficient to induce ovulation. Animal studies suggest that isoflavonoids have similar actions. For example, a low dietary dose of the isoflavonoid coumestrol suppressed ovarian cycles in adult female rats but stimulated the onset of reproductive cycles in immature females (Whitten & Naftolin, 1992; Whitten *et al.*, 1995*b*). Moreover, differing doses of isoflavones produce opposing effects on LH secretion: low doses of the isoflavone genistein enhanced LH release

Table 7.2. *Effects of isoflavones and lignans on ovarian cycles in premenopausal American and British women*

Phytoestrogen	*N*	Intake mg/kg per day	Food source	Treatment duration	Hormonal effect					Reference
					E_2	P_4	T	DHS	LH	
Isoflavones	6	0.6	Soy-protein isolate	1 month	↑	—	—	NA	↓	Cassidy *et al.*, 1994
	6	3.0	Soymilk	1 month	↓	↓	↓	↓	NA	Lu *et al.*, 1996
	24	1.0	Soy-protein isolate	6 months	↑	—	NA	NA	NA	Petrakis *et al.*, 1996
Lignans	18	~0.09	Flaxseed	3 cycles	↓ E_2:P_4	—	↑	—	—	Phipps *et al.*, 1993

Note: Hormone abbrevations: E_2: estradiol; P_4: progesterone; T: testosterone; DHS: dihydroepiandrosterone sulfate; LH: luteinizing hormone. NA: hormone was not assayed; — no significant change in hormone concentration following phytoestrogen treatment.

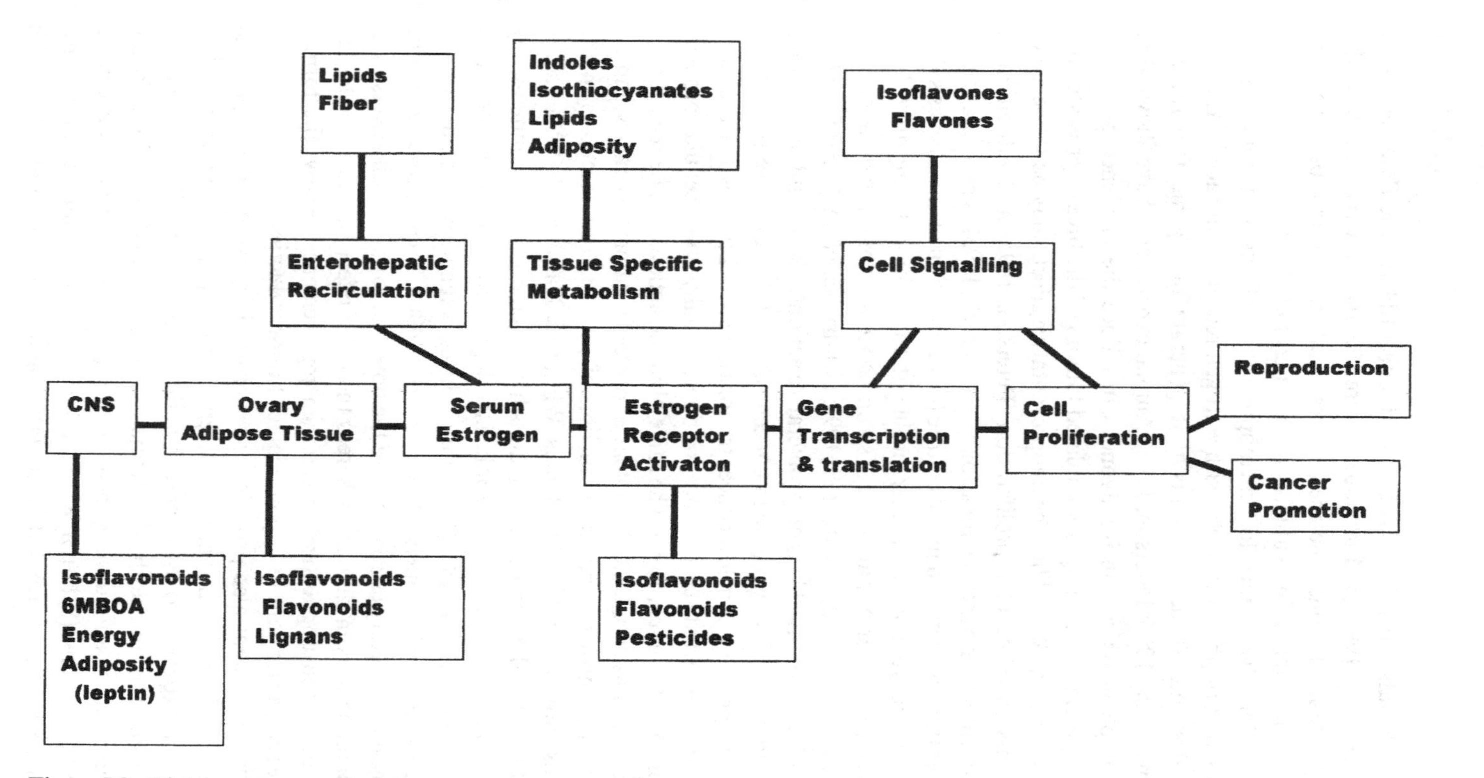

Figure 7.2. Dietary and energetic influences on estrogen action.

in female rats whereas high doses inhibited LH release (Hughes, 1988; Hughes *et al.*, 1991*a,b*). Individual differences also may play a role as the composition of isoflavone metabolites produced by gut bacteria vary considerably across individuals (Adlercreutz *et al.*, 1995).

An alternative route for chemopreventive effects is through direct interaction of phytoestrogens with estrogen receptors, intracellular signal cascades, and oxidative events that generate mutagenic free radicals (Barnes *et al.*, 1995; Peterson, 1995). Animal experiments have shown that soy supplements rich in isoflavones can reduce the incidence or rate of development of carcinogen-stimulated tumors of the breast, prostate, and colon (Barnes, 1995). These chemopreventative effects may be the result of suppression of cellular proliferation (Peterson, 1995) and inhibition of angiogenesis, or vascular invasion (Fotsis *et al.*, 1995). The isoflavonoid concentrations required in cell culture for suppression of angiogenesis (150 μM) or suppression of cellular proliferation (5–50 μM) are quite high, however, considerably in excess of serum concentrations, which range from 0.02 to – 3 μM (Adlercreutz *et al.*, 1993; Hendrich *et al.*, 1994; Franke *et al.*, 1995). Isoflavonoid effects on proliferative actions, like LH influences, vary with concentration: while high concentrations (> 2 μM) are required to suppress proliferation, low concentrations (10 nM – 2 μM), which are more similar to reported serum concentrations, actually boost cellular proliferation (Whitten *et al.*, 1997). The same biphasic actions can be seen *in vivo*. Dietary isoflavonoids can be potent estrogens in laboratory rats, stimulating uterine growth and the induction of progesterone receptors in the uterus, pituitary and hypothalamus (Whitten & Naftolin, 1992; Whitten *et al.*, 1992). Similar proliferative actions have been observed in humans: soy isoflavones stimulated breast epithelial cell proliferation in premenopausal women (Petrakis *et al.*, 1996) and induced vaginal cell maturation in post-menopausal women (Wilcox *et al.*, 1990). These variable results suggest that dietary estrogens have the potential to exert both adverse and harmful actions in humans; whereas the latter effect suggests a potential phytoestrogen application in estrogen replacement therapy, the former raises concern about their potential for tumor induction.

Phytoestrogens also provide cancer protection through developmental effects. Differences in cancer incidence across populations may reflect prenatal exposure to estrogens. In addition to their proliferative actions, estrogens can induce the differentiation of tissues and participate in the differentiation and development of the reproductive tract and central nervous system (Whitten & Naftolin, 1994). Experiments in rodents provide good evidence that most of the common isoflavonoids, along with some fungal or mycoestrogens, can organize development of tissues of the reproductive tract, breast, and central nervous system (Whitten *et al.*, 1993, 1995*a,b*, 1997; Whitten & Naftolin, 1994). These findings are remarkable in showing that dietary constituents can fundamentally alter brain development and the normal sex-specific patterns of behavior and hormone regulation, similar to the actions of xenobiotic or steroidal estrogens. In humans and other mammals, estrogen exposure early in development influences the risk of reproductive cancers in adulthood by altering the development and estrogen responsiveness of reproductive tissues (Colburn & Clement, 1992). For example, young women prenatally exposed to the synthetic estrogen diethylstilbestrol have elevated rates of vaginocervical cancer (Herbst *et al.*, 1971). Daughters of mothers with lower than normal levels of estrogen during pregnancy due to pre-eclampsia have a significantly reduced risk of breast cancer (Ekbom *et al.*, 1992) whereas dizygotic twins, whose mothers generally have higher estrogen levels, have a higher risk of breast cancer than singletons, a risk that is even higher if the female twin's sibling is a boy, perhaps due to exposure to additional androgens and estrogens secreted by her brother's testes (Hsieh *et al.*, 1992). In this way, dietary factors that raise, augment, or lower maternal estrogens during pregnancy could alter cancer risk by altering development.

Developmental exposure to exogenous estrogens can have both adverse and beneficial consequences. Whereas alterations in the neural regulation of gonadotropin secretion and sexual behavior are likely to impair adult fertility, breast duct differentiation and resistance to estrogen-induced lesions in the prostate can be chemopreventative. Dose-dependent, biphasic effects of genistein on the pituitary regulation of gonadotropin secretion suggest that dietary concentrations may help to determine the balance of adverse and beneficial effects. The possibility that developmental changes in responsiveness to estrogen are responsible for the low rates of reproductive cancers in soy-consuming populations like Japan must be considered as well as the potential role of dietary estrogens in secular changes in the timing of sexual maturation and dizygotic twinning (Whitten, 1992). A more fundamental question is why the reproductive system is so seemingly vulnerable to these exogenous estrogens and estrogen-altering components of diets.

7.4 Diet and the regulation of reproduction

The reproductive actions of dietary constituents are at least partly a response to a uniquely mammalian dilemma, the problem of supporting the high costs of lactation, when energetic costs may rise two- to threefold (Bronson, 1989). The timing of pregnancy and lactation must be adjusted to the periodicity of resources and nutritional condition; therefore, reproductive physiology must evolve in concert with dietary niche, becoming more responsive to those enviromental cues that facilitate an optimal timing of reproductive events (Sadleir, 1969; Bronson, 1989). Which proximate cues are chosen depends upon reproductive strategy. Nutrient intake and energy balance are the most immediate indices of resource availability, but are more useful proximate mechanisms for rodents and other small mammals whose lactation falls within the same season as conception than for large mammals whose cycle of reproduction spans several seasons. Of course, most animals would be expected to forego reproduction when survival is imperiled, and most do, but such dire circumstances are a less common experience for large-bodied mammals whose energy balance is buffered by large body size, body fat, and a lower relative metabolic rate (Bronson, 1989). Predictors of approaching changes in resource availability may be more useful cues for long-lived mammals. Species with a predictable seasonal patterning of food resources may utilize seasonal climatic changes to time lactation to coincide with the period of food abundance (Bronson, 1989). Climatic cues like photoperiod and rainfall regulate seasonal breeding in many species in temperate latitudes (Bronson, 1989). On the other hand, adipose tissue or work effort may be a more reliable cue for species that utilize an unpredictable food supply and rely on bodily reserves to support lactation (van Schaik & Noordwijk, 1985). The significance of work effort may be quite variable, however, as animals respond to food scarcity either by minimizing energy expenditure or by working harder to obtain food resources (Foley, 1993).

Non-nutritive factors also may serve as predictive cues by indicating adverse conditions for reproduction. The concentrations of fiber, phytoestrogens, and other chemical constituents of plants often vary markedly with plant growth phase and nutritional condition. Fiber is a structural defense that plants lay down at the close of growth and reflects lower nutrient content both because it is inversely associated with protein content and because it impedes the absorption of nutrients through its influences on gastrointestinal transit. Antimicrobial agents such as glucosinolates, indoles, and some isoflavonoids and flavones are present in highest concentrations in protein-rich parts of growing plants whereas phenolic acids and lignans that are associated with synthesis of structural defenses such as lignin and cellulose are present in highest concentrations in mature plants and plant parts (Harborne, 1979, 1982; Whitten *et al.*, 1997). These phytochemicals also can serve as reproductive cues for animal consumers. For example, many herbivorous mammals exhibit the phenomenon known as "flushing" in which a brief period of grazing on fresh pasture brings the animals into reproductive condition (Sadleir, 1969). In voles, this response has been linked to phytochemicals present in grasses that help to predict the timing of the annual growth season. The indole antimicrobial agent, 6-methoxybenzoxazolinone (6MBOA), parallels protein concentration, rising in concentration in young growing grasses and falling as the maturing plant lays down structural defenses (Sanders *et al.*, 1981). Phenolic acids form the scaffolding for lignin formation, and concentrations of hydroxycinnamic acids, *p*-coumaric and ferulic acid, increase steadily in concentration as the end of the growing season approaches (Berger *et al.*, 1977). Reproductive function in montane voles and other rodents is highly sensitive to these chemicals. 6MBOA stimulates LH secretion, initiating ovarian cycles whereas the hydroxycinnamic acids inhibit gonadal function (Berger *et al.*, 1977, 1981). Isoflavonoid concentrations increase in response to plant stress. For example, quails curtail egg laying when the isoflavone content of forage is high. As isoflavones are present in higher concentration when plant growth is limited by poor rainfall, the suppression of reproduction helps to limit reproduction when food is less abundant (Leopold *et al.*, 1976).

This evidence for mammalian reliance on environmental cues to regulate the timing of reproduction helps to explain the apparent vulnerability of the reproductive system to dietary and other environmental influences. The reproductive actions of dietary components do not simply reflect external forces that impose on reproductive systems but rather evolved responses that synchronize reproductive effort with advantageous nutritional and ecological conditions. From this perspective, the cancer literature appears curiously inverted: fat and calories subvert the normal processes of cell growth while nonnutritive components are "nutraceuticals", restoring control over runaway processes. In reality, the subversive elements are the nonnutritive factors, plant defenses that protect plants from predators and pathogens, whereas the beneficial components are the nutrients that support the maintenance of normal growth and reproduction. The inversion of these relationships in industrialized societies can be seen as an outgrowth of our hominoid heritage.

7.5 The hominoid heritage

Insights into the origin of human biological traits can be obtained through phylogenetic comparisons. Traits that are shared among closely related species are likely to represent the ancestral state whereas traits with a more restricted distribution represent more recently derived traits. Apes are the appropriate phylogenetic comparison for estimating the ancestral origins of human traits. Humans are grouped with gibbons, orangutans, gorillas, and chimpanzees in the ape superfamily Hominoidea, which is estimated to have diverged from monkeys about 30 million years ago. Recent phylogenies also group humans with other African apes in the family Hominidae (origin: 16 million years ago) and group humans with other chimpanzees in the subfamily Homininae (origin: 8 million years ago). Humans and chimpanzees differ in only 1% of their genetic material and are estimated to have diverged as recently as 6 million years ago. The following sections summarize commonalities in foraging and reproductive strategies and mechanisms found among the apes, a set of proposed ancestral traits of apes that will be termed "the hominoid heritage".

7.5.1 Food preferences

Hominoid physiology bears the indelible stamp of frugivory. All primates are omnivores, consuming an eclectic mix of plant and animal resources such as leaves, fruits, flowers, seeds, gum, insects, lizards, birds, and mammals, but by far the most common primate food is fruit (Harding, 1981), a specialization reflected in the primate inability to synthesize vitamin C (Pauling, 1970). Apes are particularly specialized frugivores, with ripe fruit accounting for 50–80% of their diets (Leighton, 1993; Wrangham *et al.*, 1996; Yamagiwa *et al.*, 1996). Even gorillas, which have been considered to be folivores, appear to prefer ripe fruit and eat large quantities of fruit whenever it is available (Tutin, 1996).

Fruits are unique plant products in that they are meant to be eaten, a sugary lure designed to give seeds a free ride in herbivore guts to distant germination sites (Janzen, 1983). Fruits utilize changes in feeding deterrents, color, and nutrient content to confine animal consumption to the optimal time for seed dispersal. Fruits grow, soften, and sweeten as they ripen, changes that are heralded by bright colors that make them more distinct against the leafy background. These changes are short-lived to limit exploitation by non-dispersing species (Janzen, 1983). Bitter or astringent chemicals like "tannins" (polyphenols that avidly bind protein and reduce protein absorption from the digestive tract) are present in high concentrations in unripe fruit but decline with fruit ripening (Janzen, 1983). Consequently, selection for detoxification mechanisms has been muted in the hominoids. Unlike Old World monkeys, apes do not possess specialized digestive apparatus for dealing with plant defenses. Whereas colobines can munch contentedly on strychnine-rich fruit or bitter leaves, we and other apes would not be so inclined.

Feeding data show that apes prefer fleshy fruits that are high in sugar and generally low in plant defenses and other nutrients (Wrangham *et al.*, 1991; Leighton, 1993; Watts, 1996). Soluble sugars make up 30% of the dry weight of fruits eaten by apes and other frugivorous primates (Wrangham *et al.*, 1991; Simmen & Sabatier, 1996). Orangutans also select for large sized fruits, and both orangutans and chimpanzees concentrate feeding effort in patches with a high density of fruit (Leighton, 1993; Wrangham *et al.*, 1996), factors that would increase the rate of caloric gain, but gorillas are less selective (Kuroda *et al.*, 1996). Plant defenses that would reduce nutrient availability like tannins are present in minimal concentrations (0–6%) in the fruits preferred by hominoids, and "alkaloids" (poisonous nitrogen-containing compounds) are generally absent (Wrangham *et al.*, 1991; Leighton, 1993). On the other hand, fat and protein intake is more limited. Lipid content is extremely low, accounting for less than 2% of fruit dry weight and less than 10% of caloric content (Wrangham *et al.*, 1991; Leighton, 1993). Protein content averages 5–8% in the diet of hominoids and other frugivorous primates (Oftedal, 1991; Wrangham *et al.*, 1991; Leighton, 1993; Simmen and Sabatier, 1996), significantly less than that of omnivorous Old World monkeys like rhesus macaques (17%; Marks *et al.*, 1988) or baboons (21%; Barton *et al.*, 1993). Moreover, although consumption of chemical deterrents is generally low, fiber content may reduce food digestibility. Although hominoids also appear to select for lower fiber content in foods, the fiber content of preferred foods is still quite high (40% neutral detergent fiber; 25–35% acid detergent fiber: Wrangham *et al.*, 1991; Leighton, 1993). Nevertheless, the hominoid diet appears to be digestible (Milton and Demment, 1988) and is likely to provide sufficient nutrients for growth and maintenance (Oftedal, 1991).

7.5.2 Perceptual mechanisms

Dietary choices entail perceptual as well as behavioral processes. In fact, sensory factors are thought to be the primary factors governing human food selection (Drewnowski, 1995a). Our perceptual sensibilities, along with those of other primates, have been molded by the frugivorous niche. Although we tend to think of brightness, succulence, and sweetness as characteristics of natural objects, they are in fact sensory responses to the adaptively significant properties of natural objects. To make subtle discriminations among fruits varying in ripeness, nutrient content, and plant defenses, frugivores must be sensitive to nuances of color, texture, and taste. For example, primates have reacquired color vision, a trait widely distributed across vertebrates such as fish, reptiles, and birds that was lost in the nocturnal origins of mammals (Goldsmith, 1990). Trichromatic color vision provides catarrhine primates with much more acute sensitivity to color than most other mammals, with most marked sensitivity in the blue-green, yellow, and yellow–orange part of the spectrum (Snodderly, 1979; Goldsmith, 1990; Jacobs, 1996). These parts of the color spectrum accord with the color changes accompanying fruit ripening; unripe fruits match the color spectrum of leaves, with spectral reflectance around 550 nm, but shift toward the long wavelength end of the spectrum, becoming yellow-orange (600 nm) or red (650 nm), as they ripen (Snodderly, 1979). Whereas bird-dispersed fruits take on red, blue, white, and purple hues as they ripen (Leighton, 1993), matching avian sensitivity to both the short and long wavelength ends of the spectrum (Goldsmith, 1990), primate-dispersed fruits taken on predominantly yellow-orange hues (Leighton, 1993). The carotenoids and flavonoids are the pigments primarily responsible for these signals of fruit ripening (Harborne, 1982). Carotenoids, with minor contributions from flavonoid flavonols, chalcones, and aurones, produce the yellow and orange colors preferred by primates whereas anthocyanidins and flavones supply the red, blue, purple, and white hues preferred by birds. Thus in selecting for fruit color, primates also indirectly select for chemical constituents.

More direct chemical selection is mediated by taste perception. Primates are acutely sensitive to sweetness, detecting sugars at concentrations as low as 3–300 mM, an adaptive response that facilitates selection for sugar content (Laska *et al.*, 1996; Simmen & Sabatier, 1996). Primates vary in taste thresholds, which appear to be related to the sugar content of the diet (Simmen & Sabatier, 1996). Sensitivities are lowest for sucrose, the most common sugar in ripe fruits, and are substantially lower than sucrose concentrations in preferred ripe fruits (Glaser, 1979; Laska *et al.*, 1996; Simmen & Sabatier, 1996). Primates that concentrate on sugar-rich fruits have higher taste thresholds than species that feed on fruits with lower sugar content (Simmen & Sabatier, 1996). For example, tamarins specializing on sugar-rich nectar and fruits with sugar concentrations as high as 500 mM have taste thresholds of 50–60 mM whereas spider monkeys consuming fruit of more varied sugar content (0.07–500 mM) have taste thresholds as low as 16 mM (Simmen & Sabatier, 1996). Chimpanzees have relatively low taste thresholds (30 mM; Hellekant *et al.*, 1996), as do humans (10 mM: Laska *et al.*, 1996), suggesting adaptation to relatively low sugar content in hominoid diets. The preferred fruits of hominoid diets are actually relatively high in sugar content (400–550 mM) but fruits eaten during periods of food scarcity are much lower (60–90 mM), only two to three times the reported taste threshold of chimpanzees, probably near the limits of palatability.

Primates are even more sensitive to phytochemical content, responding with distaste to the astringent flavors of tannins at concentrations as low as 1 mM (Critchley & Rolls, 1996) and to the bitter flavors of quinine and hydrochloric acid at concentrations as low as 5–10 mM (Hellekant *et al.*, 1997). As bitterness, sourness, and astringeny result from a variety of defensive compounds such as terpenoids, tannins, and isothiocyanates, responsiveness to these flavors is an adaptive response that protects primates from the adverse consequences of consuming these bioactive substances. Most primates respond with distaste to increasing concentrations of sour or bitter flavors (Johns, 1990), but these responses also can be modified by selection as seen in night monkeys' preference for increasing sourness, a response adapted to the sour fruits that make up an important part of their diet (Glaser & Hobi, 1986).

These sensitivities result from adaptations in taste receptors, neuroepithelial cells, and their associated neurons (Hellekant & Danilova, 1996; Nofri *et al.*, 1996). Perceptions of sweet, salty, bitter, and sour flavors are localized in specific taste receptors and discrete subsets of cranial neurons (Plata-Salaman & Scott, 1992). In addition, there are specific subsets of cortical neurons that respond to the astringent flavor of tannin along with subsets specific to the other four flavors (Critchley & Rolls, 1996). Differences among primates in sensitivity to sugar appear to be related to divergence in the structure of recognition sites on the sweetness receptor, allowing more sensitive detection of sucrose in catarrhine primates (Glaser *et al.*, 1995; Nofre & Tinti, 1996; Nofri *et al.* 1996). The hominoids appear to be more similar to humans in taste responses, and chimpanzees appear to be identical to humans in taste physiology, both in responses to sweetness modifiers and in the specificity of taste fibers (Hellekant *et al.*, 1996, 1997). In chimpanzees, the sweet subgroup is the most abundant and finely tuned neuronal subgroup and the sour subgroup is the smallest and most broadly tuned (Hellekant *et al.*, 1997). Both sour and bitter tastes are not as well distinguished without input from other fibers (Hellekant *et al.*, 1997), a response overlap that might help to weigh phytochemical load against nutrient content.

This acute sensitivity is restricted to those constituents that are a significant part of the frugivorous diet. There do not appear to be specific receptors for other nutrients like fat and protein, for example; nor is there any evidence for perception of phytoestrogens, which are rarely present in ripe fruits. Some plant constituents like fiber and fat, as well as spicy phytochemicals, may be perceived through stimulation of the trigeminal nerve and modulation of responses to other sensory receptors (Lawless, 1987; Drewnowski & Schwartz, 1990). For example, some fatty acids can prolong the stimulus of taste receptor cells in rodents, providing a potential mechanism for perception of lipid content (Gilbertson *et al.*, 1997).

7.5.3 Reproductive ecology

The most serious obstacle to reproduction for most apes is the unpredictability of fruit resources. As the fruiting cycles of different species are often highly synchronized within habitats, all frugivores must face the problem of periodic shortages in fruit (Terborgh, 1992). Ironically, unpredictability is greatest in the most aseasonal habitats like Malaysian forests where numerous species fruit in irregular but synchronous bursts of "mast fruiting" every few years. Orangutans, for example, must endure periods as long as 8–10 months in which there are no fleshy fruits (Leighton, 1993). In other tropical regions, fruiting predictability varies with the predictability of rainfall. In South American forests, a two-season climate predominates. In Africa, seasonality of rainfall varies with latitude and altitude, ranging from an alternation of two rainy and two dry seasons to a long wet and a long dry season (Terborgh, 1992). Fruit production parallels rainfall closely in two-season forests, but is less synchronous in four-season forests (Terborgh, 1992). Thus, fruit consumption by chimpanzees and gorillas is highest during the rainy season peak of fruit abundance in the two-season forests of Congo (Ndoki Forest: Kuroda *et al.*, 1996) but coincides with fruiting peaks in the long dry season and the end of the short dry season in the four-season forests of Ivory Coast (Tai: Boesch, 1996; Doran, 1997), Republic of Congo (formerly Zaire) (Kahuzi-Biega: Yamagiwa *et al.*, 1996), and Tanzania (Gombe: Goodall, 1986: 48, 233; but see Wrangham, 1977) and is highly variable, with no clear relationship to monthly rainfall, in the less seasonal forests of Kibale (Wrangham *et al.*, 1991). As in the Malayasian forests, these irregular fruiting patterns resulted in unpredictable periods of fruit scarcity (averaging 4.5 months/year) in this "unseasonal" environment (Wrangham *et al.*, 1996).

The fallback foods available to frugivores during periods of fruit scarcity vary with the patterning of fruiting and leafing. Young leaves can be important fallback foods in four-season forests where fruiting and leafing are not synchronized, but frugivores must turn to rarer resources like palm nuts and nectar or less nutritious foods like figs, pith, or mature leaves in more seasonal forests where fruiting and leafing are highly synchronized or in Malayasian forests where both leafing and fruiting are highly irregular (Terborgh & van Schaik, 1987; Terborgh, 1992). Apes occur in all of these habitats and appear to have a more generalist strategy, relying primarily on figs and piths but adding young leaves when they are available (Wrangham *et al.*, 1991, 1996; Leighton, 1993; Kuroda *et al.*, 1996).

These foraging patterns provide insight into the factors that are likely to regulate reproduction in hominoids. Due to the extreme variability in fruit production, external cues such as rainfall, temperature, or photoperiod are poor predictors of food availability, and hominoids are much more likely to rely on endogenous cues such as nutrient balance or fat stores. The nutrient composition of the diet varies substantially with fruit abundance. Fallback foods are much lower in sugar (10–12%) and higher in fiber (50% neutral detergent fiber; 35–50% acid detergent fiber) than preferred fruits although they can be higher in fat (2–4%) and protein (leaves: 17–24%; pith: 2–36%) (Wrangham *et al.*, 1991, 1996; Leighton, 1993; Kuroda *et al.*, 1996). Fluctuations in body weight are reported to parallel fruit abundance in both chimpanzees and orangutans, suggesting that apes may rely on fat stores to sustain them through lengthy periods of food scarcity (Wrangham, 1977; Leighton, 1993). This strategy is supported by a uniquely primate fat depot, the abdominal paunch (Pond, 1997), and a pattern of energy minimization during fruit scarcity and gluttony during periods of fruit abundance. For example, both chimpanzees (Wrangham, 1977; Doran, 1997) and gorillas (Tutin, 1996; Watts, 1996) minimize energetic expense by reducing daily travel when fruit is scarce, and orangutan fat storage varies markedly with fruit availability (Leighton, 1993; Knott, 1997). Thus dietary cues such as caloric or sugar intake or low fiber intake are likely to be good predictors of fruit abundance for hominoids as are endogenous cues such as energy balance and adipose tissue. Although reproductive ecology has received much less attention than other topics in studies of apes, there is recent evidence that annual variations in energy balance and adiposity are associated with changes in ovarian activity. The incidence of conceptions and the onset of postpartum and adolescent cycles in chimpanzees at Gombe and Mahale is higher during the late dry season, a time of increased body weight at Mahale (Nishida *et al.*, 1990; Wallis, 1995), and ovarian activity appears to be enhanced in orangutans during periods of fruit abundance (Knott, 1997). The links between foraging pattern in relation to food abundance and reproductive sensitivity to alternating "feast or famine" regimens can be termed a hominoid reproductive strategy (Figure 7.3), a related set of behavioral, energetic, and endocrine responses to nutrient balance that may be an important part of the evolutionary heritage we share with other hominoids.

7.6 Human food choices

7.6.1 Perceptual biases

Global colonization and adaptation to local ecosystems have resulted in a much broader array of diets across human populations, but it could be argued that humans have maintained the perceptual bases for food choice found in our hominoid ancestors. Even though fruit is no longer the predominant food staple in human diets, we have maintained preferences for nutrient and phytochemical content that are similar to those of frugivorous apes. These preferences are mediated primarily by taste and appearance and reflect our ancestral biochemical defenses. Although human dietary practices have resulted in some biological adaptations (Jackson, 1991), the relatively short history of human domestication makes it unlikely that our physiological ability to process food chemicals has diverged markedly from that of our hominoid ancestors who probably were poorly adapted for deactivation of pharmacological agents.

The hominoid orientation to color as an index of palatability is reflected in the modern propensity to dye and polish fruit and to add color to colorless food products (Faber & Armelagos, 1980), and the predominance of oranges and yellows in our fruits, vegetables, and processed foods. Humans also are acutely sensitive to sugar, more so, in fact, than most other primates (Laska *et al.*, 1996). Although this heightened sensitivity is often taken as evidence of our "sweet tooth", it actually enhances the palatability of foods low in sucrose, expanding the range of potential fallback foods. A "fat tooth" has been proposed as an alternative human craving that reflects human preferences for mixtures of fat and sugar (Drewnowski, 1995*b*). The food concentrations of fat (20%) and sugar (8%) preferred by most adults (Drewnowski, 1995*b*) appear at first glance to be a uniquely human preference because these concentrations are so unlike the high sugar, low fat diets of hominoids. Humans do not appear to have any physiological mechanisms for regulating fat intake (Drewnowski, 1997), so preference for high fat may result from other hominoid preferences. Humans are actually rather poor at rating fat content, responding primarily to texture, which acts as a sweetness enhancer (Drewnowski &

Schwartz, 1990). Thus it is conceivable that the perceived smoothness of fatty foods may reflect a preference for foods that are lower in fiber. These modifying effects of food texture might have provided a mechanism for weighing the relative concentrations of fiber and fat against the low carbohydrate content of potential fallback foods.

Human responses to bitter flavors, on the other hand, are governed by "the yuck factor" (Fackelmann, 1997), an inherited polymorphism familiar to generations of anthropology students through PTC tasting. About 25% of the population of the USA are homozygous supertasters, who experience phenylthiocarbamide (PTC) and 6-*n*-propylthiouracil (PROP) as incredibly bitter (Bartoshuk *et al.*, 1996). Another 50% of the population are heterozygous for the taster gene and find the bitter flavors mildly adversive. The remaining 25% lack the gene and the ability to taste the bitter flavor. The degree of aversion is related to the threshold of detection of PROP; non-tasters have a threshold of more than 180 μM whereas supertasters can detect PROP at concentrations below 32 μM (Drewnowski *et al.*, 1997). The taster allele also confers enhanced sensitivity to other bitter substances, an ability that would facilitate the detection of potentially harmful phytochemicals. For example, like PTC, the naturally occurring thioureas, thiocynates, and isothiocyanates, bitter-tasting substances that give cruciferous vegetables their distinctive flavor, inhibit thyroid hormone metabolism, resulting in goiter. Populations where these substances are present in dietary staples, such as the Aymara Indians of Bolivia and other Andean populations in Ecuador, have higher frequencies of the taster allele (Greene 1974; Johns 1990). In modern industrialized societies, on the other hand, the taster gene is maladaptive because it causes supertasters to reject bitter-tasting foods containing chemopreventative agents like isothiocyanates (Drewanoski & Rock, 1995). It is less clear what selective factors might have produced the current high frequencies of the non-taster gene. Tasters also are more sensitive to a variety of food components besides bitter flavors, including sweet, spicy, and fat flavors (Tepper & Nurse, 1997), and selection for tolerance of these factors in foods may have favored the non-taster allele. For example reduced sensitivity to PTC is associated with consumption of bitter, niacin-rich coffee in Yucatan (Davis, 1978).

7.6.2 Foraging and plant domestication

These preferences have guided the selection and domestication of human food plants. Like other hominoids, human foragers select foods that are high in carbohydrates and low in fiber, but their choices are also influenced by their more organized division of labor and food sharing, food collection and storage, and more extensive food processing. Fruits remain a significant proportion of the diet, but hunter–gatherers add the seeds, roots, and tubers of herbaceous plants, items that can be more readily collected, stored, and cultivated than the fruits of tropical hardwoods (Lee, 1979; Harris, 1987). Although the seeds eaten by hunter–gatherers are not as high in carbohydrate content (25–55%) as their fruits (55–80%), their tubers are relatively starchy (50–60%) even though only a few are sweet (Lee, 1979; Vincent, 1984). Although the average fat content of fruits and roots is as low as typical ape foods, seeds are much higher in fat and form an important part of the hunter–gatherer diet (Table 7.3). Fiber content is lower than that found in most ape foods, averaging less than 10% (range 2–30%) (Lee, 1979; Vincent, 1984; Eaton *et al.*, 1996). For example, the fiber intake of modern and prehistoric hunter-gatherers is estimated at 85–100 g/day, or 1–1.5 g/kg body weight (Kliks, 1978; Eaton *et al.*, 1988, 1996) whereas the fiber intake of chimpanzees may be as high as 200 g/day or 4 g/kg (Milton, 1993). Little is known about concentrations of other phytochemicals in forager diets, but seeds or tubers are often better protected chemically because of their role as storage organs (Southgate, 1991). It is likely that phytoestrogens are present, since major tubers of the Hadza and !Kung (Lee, 1979; Vincent, 1984) are members of the legume family noted for high isoflavonoid content (Whitten *et al.*, 1997). Overall, human forager foods tend to be more energy-dense than ape foods (Table 7.4), primarily due to higher fat and lower fiber content.

These food traits were elaborated and magnified in the domestication of plants. Domestication has increased the size, energy content, and palatability of plant foods (Johns, 1990; Smith, 1995). As humans took control of the reproductive cycles of these plants, they markedly altered the adaptive landscape of domesticants, changing plant defensive and reproductive strategies, with consequent changes in plant morphology and phytochemistry. These changes are particularly evident in cereal crops where reductions in the depth of the protective seed coat and increases in seed size are markers of domestication (Smith, 1995). The morphological hallmarks of domestication were accompanied by phytochemical changes. The fiber content of cultivated fruits and vegetables is low, averaging 9–20 % (USDA, 1997). Most cultivars also contain lower concentrations of tannins, saponins, glycoalkaloids, and cucurbitacins than their wild relatives (Johns, 1990; Chikwendu & Okezie, 1989). Soaking, crushing and grating, cooking, and fermentation, further reduce the content of perceptible plant defenses (Stahl, 1989; Southgate, 1991).

Table 7.3. *Lipid content of primate foods*

	Lipid concentration % of dry weight (% calories as fat)					
	Fruit	Figs	Seeds	Leaves	Pith	Storage organs
Baboons			5–12 (28–36%)			1–7 (51–72%)
Orangutan	1.7 (6%)	0 (0%)	1 (3%)			
African apes	1.7 (8%)	3.4 (23%)		2.6 (22%)	1.6 (16%)	
Chimpanzee	0 (0%)	3.5 (28%)		0.8 (6%)	0.8 (9%)	
Human foragers	1.0 (3%)		3.9, 43.4 (62%)			0.5–4.8 (2–14%)

Note: Blank cells reflect the absence or minimal representation of a food type in the diet.
From Lee, 1979; Vincent, 1984; Eaton *et al.*, 1988, 1996; Cane, 1989; Whitten *et al.*, 1991; Wrangham *et al.*, 1991; Leighton, 1993.

Table 7.4. *Energy density (MJ/kg dry weight) of primate foods*

Food	Baboon	Orangutan	African ape	Chimpanzee	Human foragers	Human farmers
Fruit		12.4	8.4	6.5	8.8	14–15
Fig fruit		7.4	7.4	4.7		
Leaves			4.4	4.8		12–14
Pith	19		3.8	3.5		
Seeds, grains	17–21	14.1			15–23	16–100
Tubers	18				2–14	12–18

Note: Blank cells reflect the absence or minimal representation of a food type in the diet.
Energy contents estimated from food composition data of Hamilton *et al.*, 1978; Lee, 1979; Vincent, 1984; Whitten *et al.*, 1991; Wrangham *et al.*, 1991; Leighton, 1993; USDA, 1997.

At the same time, reductions in the physical and chemical defenses of cultivars make them more vulnerable to competitors, predators, and pathogens, and more dependent on human intervention for survival and reproduction. Human selection for crop hardiness and storage capacity has inadvertently increased our exposure to chemical defenses that our perceptual systems are ill-equipped to detect. For example, many of the

Table 7.5. *Phytoestrogens in domesticated food staples*

Cultivar	Phytoestrogen
Near East	
Pea	Daidzein, biochanin A, coumestrol
Bean	Genistein, biochanin A, coumestrol
Chickpea	Isoflavone
Flax	Lignans
Wheat	Lignans, tricin
Europe	
Wheat	Lignans, tricin
Africa	
Pearl millet	Daidzein
Asia	
Soybean	Genistein, daidzein, biochanin A
SE Asia	
Yam	Diosgenin, yamogenin
North America	
Sunflower	Coumestrol
Chenopod	Contraceptive substance
MesoAmerica	
Bean	Genistein, biochanin A, coumestrol
Andes	
Quinua	Contraceptive substance
Anu	Isothiocyanates
Maca	Isothiocyanates

plant families domesticated by humans are noted for their phytoestrogen content (Farnsworth *et al.*, 1975; Harris & Hillman, 1989; Table 7.5), which may reflect the benefits of the fungicidal actions of many phytoestrogens (Whitten & Naftolin, 1991; Whitten *et al.*, 1997) for food storage and perhaps medicinal uses (Etkin, 1994). Modern industrial agriculture solves the problem of plant protection in a different way, through the application of synthetic pesticides and herbicides (Janzen, 1983). Ironically, many of these pesticides have endocrine actions themselves (Colburn & Clement, 1992; Davis *et al.*, 1993).

7.7 Conclusions

This chapter provides a new perspective on the advent and resilience of "Western diseases" in modern industrialized societies. The ill health of the modern age is often portrayed as a result of a mismatch between Stone Age biology and twentieth century cuisine, a situation that can be remedied by emulating Paleolithic diet and activity patterns (Eaton *et al.*, 1988). An alternative model proposes an ancestral behavioral repertoire as the biological component outpaced by modern food resources (Hamilton, 1987). This chapter has taken a different approach, focusing on the ways in which food choices and predispositions create both environments and biologies (Figure 7.3). From this perspective, our twentieth century environments are as much a construction of our hominoid perceptions and predispositions as our Paleolithic environments. Humans, like good frugivores, select diets that are high in carbohydrates and energy and low in phytochemical content. We have gone far beyond our primate relatives in our ability to optimize dietary content through food processing, plant domestication, and market-based exchange. Healthy as the hunter–gatherer diet might be in comparison to our own, it mirrors the diets of industrialized societies in the emphasis on energy-dense, lipid-rich, and fiber-poor foods. Domestication has magnified these properties, and industrial food production has concentrated them to produce a dietary composition that is the ideal outcome of hominoid preferences. These delectable treats are gooey, moist, soft, and sweet, a carbohydrate- and fat-rich feast of which any ape would heartily approve. Our twentieth century lifestyles are not mismatched to our evolutionary heritage; they are the ideal outcome of that heritage. The catch is that our reproductive physiology is adapted not to our "optimal diet" of steak and cake but to the bitter, fibrous, and lean diet of hominoid hard times. Just as our perceptual systems are acutely tuned to detect sugar, our reproductive systems appear to be attuned to dietary factors that reflect resource availability such as fiber, fat, and phytoestrogens. Our sensitivity to these environmental factors also makes us more vulnerable to the chemical residue of industrial society, which affects reproductive processes through many of the same pathways by which phytochemicals and nutrients influence estrogen metabolism and action (Davis *et al.*, 1993).

Cancer rates have proven surprisingly resistant to the war waged in research laboratories and public awareness campaigns over the past 25 years (Beardsley, 1994). This apparent lack of progress reflects past emphases on treatment over prevention, unidimensional solutions, and a promotion of a "healthy lifestyle" as merely a product of individual choice. A hominoid perspective suggests that winning the war will require a more holistic emphasis on the multiple pathways by which environments and predispositions influence development, endocrine function, and behavior.

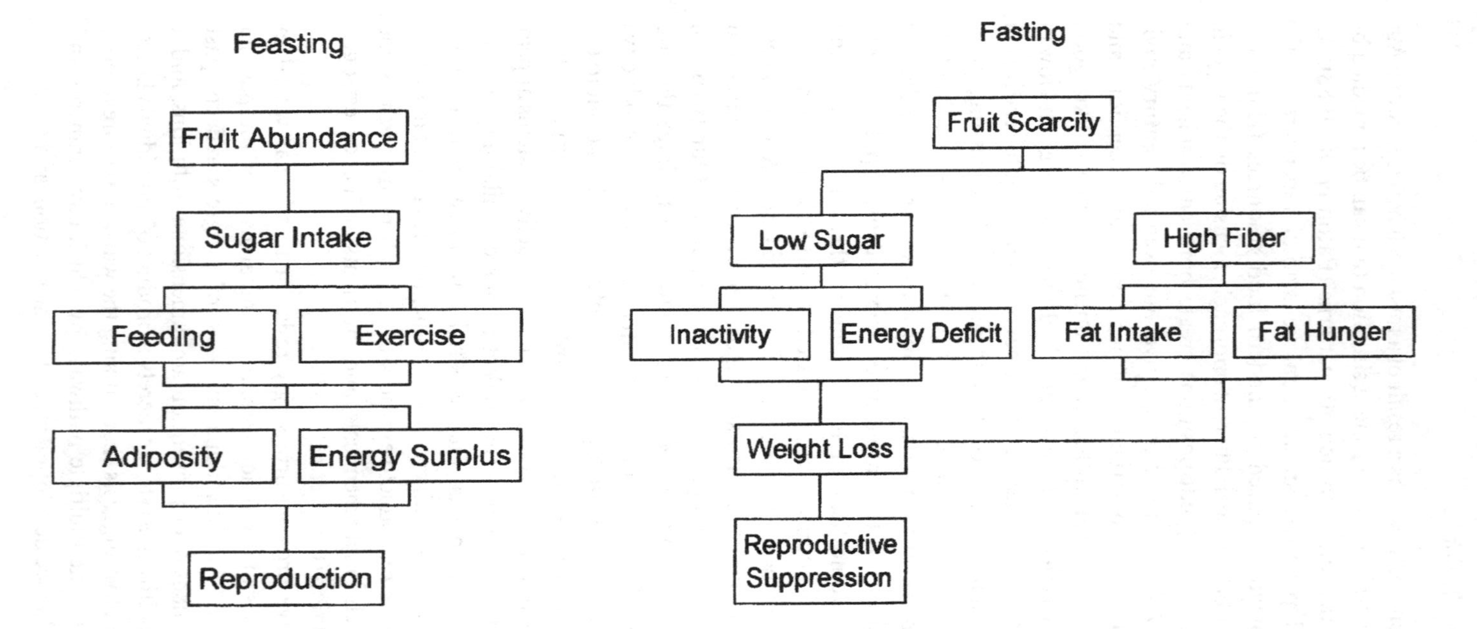

Figure 7.3. Outline of the proposed hominoid reproductive strategy, depicting behavioral and reproductive responses to feast and famine conditions.

7.8 References

Adlercretutz, H. (1990). Western diet and western diseases: some hormonal and biochemical mechanisms and associations. *Scandinavian Journal of Clinical Laboratory Investigation*, **50**, Suppl. 201, 3–23.

Adlercreutz, H., Goldin, B. R., Gorbach, S. L., Höckerstedt, K. A. V., Watanabe, S., Hämälainen, E. S., Markkanen, M. H., Mäkelä, T. H., Wähäla, K.T., Hase, T. A., & Fotsis, T. (1995). Soybean phytoestrogen intake and cancer risk. *Journal of Nutrition*, **125**, (Suppl. 3S), 757S–70S.

Adlercreutz, H., Markkanen, H., & Watanabe, S. (1993). Plasma concentrations of phyto-oestrogens in Japanese men. *Lancet*, **342**, 1209–10.

Bagga, D., Ashley, J. M., Geffrey, S. P., Wang, H. J., Barnard, R. J., Korenman, S., & Heber, D. (1995). Effects of a very low fat, high fiber diet on serum hormones and menstrual function. Implications for breast cancer prevention. *Cancer*, **76**, 2491–6.

Barnes, S. (1995). Effect of genistein on *in vitro* and *in vivo* models of cancer. *Journal of Nutrition*, **125** (Suppl. 3S), 777S–83S.

Barnes, S., Peterson, T. G., & Loward, L. (1995). Rationale for the use of genistein-containing soy matrices in chemoprevention trials for breast and prostate cancer. *Journal of Cell Biochemistry*, Suppl., **22**, 181–7.

Barton, R. A., Whiten, A., Byrne, R. W., & English, M. (1993). Chemical composition of baboon plant foods: implications for the interpretation of intra- and interspecific differences in diet. *Folia Primatol*, **61**, 1–20.

Bartoshuk, L. M., Duffy, V. B., Reed, D., & Williams, A. (1996). Supertasting, earaches and head injury: Genetics and pathology alter our taste worlds. *Neuroscience and Biobehavioral Review*, **20**, 79–87.

Beardsley, T. (1994). A war not won. *Scientific American*, **270**, 130–8.

Berger, P. J., Sanders, E. H., Gardner, P. D., & Negus, N. C. (1977). Phenolic plant compounds functioning as reproductive inhibitors in *Microtus montanus*. *Science*, **195**, 575–7.

Berger, P. J., Negus, N. C., Sanders, E. H., & Gardner, P. D. (1981). Chemical triggering of reproduction in *Microtus montanus*. *Science*, **214**, 69–70.

Boesch, C. (1996). Social grouping in Tai chimpanzees. In *Great Ape Societies*, ed. W. C. McGrew, L. F. Marchant, & T. Nishida, pp. 101–13. London, UK: Cambridge University Press.

Bradlow, H. L., Michnovicz, J. J., Telang, N. T., & Osborne, M. P. (1991). Effect of dietary indole-3-carbinol on estradiol metabolism and spontaneous mammary tumors in mice. *Carcinogenesis*, **12**, 1571–4.

Bronson, F. (1989). *Mammalian Reproductive Biology*. Chicago, IL: University of Chicago Press.

Cane, S. (1989). Australian aboriginal seed grinding and its archaeological record: a case study from the Western Desert. In *Foraging and Farming: The Evolution of Plant Exploitation*, ed. D. R. Harris, & G. C. Hillman, pp. 99–119. London: Unwin Hyman.

Cassidy, A., Bingham, S., & Setchell, K. D. R. (1994). Biological effects of a diet of soy protein rich in isoflavones on the menstrual cycle of premenopausal women. *American Journal of Clinical Nutrition*, **60**, 333–40.

Chikwendu, V. E., & Okezie, C. E. A. (1989). Factors responsible for the ennoblement of African yams: inferences from experiments in yam domestication. In *Foraging and Farming: The Evolution of Plant Exploitation*. ed. D. R. Harris, & G. C. Hillman, pp. 335–57. London: Unwin Hyman.

Cohen, M. N., & Armelagos, G. J., eds. (1984). *Paleopathology at the Origins of*

Agriculture. Orlando FL: Academic Press.
Colburn, T., Clement, C., eds. (1992). *Chemically Induced Alterations in Sexual and Functional Development: The Wildlife-Human Connection*. Princeton, NJ: Princeton Scientific Publishing.
Critchley, H. D., & Rolls, E. T. (1996). Responses of primate taste cortex neurons to the astringent tastant tannic acid. *Chemical Senses*, **21**, 135–45.
Davis, D. L., Bradlow, H. L., Wolff, M., Woodrufff, T., Hoel, D. G., & Anton-Culver, H. A. (1993). Medical hypothesis: Xenoestrogens as preventable causes of breast cancer. *Environmental Health Perspectives*, **101**, 372–7.
Davis, R. G. (1978). Increased bitter taste detection thresholds in Yucatan inhabitants related to coffee as a dietary source of niacin. *Chemical Senses Flavours*, **3**, 423–9.
Doran, D. (1997). Influence of seasonality on activity patterns, feeding behavior, ranging, and grouping patterns in Tai chimpanzees. *International Journal of Primatology*, **18**, 183–206.
Dorgan, J. F., Reichman, M. E., Judd, J. T., Brown, C., Longcope, C., Schatzkin, A., Forman, M., Campbell, W. S., Franz, C., Kahle, L., & Taylor, P. R. (1996). Relation of energy, fat, and fiber intakes to plasma concentrations of estrogens and androgens in premenopausal women. *American Journal of Nutrition*, **64**, 25–31.
Drewnowski, A. (1995*a*). Energy intake and sensory properties of food. *American Journal of Clinical Nutrition*, **62**, (Suppl. 5), 1081S–5S.
Drewnowski, A. (1995*b*). Sensory perceptions and individual preferences for sugar and fat. In *Proceedings from Health and Pleasure at the Table*, ed. L. Dube, J. Le Bel, C. Tougas, & V. Troche, pp. 211–26. Montreal, Canada. Enjeux actuel du marketing dans l'alimentation.
Drewnowski, A. (1997). Why do we like fat? *Journal of the American Diet Association*, **97**, Suppl. 7, S58–62.
Drewnowski, A., & Rock, C. L. (1995). The influence of genetic taste markers on food acceptance. *American Journal of Clinical Nutrition*, **62**, 506–11.
Drewnowski, A., & Schwartz, M. (1990). Invisible fats: sensory assessment of sugar/fat mixtures. *Appetite*, **14**, 203–17.
Drewnowski, A., Henderson, S. A., & Shore, A. B. (1997). Genetic sensitivity to 6-n-propylthiouracil (PROP) and hedonic responses to bitter and sweet tastes. *Chemical Senses*, **22**, 27–37.
Durham, W. H. (1991). *Coevolution Culture Genes and Human Diversity*. Stanford, CA: Stanford University Press.
Eastwood, M. A., Brydon, W. G., & Anderson, D. M. W. (1986). The effect of the poly saccharide composition and structure of dietary fibers on cecal fermentation and fecal excretion. *American Journal of Clinical Nutrition*, **44**, 51–5.
Eaton, S. B., Eaton, III S. B., Konner, M. J., & Shostak, M. (1996). An evolutionary perspective enhances understanding of human nutritional requirements. *Journal of Nutrition*, **126**, 1732–40.
Eaton, S. B., Shostak, M., & Konner, M. (1988). *The Paleolithic Prescription*. New York: Harper & Row.
Ekbom, A., Trichopoulous, D., Adami, H. O., Hsieh, C. C., & Lan, S. J. (1992). Evidence of prenatal influences on breast cancer risk. *Lancet*, **340**, 1015–19.
Ellison, P. T., Lipson, S. F., O'Rourke, M. T., Bentley, G. R., Harrigan, A. M., Panter-Brick, C., & Vitzthum, V. J. (1993). Population variation in ovarian

function. *Lancet*, **342**, 433–4.
Etkin, N. L., ed. (1994). *Eating on the Wild Side*. Tucson AZ: University of Arizona Press.
Faber, P., & Armelagos, G. (1980). *The Anthropology of Eating*. Boston, MA: Houghton Mifflin.
Fackelmann, K. (1997). The bitter truth. *Science News*, **152**, 24–5.
Farnsworth, N. R., Bingel, A. S., Cordell, G. A., Crane, F. A., & Fong, H. H. S. (1975). Potential value of plants as sources of new antifertility agents II. *Journal of Pharmacological Sciences*, **64**, 717–54.
Feng, W., Marshall, R., Lewis-Barned, N. J., & Goulding, A. (1993). Low follicular oestrogen levels in New Zealand women consuming high fibre diets: A risk factor for osteopenia? *NZ Medical Journal*, **106**, 419–22.
Foley, R. A. (1993). The influence of seasonality on hominid evolution. In *Seasonality and Human Ecology*, ed. S. J. Ulijaszek, & S. S. Strickland, pp. 17–37. Cambridge, UK: Cambridge University Press.
Fotsis, T., Pepper, M., Adlercreutz, H., Hase, T., Montesano, R., & Schweigerer, L. (1995). Genistein, a dietary ingested isoflavonoid, inhibits cell proliferation and *in vitro* angiogenesis. *Journal of Nutrition*, **125** (Suppl. 3S), 790S–7S.
Franke, A. A., Custer, L. J., Cerna, C. M., & Narala, K. (1995). Rapid HPLC analysis of dietary phytoestrogens from legumes and from human urine. *Proceedings of the Society for Experimental Biology Medicine*, **208**, 18–26.
Gilbertson, T. A., Fontenot, D. T., Liu, L., Zhang, H., & Monroe, W. T. (1997). Fatty acid modulaton of K + channels in taste receptor cells: gustatory cues for dietary fat. *American Journal of Physiology*, **272**, (4 Pt 1), C1203–10.
Glaser, D. (1979). Gustatory preference behavior in primates. In *Preference Behavior and Chemoreception*, ed. J. H. A. Kroeze, pp. 51–61. London, UK: IRL Press.
Glaser, D., & Hobi, G. (1986). Taste responses in primates to citric and acetic acid. *International Journal of Primatology*, **6**, 395–8.
Glaser, D., Tinti, J. M., & Nofre, C. (1995). Evolution of the sweetness receptor in primates. 1. Why does alitame taste sweet in all prosimians and simians, and aspartame only in Old-World simians. *Chemical Senses*, **20**, 573–84.
Goldin, B. R., Adlercretuz, H., Gorbach, S. L., Wood, M. N., Dwyer, J. T., Conlon, T., Bohn, E., & Gershoff, S. N. (1986). The relationship between estrogen levels and diets of Causcasian American and Oriental immigrant women. *American Journal of Clinical Nutrition*, **44**, 945–53.
Goldin, B. R., Woods, M. N., Spiegelman, D. L., Longcope, C., Morrill-LaBrode, A., Dwyer, J. T., Gualtieri, L. J., Hertzmark, E., & Gorbach, S. L. (1994). The effect of dietary fat and fiber on serum estrogen concentrations in women under controlled dietary conditions. *Cancer*, **74** (Suppl. 3), 1125–31.
Goldsmith, T. H. (1990). Optimization, constraint, and history in the evolution of eyes. *Quarterly Review of Biology*, **65**, 281–322.
Goodall, J. (1986). *The Chimpanzees of Gombe. Patterns of Behavior*. Cambridge, MA: Belknap Press of Harvard University Press.
Gorbach, S. L., & Goldin, B. R. (1987). Diet and the enterohepatic recycling of estrogens. *Preventive Medicine*, **16**, 525.
Greene, L. S. (1974). Physical growth and development, neurological maturation, and behavioral functioning in two Ecuadorian Andean communities in which goiter is endemic. *American Journal of Physical Anthropology*, **41**, 139–52.
Greenwald, P. (1989). Principles of cancer prevention: Diet and nutrition. In *Cancer: Principles and Practice of Oncology*, third edn., ed. V. T. DeVita Jr, S.

Hellman, & S. A. Rosenberg, pp. 167–180. Philadelphia, PA: J. B. Lippincott.

Hagerty, M. A., Howie, B. J., Tan, S., & Shultz, T. D. (1988). Effect of low- and high-fat intakes on the hormonal milieu of pregnancy. *American Journal of Clinical Nutrition*, **47**, 653–9.

Hamilton, W. J. III, Buskirk, R. E., & Buskirk, W. H. (1978). Omnivory and utilization of food resources by chacma baboons, *Papio ursinus*. *American Naturalist*, **112**, 911–24.

Hamilton, W. J. III. (1987). Omnivorous primate diets and human overconsumption of meat. In *Food and Evolution Toward a Theory of Human Food Habits*, ed. M. Harris, & E. B. Ross, pp. 117–32. Philadelphia, PA: Temple University Press.

Harborne, J. B. (1979). Flavonoid pigments. In *Herbivores: Their Interaction with Secondary Plant Chemicals*, ed. G. A. Rosenthal, & D. H. Janzen, pp. 619–55. New York: Academic Press.

Harborne, J. B. (1982). *Introduction to Ecological Biochemistry*. 2nd edn. New York: Academic Press.

Harding, R. S. O. (1981). A order of omnivores: Nonhuman primate diets in the wild. In *Omnivorous Primates Gathering and Hunting in Human Evolution*, ed. R. S. O. Harding, pp. 191–214. New York: Teleki G. Columbia University Press.

Harris, D. R. (1987). Aboriginal subsistence in a tropical rain forest environment: Food procurement, cannibalism, and population regulation in Northeastern Australia. In *Food and Evolution Toward a Theory of Human Food Habits*, ed. M. Harris, & E. B. Ross, pp. 357–385. Philadelphia, PA: Temple University Press.

Harris, D. R., & Hillman, G. C., eds. (1989). *Foraging and Farming. The Evolution of Plant Exploitation*. London: Unwin Hyman.

Hellekant, G., & Danilova, V. (1996). Species differences toward sweeteners. *Food Chemistry*, **56**, 322–8.

Hellekant, G., Ninomiya, Y., & Danilova, V. (1997). Taste in chimpanzees II: Single chorda tympani fibers. *Physiology of Behavior*, **61**, 829–91.

Hellekant, G., Ninomiya, Y., DuBois, G. E., Danilova, V., & Roberts, T. W. (1996). Taste in chimpanzee: I. the summated response to sweeteners and the effect of gymnemic acid. *Physiology of Behavior*, **60**, 469–79.

Henderson, B. E., Ross, R. K., & Pke, M. C. (1993). Hormonal chemoprevention of cancer in women. *Science*, **259**, 633–8.

Hendrich, S., Xu, X., Wang, H. J., & Murphy, P. A. (1994). Neither diet selection nor type of soy food significantly affect bioavailability of isoflavones fed in a single meal to young adult females. *Journal of Nutrition*, **125**, Suppl 3S, 805S–6S.

Herbst, A., Ulfelder, H., & Poskanzer, D. (1971). Adenocarcinoma of the vaginal: association of maternal stilbesterol therapy and tumor appearance in young women. *New England Journal of Medicine*, **284**, 878–81.

Howe, G. R., Hirohata, T., Hislop, T. G., Iscovich, J. M., Yuan, J. M., Katsouyanni, K., Lubin, F., Marubini, E., Modan, B., & Rohan, U. (1990). Dietary factors and risk of breast cancer: combined analysis of 12 case-control studies. *Journal of the National Cancer Institute*, **82**, 561–9.

Hsieh, C-C., Lan, S-J., Ekbom, A., Petridou, E., Adami, H-O., & Trichopoulos, D. (1992). Twin membership and breast cancer risk. *American Journal of Epidemiology*, **136**, 1321–6.

Hughes, C. L. (1988). Effects of phytoestrogens on GnRH-induced luteinizing hormone secretion in ovariectomized rats. *Reproductive Toxicology*, **1**,

179–81.

Hughes, C. L., Chakinala, M. M., Reece, S. G., Miller, R.N., Schomberg, D. W., & Basham, K. B. (1991*b*). Acute and subacute effects of naturally occurring estrogens on luteinizing hormone secretion in the ovariectomized rat: part 2. *Reproductive Toxicology*. **5**, 133–7.

Hughes, C. L., Kaldas, R. S., Weisinger, A. S., McCants, C. E., & Basham, K. B. (1991*a*). Acute and subacute effects of naturally occurring estrogens on luteinizing hormone secretion in the ovariectomized rat: Part 1. *Reproductive Toxicology*, **5**, 127–32.

Jackson, F. L. C. (1991). Secondary compounds in plants (allelochemicals) as promotors of human biological variability. *Annual Reviews in Anthropology*, **20**, 506–46.

Jacobs, G. H. (1996). Primate photopigments and color vision. *Proceedings of the National Academy of Sciences, USA*, **93**, 577–81.

Jacobs, L. R. (1986). Modification of experimental colon carcinogenesis by dietary fibers. *Advances in Experimental Medical Biology*, **206**, 105–18.

Janzen, D. H. (1983). Dispersal of seeds by vertebrate guts. In *Coevolution*, ed. D. J. Futuyma, & M. Slatkin, pp. 232–62. Sunderland, MA: Sinauer Associates Inc.

Johns, T. (1990). *With Bitter Herbs They Shall Eat It: Chemical Ecology and the Origins of Human Diet and Medicine*. Tucson, AZ: University of Arizona Press.

Kaneda, N., Nagata, C., Kabuuto, M., & Shimizu, H. (1997). Fat and fiber intakes in relation to serum estrogen concentration in premenopausal Japanese women. *Nutrition and Cancer*, **27**, 279–83.

Key, T. J. A., Chen, J., Wang, D.Y., Pike, M. C., & Boreham J. (1990). Sex hormones in women in rural China and in Britain. *British Journal of Cancer*, **62**, 631–6.

Kliks, M. (1978). Paleodietetics: A review of the role of dietary fiber in preagricultural human diets. In *Topics in Dietary Fiber Research*, ed. G. A. Spiller, R. J. Amen, pp. 181–202. New York: Plenum Press.

Knott, C. D. (1997). Monitoring health status of wild orangutans through field analysis of urine. *American Journal of Physical Anthropology*, Suppl. **22**, 139–40.

Kuroda, S., Nishihara, T., Suzuki, S., & Oko, R. A. (1996). Sympatric chimpanzees and gorillas in the Ndoki Forest, Congo. In *Great Ape Societies*, ed. W. C. McGrew, L. F. Marchant, & T. Nishida, pp. 71–81. London, UK: Cambridge University Press.

Laska, M., Sanchez, E. C., Rivera, J. A. R., & Luna, E. R. (1996). Gustatory thresholds for food-associated sugars in the spider monkey (*Ateles geoffroyi*). American Journal of Primatology, **39**, 189–93.

Lawless, H. T. (1987). Gustatory psychophysics. In *Neurobiology of Taste and Smell*, ed. H. E., Finger, & W. L. Silver, pp. 401–20. New York: John Wiley.

Lee, R. B. (1979). *The Kung San Men, Women, and Work in a Foraging Society*. London, UK: Cambridge University Press.

Leighton, M. (1993). Modeling dietary selectivity by Bornean orangutans: evidence for integration of multiple criteria in fruit selection. *International Journal of Primatology*, **14**, 257–313.

Leopold, A. S., Erwin, M., & Oh, J. (1976). Phytoestrogens: Adverse effects on reproduction in California quail. *Science*, **191**, 98–100.

Lu, L-J. W., Anderson, K. E., Grady, J. J., & Nagamani, M. (1996). Effects of soya consumption for one month on steroid hormones in premenopausal women: Implications for breast cancer risk reduction. *Cancer Epidemiology Biomarkers Prevention*, **5**, 63–70.

Marks, D. L., Swain, T., Goldstein, S., Richard, A., & Leighton, M. (1988). Chemical correlates of rhesus monkey food choice: The influence of hydrolyzable tannins. *Journal of Chemical Ecology*, **14**, 213–35.

Marshall, E. (1993). Search for a killer: Focus shifts from fat to hormones in special report and breast cancer. *Science*, **259**, 618–21.

Messina, M., Persky, V., Setchell, K. D. R., & Barnes, S. (1994). Soy intake and cancer risk: A review of the *in vitro* and *in vivo* data. *Nutrition and Cancer*, **21**, 113–31.

Milton, K. (1993). Diet and primate evolution. *Scientific American*, **269** (Aug), 86–93.

Milton, K., & Demment, M. W. (1988). Digestion and passage kinetics of chimpanzees fed high and low fiber diets and comparison with human data. *Journal of Nutrition*, **118**, 1082–8.

Mousavi, Y., & Adlercreutz, H. (1992). Enterolactone and estradiol inhibit each other's proliferative effect on MCF-7 breast cancer cells in culture. *Journal of Steroid Biochemistry*, **41**, 615–19.

Nicholls, J., Lasley, B. L., Gold, E. B., Nakajima, S. T., & Schneeman, B. O. (1995). Phytoestrogens in soy and changes in pituitary response to GnRH challenge tests in women. *Journal of Nutrition*, **125**, 803S.

Nishida, T., Takasaki, H., & Takahata, Y. (1990). Demography and reproductive profiles. In *The Chimpanzees of Mahale Mountains: Sexual and Life History Strategies*, ed. T. Nishida, pp. 63–98. Tokyo, Japan: University of Tokyo Press.

Nofre, C., & Tinti, J. M. (1996). Sweetness reception in man: The multipoint attachment theory. *Food Chemistry*, **56**, 263–74.

Nofre, C., Tinti, J. M., & Glaser, D. (1996). Evolution of the sweetness receptor in primates. 2. Gustatory responses of nonhuman primates to 9 compounds known to be sweet in man. *Chem Senses*, **21**, 747–62.

Oftedal, O. T. (1991). The nutrient consequences of foraging in primates: the relationship of nutrient intakes to nutrient requirements. *Philosophical Transactions of the Royal Society of London B*, **334**, 161–70.

Pauling, L. (1970). Evolution and the need for ascorbic acid. *Proceedings of the National Academy Sciences, USA*, **67**, 1643–8.

Peterson, G. (1995). Evaluation of the biochemical targets of genistein in tumor cells. *Journal of Nutrition*, **125** (3S), 784S–9S.

Petrakis, N. L., Barnes, S., King, E. B., Lowenstein, J., Wiencke, J., Lee, M. M., Miike, R., Kirk, M., & Coward, L. (1996). Stimulatory influence of soy protein isolate on breast secretion in pre- and post-menopausal women. *Cancer Epidemiol Biomarkers Prev*, **5**, 785–94.

Phipps, W. R., Martini, M. C., Lampe, J. W., Slavin, J. L., & Kurzer, M. S. (1993). Effect of flax seed ingestion on the menstrual cycle. *Journal of Clinical Endocrinology and Metabolism*, **77**, 1215–19.

Plata-Salaman, C. R., & Scott, T. R. (1992). Taste neurons in the cortex of the alert cynomologous monkey. *Brain Research Bulletin*, **28**, 333–6.

Pond, C. M. (1997). The biological origins of adipose tissue in humans. In *The Evolving Female: A Life-History Perspective*, ed. M. E. Morbeck, A. Galloway, & A. L. Wihlman, pp. 147–62. Princeton, NJ: Princeton University Press.

Rindos, D. (1989). Darwinism and its role in the explanation of domestication. In *Foraging and Farming: The Evolution of Plant Exploitation*, ed. D. R. Harris, & G. C. Hillman, PP. 27–41. London: Unwin Hyman.

Rose, D. P., Boyar, A. P., Cohen, C., & Strong, L. E. (1987). Effect of a low-fat diet on hormone levels in women with cystic breast disease. I. Serum steroids and gonadotropins. *Journal of the National Cancer Institute*, **78**, 623–6.

Rose, D. P., Goldman, M., Connolly, J. M., & Strong, L. E. (1991). High-fiber diet reduces serum estrogen concentrations in premenopausal women. *American Journal of Clinical Nutrition*, **54**, 520–5.

Sadleir, R. M. F. S. (1969). *The Ecology of Reproduction in Wild and Domestic Mammals*. London, UK: Methuen.

Sanders, E. H., Gardner, P.D., Berger, P. J., & Negus, N. C. (1981). 6-Methoxybenzoxazolinone: a plant derivative that stimulates reproduction in *Microtus montanus*. *Science*, **214**, 67–9.

Schaefer, E. J., Lamon-Fava, S., Spiegelman, D., Dwyer, J. T., Lichtenstein, A. H., McNamara, J. R., Goldin, B. R., Woods, M. N., Morrill-LaBrode, A., & Hertzmarke, E. (1995). Changes in lipoprotein concentration and composition in response to a low-fat, high-fiber diet are associated with changes in serum estrogen concentrations in premenopausal women. *Metabolism*, **44**, 749–56.

Setchell, K. D. R., Lawson, A. M., Mitchell, F. L., Adlercreutz, H., Kirk, D. N., & Axelson, M. (1984). Lignans in man and animal species. *Nature*, **287**, 740–2.

Shimizu, H., Ross, R. K., Bernstein, L., Pike, M. C., & Henderson, B. E. (1990). Serum oestrogen levels in postmenopausal women: comparison of American whites and Japanese in Japan. *British Journal of Cancer*, **62**, 451–3.

Shimizu, H., Ross, R. K., Bernstein, L., Yatani, R., Henderson, B. E., & Mack, T. M. (1991). Cancer of the prostate and breast among Japanese and white immigrants in Los Angeles County. *British Journal of Cancer*, **63**, 963–6.

Simmen, B., & Sabatier, D. (1996). Diets of some French Guianan primates: Food composition and food choices. *International Journal of Primatology*, **17**, 661–93.

Slavin, J. L. (1987). Dietary fiber: Classification, chemical analyses, and food sources. *Journal of the American Dietary Association*, **87**, 1164–71.

Smith, B. D. (1995). *The Emergence of Agriculture*. New York: W. H. Freeman & Co. Scientific American Library.

Snodderly, D. M. (1979). Visual discriminations encountered in food foraging by a neotropical primate: Implications for the evolution of color vision. In *The Behavioral Significance of Color*, ed. E. H. Burtt, Jr, pp. 237–79. New York: Garland STPM Press.

Southgate, D. A. T. (1991). Nature and variability of human food consumption. *Philosophical Transactions of the Royal Society of London B*, **334**, 281–8.

Stahl, A. B. (1989). Plant-food processing: implications for dietary quality. In *Foraging and Farming: The Evolution of Plant Exploitation*, ed. D. R. Harris, & G. C. Hillman, pp. 171–94.London: Unwin Hyman.

Steinmetz, K. A., & Potter, J. D. (1991). Vegetables, fruit, and cancer. II. Mechanisms. *Cancer Causes and Control*, **2**, 427–41.

Telang, N. T., Suto, A., Wong, G. Y., Osborne, M. P., & Bradlow, H. L. (1992). Induction by estrogen metabolite 16α-hydroxyestrone of genotoxic damage and aberrant proliferation in mouse mammary epithelial cells. *Journal of the National Cancer Institute*, **84**, 634–8.

Tepper, B. J., & Nurse, R. J. (1997). Fat perception is related to PROP taster status. *Physiology and Behavior*, **61**, 949–54.
Terborgh, J. (1992). *Diversity and The Tropical Rain Forest*. New York: W. H. Freeman and Co.
Terborgh, J., & van Schaik, K. (1987). In *Organization of Communities Past and Present*, ed. J. H. R. Gee, & P. S. Giller. London: Blackwell.
Thompson, L. U. (1994). Antioxidants and hormone-mediated health benefits of whole grains. *Critical Reviews of Food Science and Nutrition*, **34**, 473–97.
Trowell, H. C., & Burkitt, D. P. (1983). *Western Diseases: Their Emergence and Prevention*. London: Edward Arnold.
Tutin, C. E. G. (1996). Ranging and social structure of lowland gorillas in the Lope Reserve, Gabon. In *Great Ape Societies*, ed. W. C. McGrew, L. F. Marchant, & T. Nishida, pp. 58–70.London, UK: Cambridge University Press.
United States Department of Agriculture, Agricultural Research Service. (1997). USDA nutrient database for standard reference. Release 11–1, Nutrient Data Laboratory. Home page, http://www.nal.usda.gov/fnic/foodcomp.
van Schaik, C. P., & Noordwijk, M. A. (1985). Interannual variability in fruit abundance and the reproductive seasonality in Sumatran Long-tailed macaques (*Macaca fascicularis*). *Journal of Zoology. Lond A*, **206**, 533–49.
van Thiel, D. H., Galvao-Teles, A., Monteiro, E., Rosenblum, E. R., & Gavaler, J. S. (1991). The phytoestrogens present in de-ethanolized bourbon are biologically active: A preliminary study in postmenopausal women. *Alc Clin Exp Res*, **15**, 822–3.
Vincent, A. S. (1984). Plant foods in savanna environments: a preliminary report of tubers eaten by the Hadza of northern Tanzania. *World Archaeology*, **17**, 131–48.
Wallis, J. (1995). Seasonal influence on reproduction in chimpanzees of Gombe National Park. *International Journal of Primatology*, **16**, 435–51.
Wattenberg, L. W. (1992). Inhibition of chemical carcinogenesis by minor dietary constituents. *Cancer Research*, **52**, 2085–91.
Watts, D. (1996). Comparative socio-ecology of gorillas. In *Great Ape Societies*, ed. W. C. McGrew, L. F. Marchant, & T. Nishida, pp. 16–28. London UK: Cambridge University Press.
Whitten, A., Byrne, R. W., Barton, R. A., Waterman, P. G., & Henzi, S. P. (1991). Dietary and foraging strategies of baboons. *Philosophical Transactions of the Royal Society of London*, **334**, 187–97.
Whitten, P. L. (1992). Chemical revolution to sexual revolution: Historical changes in human reproductive development. In *Chemically-Induced Alterations in Sexual and Functional Development: The Wildlife/Human Connection*, ed. T. Colburn, & C. Clement, pp. 311–34. Princeton NJ: Princeton Scientific.
Whitten, P. L., Lewis, C., & Naftolin, F. (1993). A phytoestrogen diet induces the premature anovulatory syndrome in lactationally exposed female rats. *Biology of Reproduction*, **49**, 1117–21.
Whitten, P. L., Kudo, S., & Okubo, K. (1997). Isoflavonoids. In *Handbook of Plant and Fungal Toxicants*. Boca Ratan, FL: CRC Press.
Whitten, P. L., Lewis, C., Russell, E., & Naftolin, F. (1995*a*). Phytoestrogen influences on the development of behavior and gonadotropin function. *Proceedings of the Society for Experimental Biology and Medicine*, **208**, 82–6.
Whitten, P. L., Lewis, C., Russell, E., & Naftolin, F. (1995*b*). Potential adverse

effects of phytoestrogens. *Journal of Nutrition*, **125**, 771S–6S.
Whitten, P. L., & Naftolin, F. (1991). Dietary estrogens: A biologically active background for estrogen action. In *The New Biology of Steroid Hormones*, ed. R. Hochberg, & F. Naftolin, pp. 155–67. New York: Raven Press.
Whitten, P. L., & Naftolin, F. (1992). Effects of a phytoestrogen diet on estrogen-dependent reproductive processes in immature female rats. *Steroids*, **57**, 56–61.
Whitten, P. L., & Naftolin, F. (1994). Xenoestrogens and neuroendocrine development. In *Prenatal Exposure to Toxicants: Developmental Consequences*, ed. H. L. Needleman, & D. Bellinger, pp. 268–93. Johns Hopkins University Press.
Whitten, P. L., Russell, E., & Naftolin, F. (1992). Effects of a normal, human-concentration, phytoestrogen diet on rat uterine growth. *Steroids*, **57**, 98–106.
Whitten, P. L., Russell, E., & Naftolin, F. (1994). Effects of phytoestrogen diets on estradiol action in the rat uterus. *Steroids*, **59**, 443–9.
Wilcox, G., Wahlqvist, M. L., Burger, H. G., & Medley, G. (1990). Oestrogenic effects of plant foods in postmenopausal women. *British Medical Journal*, **301**, 905–6.
Willett, W. C., Stampfer, M. J., Colditz, G. A., Rosner, B. A., Hennekens, C. H., & Speizer, F. E. (1987). Dietary fat and the risk of breast cancer. *New England Journal of Medicine*, **315**, 22–8.
Woods, M. N., Gorbach, S. L., Longcope, C., Goldin, B. R., Dwyer, J. T., & Morril-LaBrode, A. (1989). Low-fat, high-fiber diet and serum estrone sulfate in premenopausal women. *American Journal of Clinical Nutrition*, **49**, 1179–83.
Woods, M. N., Barnett, J. B., Spiegelman, D., Trail, N., Hertzmark, E., Loncope, C., & Gorbach, S. L. (1996). Hormone levels during dietary changes in premenopausal African–American women. *Journal of the National Cancer Institute*, **88**, 1369–74.
Wrangham, R. W. (1977). Feeding behaviour of chimpanzees in Gombe National Park, Tanzania. In: *Primate Ecology*, ed. T. H. Clutton-Brock, pp. 504–38. London: Academic Press.
Wrangham, R. W., Chapman, C. A., Clark-Arcadi, A. P., & Isabirye-Basuta, G. (1996). Social ecology of Kanyawara chimpanzees: implications for understanding the costs of great ape groups. In *Great Ape Societies*, ed. W. C. McGrew, L. F. Marchant, & T. Nishida, pp. 45–57. London, UK: Cambridge University Press.
Wrangham, R. W., Conklin, N. L., Chapman, C. A., & Hunt, K. D. (1991). The significance of fibrous foods for Kibale Forest chimpanzees. *Philosophical Transactions of the Royal Society of London, B*, **334**, 171–8.
Yamagiwa, J., Marushashi, T., Yumoto, T., & Mwanza, N. (1996). Dietary and ranging overlap in sympatric gorillas and chimpanzees in Kahuzi-Biega National Park, Zaire. In *Great Ape Societies*, ed. W. C. McGrew, L. F. Marchant, & T. Nishida, pp. 82–98. London, UK: Cambridge University Press.
Yu, H., Harris, R. E., Gao, Y. T., Gao, R. N., & Wynder, E. L. (1991). Comparative epidemiology of cancers of the colon, rectum, prostate and breast in Shanghai, China versus the United States. International Journal of Epidemiology, **20**, 76–81.

8

Modernization, psychosocial factors, insulin, and cardiovascular health

STEPHEN T. McGARVEY

8.1 Introduction

The health effects of the alteration of traditional occupations, physical activity, and diets by exposure to modern ways of life have been described in many cross-national, cross-sectional, and longitudinal studies. The energetic and metabolic consequences of modernization, such as adiposity in adults and its role in cardiovascular disease (CVD) risk factors, have been the focus of most modernization studies. Fewer studies have been able to measure behavioral, attitudinal, and emotional factors, their physiological concomitants, and their associations with health and disease.

The purpose of this chapter is to describe the role of hormones associated with energy metabolism and the sympathetic nervous system (SNS) as they relate to changes in cardiovascular health and disease with modernization. The emphasis is on insulin, and its role in adiposity and psychophysiological stress in modernizing groups. CVD-related outcomes discussed here will largely be risk factors such as blood pressure and hypertension, lipid and lipoprotein levels, adiposity and catecholamine levels. Potential interactions between positive energy balance and psychophysiological arousal will be discussed because such interrelationships may elevate CVD risk in groups experiencing social and economic change. This unitary model of insulin's several effects on the two physiological and psychological systems will be exemplified by describing work among modernizing Samoans and how future studies could benefit from this integrative focus on hormones and lifestyle changes.

Before considering the biological responses to modernization and the interacting role of insulin and catecholamines in cardiovascular health and disease, concepts and operational definitions of modernization, lifestyle, and psychosocial stress are described.

8.2 Concepts and aspects of modernization

Modernization is a poorly defined and heterogeneous process. The common characteristics of this general process involve changes in energetic, economic, socio-political, and ideological domains of pre-industrial communities as a result of interactions with industrial nation states. Specific changes in traditional communities are dependent on the history, political economy, ecology, and culture of the interacting societies. From the perspective of human population biology, economic modernization has been defined as the interaction of relatively simple technological and local socio-economic systems with industrial economic systems along with their national and international market and political factors (McGarvey *et al.*, 1989).

The difficulty in positing a specific and concrete definition of modernization must remind researchers and readers to be very cautious in making overly broad generalizations. For human population biology, modernization is useful only as a general term which must be followed by attempts to measure specific exposures in the altered economic, socio-political, and psychosocial domains, and their effects on human biology and biocultural adaptation. The first goals of such research should be to describe the context of economic and social change in the study population, to specify the concrete exposures of individuals in the particular modernizing population, and to hypothesize detectable associations between exposures and health outcomes in specific subgroups.

In addition to modernization, several other terms have also been used to describe changes in the way of life of formerly traditional communities, including Westernization, Americanization, development, and acculturation. All have been criticized for explicit or implicit unilineal progressive assumptions. That is, traditional societies will change and develop along a path of sequences with a narrow range of possibilities towards the late twentieth century industrial society model (Howard, 1986). This critique certainly applies to the first three terms, and newer concepts such as Howard's (1986) cultural diversification are recognitions of the increased heterogeneity that occurs during economic change.

In the following overview of modernization, I divide its influence into first, energetic and socio-economic aspects, and second, psychosocial aspects.

8.2.1 Energetic and socio-economic aspects

A substantial body of work in biological anthropology examines socio-economic changes in small isolated groups which lead to an overall economic, energetic and nutritional enrichment, for example, in Pacific island groups (Baker *et al.*, 1986; Friedlaender, 1987; Wessen *et al.*, 1992). Conversely, such changes can also lead to loss of local control of land, cash cropping for export, reduction in diversity of food stuffs, and greater economic fragility and malnutrition, for example in Brazil (Gross and Underwood, 1971), Jamaica (Marchione & Prior, 1980), Mexico (Dewey, 1981), Indonesia and other countries (Franke, 1987).

There is also increasing heterogeneity in socio-economic responses to economic development within large societies leading to greater social stratification and differences in nutritional and health status (Frenk *et al.*, 1991). Increased malnutrition and child growth have been described in industrial countries as a result of the level of poverty and its expansion, for example in the USA (Crooks, 1995) and China (Shen *et al.*, 1996).

Thus modernization can lead to economic and nutritional impoverishment, as well as to enrichment. The search for general rules about the pace and direction of modernization's impact on human biology on all populations is fruitless to this author. Regional ecological, historical and political influences must be understood to reconstruct why enrichment or impoverishment or both occurred in different areas as a response to modernization forces. Anthropology can offer an ecological and holistic perspective to help with such an endeavour.

Regardless of the heterogeneous processes of modernization in diverse populations, the individual hallmarks of these changes are increased work for wage labor and decreased subsistence work. Many studies emphasize the changes in energetic variables during modernization and report on increased adult and child weight, adiposity, increased dietary intake of calories and fats, and reduced physical activity (WHO, 1990). An earlier review discussed such results and concluded that increases in adiposity, obesity, non-insulin-dependent diabetes mellitus (NIDDM), adult blood pressure (BP), and cardiovascular disease (CVD) are generally seen with modernization (McGarvey *et al.*, 1989). Diet and activity behavior changes, leading to a positive energetic balance and increased adiposity, appeared to be the major, but clearly not exclusive, forces in changing patterns of adult CVD morbidity and mortality. Additional studies continue to document such changes in small Pacific Island societies (Wessen *et al.*, 1992; Hodge *et al.*, 1993; McGarvey *et al.*, 1993; Gershater & McGarvey, 1995; Hodge *et al.*, 1994), West Africa (Kaufman *et al.*, 1996), China (Leung *et al.*, 1996), and minority ethnic groups in the US (Ernst and Harlan, 1991).

8.2.2 Psychosocial aspects

The above conceptual and operational definitions of modernization, which emphasize energetic and economic factors, excluded the crucial psychosocial changes stemming from alterations in family, household, and community social interactions. Such factors were excluded purposefully, despite awareness of classic formulations about social change, stress and health (Scotch, 1963; Henry & Cassel, 1969; Cassel, 1976; Ostfeld and D'Atri, 1977), and despite work done in parallel on psychosocial influences, physiological stress, and CVD risk factors (Harrison *et al.*, 1981; Brown, 1982; Dressler, 1982; Jenner *et al.*, 1982; James *et al.*, 1985; McGarvey & Schendel, 1986). Without considering psychosocial factors, studies which focused on changes in diet, physical activity, and energy balance had limited value for generalizing about the overall effects of modernization on CVD risk factors and the potential interactions between energetic and psychosocial domains.

Recent psychosocial studies in modernizing groups have sought to operationalize concepts such as stress, acculturation, coping styles, lifestyle and lifestyle incongruity, prestige enhancement and status inconsistency, and found significant influences on CVD risk factors (Harrison, 1995; Chin-Hong & McGarvey, 1996; Dressler, 1985, 1995; James *et al.*, 1989). Behaviors such as alcohol consumption, cigarette smoking, and sleep patterns have also been studied in modernizing groups for their interactions with psychosocial and physiological stress factors and their independent association with CVD risk (James *et al.*, 1989).

Acculturation

Acculturation can be referred to generally as the process of culture change during the same general macro-level processes, but can be defined as the adoption of attitudes, values, beliefs, assumptions, and cultural practices characteristic of the economically dominant society, or the demographically dominant society in the case of ethnic minorities and migrants. The adoption of non-traditional attitudes, values, and practices can lead to increased stress from the interaction of traditional expectations and new social demands. Acculturation also might be adaptive as individuals begin to assimilate materially and socially by decreasing the putative dissonance between the roles and expectations they learned and those they begin to practice and aspire to. Psychological processes become important in acculturation studies, and bioanthropologists can benefit from the methodological rigor of psychometric research. Acculturation has been linked to health status in studies of ethnic minorities in developed nations, although measures and findings are quite specific to the population studied (Marmot & Syme, 1976; Hazuda *et al.*, 1988).

Lifestyle incongruity

Lifestyle has become recently a key concept in social science and health studies, particularly in anthropology, through the work of Dressler (1990, 1991, 1995). The theoretical definition relates to changes in the consumption of material goods, and its association with attempts to achieve higher social rank and enhanced prestige, thus, is well suited for modernization and CVD studies. As economic changes and stratification occur, so too does social differentiation. This increasing social heterogeneity is often characterized by adoption of non-traditional ways of life which provide external evidence to others of participation in the more prestigious "modern" ways of life. Outward material displays in order to raise prestige rankings can include purchased goods from outside, such as motor vehicles, televisions, refrigerators, and non-traditional clothing and furnishings. Attempts to raise prestige by social displays that are incongruous with occupational or educational level may produce "uncertainty and vigilance in social interaction" (Dressler, 1995), because individuals are not being treated in daily interpersonal routines as if they were of the higher projected status. The common phrase "living beyond one's means" is analogous to aspects of lifestyle incongruity, implying the psychological burden of maintaining a costly material way of life. Thus, lifestyle incongruity is similar to other stressful conditions which can cause psychophysiological arousal, increased SNS activity, and if present chronically, elevated blood pressure, heart rate, and lipid levels (Henry & Cassel, 1969).

Ideas about lifestyle processes were initially described by social scientists with now familiar terms such as conspicuous consumption, prestige enhancement, and the leisure class (Veblen, 1899; Weber, 1946). Recent scholarship in cultural anthropology on development and modernization has also considered these phenomena with analyses of the commodification of consumption, cultural hegemony, and critiques of the consumer society (Rutz & Orlove, 1989). Unfortunately, the meaning of lifestyle has been debased in contemporary English to a very general interpretation and even in public health the expression "healthy lifestyle" has come to connote avoiding smoking, engaging in exercise, moderating alcohol use, and reducing stress. The term remains useful because of its link to classic social science scholarship on social status and prestige changes with economic development.

Lifestyle incongruity has been operationalized as the difference in the standardized rank distributions of: (1) a scale of household material goods and media-oriented behaviors such as reading newspapers, and (2) occupation or education (Dressler, 1990). Lifestyle incongruity was significantly related to blood pressure and lipid levels in several cross-sectional studies of modernizing societies, including a Southern USA black community (Dressler, 1990), Brazil (Dressler *et al.*, 1991), and Western Samoa (Chin-Hong & McGarvey, 1996). Further studies are necessary to replicate the findings of an association between lifestyle incongruity and CVD risk factor, their longitudinal predictability, and to elucidate proposed mediating mechanisms. The challenge remains to measure the ethnic group, sex, age, and cohort specificity of lifestyle incongruity, to contextualize explanations about these measures, and specify the hypothesized associations with CVD risk factors.

As a result of the social and economic changes with modernization, individuals develop an almost entirely different set of activities, interests and values than their parents. In some cases, the activities of adults are very different from those they were expected and enculturated to pursue when children and adolescents. These processes are tapped by other concepts such as social mobility and socio-cultural incongruity explored earlier by psychosocial epidemiologists (Syme *et al.*, 1965, Smith, 1967). Related concepts such as status inconsistency and social consistency have been related to CVD risk factors in modernizing groups (McGarvey & Schendel, 1986; Dressler, 1988). All these concepts share the common theme of excessive psychosocial costs to individuals who have a mismatch or lack of congruence in social and economic domains such as between their education and occupation, or their own and parents' education or occupation.

Individual coping

Psychological *coping*, namely the cognitive, emotional, and behavioral processes of evaluating the salience of stressors, and adapting to them, is another important source of variation in understanding who is vulnerable to physiological arousal in the context of modernization. Some individuals buffer their arousal by assigning little value to certain stressors. Some individuals are able to actively find solutions and lessen the burden of perceived stressors. Others remain aroused, unable to solve the perceived problems, but with heightened emotions including anger, anxiety, depression, and fear. Some individuals deny any salience or even presence of certain stressors. There is clearly a spectrum of coping responses and concomitant emotions to everyday hassles and chronic life stressors. Detailed experience and knowledge about cultural and psychological factors affecting individuals within study populations is required for such research.

In St. Lucia, Dressler (1982) showed the apparent success of denial as a coping pattern in reducing stress due to lifestyle incongruity. Influenced by a series of studies on anger and blood pressure in Black Americans (Harburg *et al.*, 1979; Gentry *et al.*, 1982), I and colleagues have been working on anger expression and health in modernizing Samoa (McGarvey & Schendel, 1986; Steele & McGarvey, 1996, 1997), as described below in the section on Samoan research.

Social support

Social support can buffer stressors through both problem solving and emotional sharing of stressful events. This concept has been applied to some modernizing populations, for example St. Lucia and American Samoa (Dressler, 1982; Hanna *et al.*, 1986). It is difficult to assess social support. Thus studies of family and community networks need to carefully evaluate the specific type and amount of social support rendered to modernizing individuals, not just the size of a network of family, friends and others. Studies of social support and individual coping require long-term familiarity and detailed ethnographic knowledge of study populations. Detailed knowledge of ethnographic studies is essential. Increasingly, such studies should be done with the close collaboration of individuals who originated from or currently reside in the local populations. This will reduce the temptation to use off-the-shelf interview schedules of dubious validity in novel situations. It may increase the sophistication of the contextual understanding of exposures to the concrete psychosocial burdens resulting from economic and social changes of modernization in specific societies.

8.3 The role of insulin

The hormone insulin plays crucial roles in several physiological systems, especially in glucose disposal and regulation of blood glucose levels. This primary role of insulin is well known through many intensive investigations of hyperglycemia, diabetes, energy balance and adiposity and their metabolic causes and consequences. In brief, insulin is secreted by the beta cells of the pancreatic islets of Langerhans in response to elevated blood glucose levels, after food intake. Insulin promotes the storage of glucose in various organs, muscles and adipose tissue. In some individuals and conditions there is a reduction in insulin's ability to clear glucose, referred to as insulin resistance. Insulin resistance is thought to be due to a loss of sensitivity to insulin by cell membrane receptors, resulting in compensatory hyperinsulinemia to effect the same level of glucose transport. Hyperinsulinemia itself can also be a primary cause of the development of insulin resistance stemming from increased insulin secretion in response to glucose. Several studies have shown familial and genetic bases for disordered pulsatile insulin secretion in those with untreated NIDDM and first degree relatives of those with NIDDM (Polonsky *et al.*, 1988; O'Rahilly *et al.*, 1988). In addition, insulin resistance has been attributed to post-receptor defects in glucose uptake and increased free fatty acids which attenuate skeletal muscle glucose uptake and hepatic insulin clearance.

In the past 10–15 years the effects of insulin levels on sympathetic nervous system (SNS) activity and energy balance have been demonstrated. In the next sections, insulin's influence on these systems will be summarized followed by an overview of insulin's importance in studies of health among modernizing societies. The summary of insulin's activities described below relies not only on the primary studies but on the syntheses of Landsberg (1986, 1990), Björnthorp (1988, 1991), and Defronzo and Ferrannini (1991).

8.3.1 Insulin and the sympathetic nervous system

Increased SNS activity has direct effects in the short-term on cardiovascular functioning such as increased heart rate, cardiac output, and increased peripheral vascular resistance, resulting in an increased blood pressure. Chronic stimulation of the SNS has been associated with hypertension, damage to blood vessel endothelium and mobilization of lipids, and several other aspects of cardiovascular health (Henry & Stephens, 1977). One of the most important putative sources of SNS stimulation in modernizing groups is the many potential psychosocial exposures discussed above.

A series of studies have indicated that increased insulin levels lead independently to increased SNS activity (Rowe *et al.*, 1981; Landsberg & Young, 1985*a*). Insulin administration to normal and diabetic individuals increases levels of norepinephrine (noradrenaline). This experimental hyperinsulinemia increases also blood pressure and oxygen consumption.

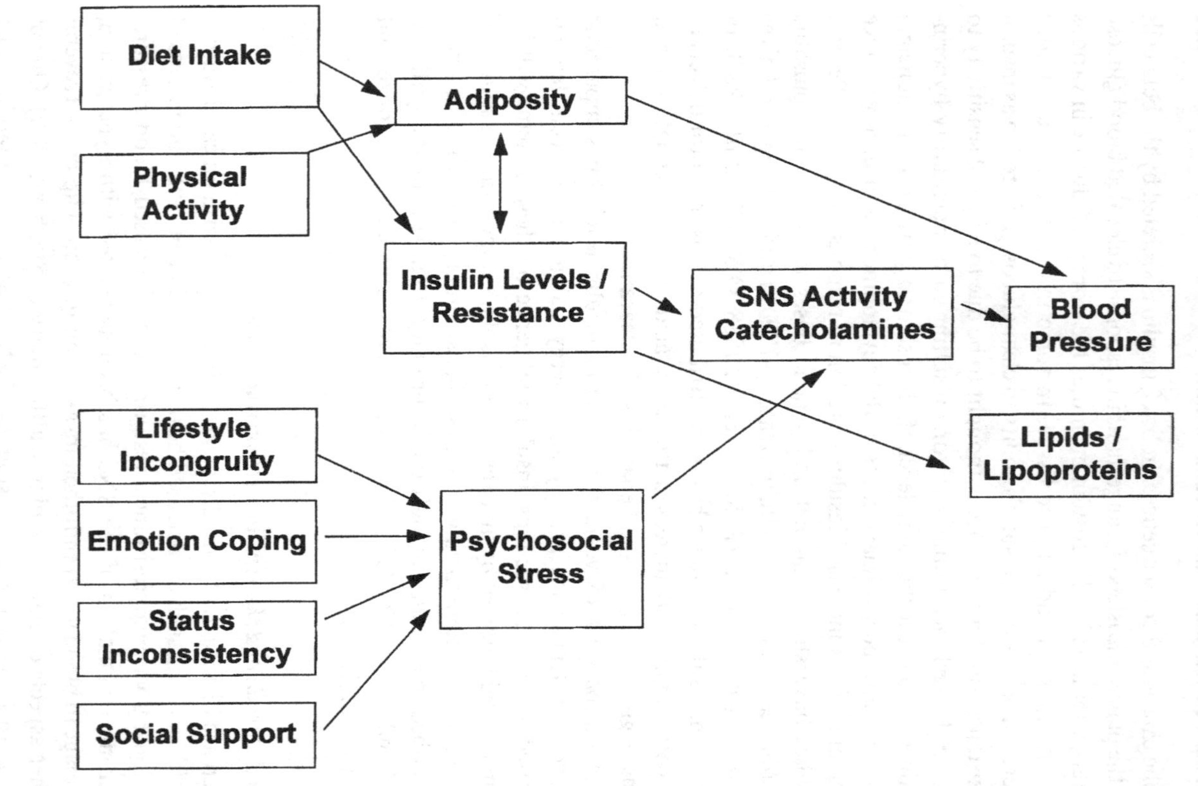

Figure 8.1. Interactions of adiposity, insulin, psychosocial stress, the sympathetic nervous system, and cardiovascular disease risk factors.

The stimulation of SNS activity by insulin is not sensitive to blood glucose levels in either obese or lean subjects. This suggests that there is no downregulation of insulin sensitivity in the centrally mediated SNS activity control centers. Whereas the obese are resistant to insulin's effect on glucose, because of loss of insulin sensitivity in muscles, they retain sensitivity to SNS stimulation by insulin. The hyperinsulinemia concomitant with being overweight apparently leads to increased SNS stimulation, resulting in increased levels of epinephrine, norepinephrine, and cortisol, among other hormones.

These findings and interpretations have important implications for modernizing populations. SNS activity may be increased in both adults and children due solely to the changes in insulin from increased adiposity, stemming from a positive balance between energy consumed and expended. Additional increases in SNS activity through mechanisms of psychophysiological arousal could result in further increases in CVD risk factors.

8.3.2 Insulin, sympathetic nervous system, and energy balance

There is a growing consensus on the interdependence between diet, metabolic rate, insulin, and SNS activity (Figure 8.1). Changes in dietary intake are related to activity of the SNS. Fasting decreases sympathetic outflow, and overfeeding increases it (Young & Landsberg, 1977; Landsberg & Young, 1985*b*). Increased intake of fats and some carbohydrates also leads to increased SNS activity.

Insulin stimulated SNS activity influences the regulation of energy balance through its effects on metabolic rate and thermogenesis (Landsberg, 1986). Catecholamines, i.e., norepinephrine and epinephrine, increase metabolic rate, presumably through brown adipose tissue activity. Insulin has been shown to lead to a SNS mediated rise in metabolic rate (Katzeff *et al.*, 1986; Acheson *et al.*, 1983).

There appears to be an adaptive process by which the decreased SNS activity of fasting conserves energy at a time of low intake because this leads to reduced thermogenesis. Similarly, increased SNS with raised caloric or fat intake stimulates thermogenesis in the form of heat, at a caloric cost, which attempts to restore energy balance altered by overfeeding. Thus, the hyperinsulinemia of obesity appears to increase SNS activity and thermogenesis, which limits weight gain. This suggests the possibility that insulin sensitivity, resistance and hyperinsulinemia are responsive mechanisms that limit further weight gain. Also local cellular insulin resistance may attenuate further fat deposition into adipocytes, limiting the fat cell mass that would otherwise accrue. The epidemiological evidence regarding persistent weight gain and adiposity in modernizing adults suggests that this compensatory thermogenesis mechanism is not sufficient to regulate large, chronic excesses of energy intake over expenditure.

Positive energy balance over time may be a better explanation at the population level than the perspective that obese individuals have a genetically based defect in dietary thermogenesis (Laville *et al.*, 1993). There are elegant demonstrations of substantial individual variation in energy intakes with maintenance of energy balance and expenditure, and thus inferentially of genetic variability in metabolic rate and energy expenditure (Prentice *et al.*, 1985; Ravussin *et al.*, 1986). Thermogenesis, or metabolic rate, appears to be highly heritable (Bogardus *et al.*, 1986). Clearly, a combined biobehavioral perspective which posits both a partial role for the genetic control of insulin mediated diet-induced thermogenesis and a role for socio-culturally mediated behaviors affecting differences in energy expenditure and intake is more fruitful for studies in modernizing groups.

8.3.3 Insulin and adiposity

Adiposity and insulin levels are positively associated in virtually every study of healthy individuals, including children, adolescents, and adults. Adiposity appears to be both a by-product of the role of insulin in energy balance and metabolic activities, as described above, and a contributor to insulin regulation. Insulin levels and glucose response are higher in those with relatively more abdominal adiposity than femoral or gluteal adiposity (Krotkiewski *et al.*, 1983; Rebuffe-Scrive *et al.*, 1985). Abdominal adiposity is marked by large fat cells, increased lipolysis and a blunted anti-lipolytic influence of insulin (Bouchard, 1994).

There is a debate about the causal direction of relationships between insulin levels and resistance on the one hand and adipose tissue or weight changes on the other. As for insulin levels and adiposity, cross-sectional studies have related insulin resistance and adiposity; weight gain increases while weight loss decreases insulin resistance (Swinburn *et al.*, 1991; Sims, 1976). Longitudinal investigations show that insulin resistance is not simply related to gains in weight or adipose tissue. Several recent studies show that insulin resistance attenuates weight gain rather than accelerates it. In Pima Indians (Swinburn *et al.*, 1991), Mexican Americans (Valdez *et al.*, 1994), and a mixed ethnic population of Hispanic Americans in San Luis Valley, Colorado (Hoag *et al.*, 1995), those with the lowest insulin resistance or lowest insulin levels were those with the greatest weight gain. Although this seems paradoxical given the positive correlation between adipose tissue and insulin levels, these longitudinal studies suggest the adaptive scenario described above for insulin, SNS, and energy balance (Figure 8.1).

Those who are already overweight or in positive energy balance have hyperinsulinemia which then leads to insulin resistance and increased thermogenesis, both factors leading to slower weight gain and partial restoration of energy balance. We would not expect insulin levels to have the same dose–response effect on increasing fat, or weight, among those with existing hyperinsulinemia and insulin resistance. Among those with low insulin levels and the lowest insulin resistance, glucose disposal into adipocytes would be greatest per unit insulin, leading to the findings described above. Clearly, the underlying mechanisms require much more study, but there appears to be an exponential decaying relation between insulin levels as well as insulin resistance, and weight gain. At low insulin levels there is a steep positive slope between insulin and weight gain, and as insulin levels and resistance increase, weight gain slows down. At the highest insulin levels, among normoglycemic individuals, there may be little *longitudinal* relationship between insulin and weight gain. There would be a positive cross-sectional correlation between insulin levels (or resistance) and weight (or adiposity), as described above, due to the presence of both high insulin levels and high body fat. Among hyperinsulinemic individuals in modernizing societies, the compensatory SNS activation, increased metabolic rate and thermogenesis are all mechanisms attempting to reduce weight gain, but these mechanisms may be inadequate in the face of increasing positive energy balance due to excess caloric intake, decreased activity expenditure, or both. The compensatory mechanisms discussed above may only be detectable with very frequent measurement intervals and only among those early in the natural history of hyperinsulinemia and adiposity.

8.3.4 Insulin, blood pressure, and lipids

Insulin levels and resistance have been related to blood pressure and hypertension in a variety of groups including representative population samples, healthy but overweight individuals and young hypertensives (Modan *et al.*, 1985; Fournier *et al.*, 1986). Among both the obese Pima Indians and Nauruans, insulin tends not to be related to blood pressure, probably because of the prevalence of disorders of established hyperinsulinemia (Saad *et al.*, 1991; Collins *et al.*, 1990). In a study of insulin and blood pressure in Pacific populations, there was a positive association in Western Samoans, especially among those with the lowest glucose levels, but no association among the very overweight Nauruans (Collins *et al.*, 1990).

Several mechanisms have been described to support a causal role for insulin levels and resistance. These include increased renal reabsorption of sodium due to hyperinsulinemia (DeFronzo, 1981), and the increased SNS activity due to hyperinsulinemia which raises cardiac output and peripheral vasoconstriction, as described above.

Insulin levels also consistently influence lipid and lipoprotein levels (DeFronzo and Ferrannini, 1991; Stalder *et al.*, 1981). Hyperinsulinemia increases hepatic very-low-density lipoprotein (VLDL) cholesterol synthesis which leads to elevations in VLDL cholesterol and triglycerides, and eventually to increased low density lipoprotein (LDL) and total cholesterol levels (Olefsky *et al.*, 1974; Reaven, 1988). Both LDL and VLDL levels have been related to risk of heart disease. Insulin stimulates lipoprotein lipase levels which also increases triglyceride levels (DeFronzo & Ferrannini, 1991). These processes are pronounced in those with abdominal adiposity, probably due to their increased hyperinsulinemic responses.

Insulin levels have also been positively associated prospectively with incidence of CVD itself, not just elevated risk factors such as blood pressure and lipids (Pyörälä 1979; Ducimetiere *et al.*, 1980; Welborn & Wearne, 1979). Thus, hyperinsulinemia has consequences for CVD regardless of the exact mechanisms.

Recent research also indicates that insulin plays an important role in ovarian steroid production in women, including healthy women and those with polycystic ovarian disease (Willis *et al.*, 1996). It appears that insulin's effects on ovarian steroids is not characterized by insulin resistance and that hyperinsulinemia may potentially contribute to hyperandrogenism and increased disease risk (Nahum *et al.*, 1995; Gates *et al.*, 1996).

8.4 Implications for modernizing populations

In modernizing populations, changes in occupation, diet choices, and activity patterns are likely to lead to positive energy balance and adiposity (Figure 8.1). Psychosocial stress also tends to increase with new social interactions, altered psychological relevance of stimuli, efforts to obtain novel prestige items, and greater work demands. The above review strongly indicates that insulin levels will increase concomitantly with adiposity, and that hyperinsulinemia will likely increase SNS activity. In addition, psychophysiological arousal will elevate SNS activity. Hyperinsulinemia will have deleterious effects on glucose regulation, blood pressure, lipid and lipoprotein levels. SNS activity will also increase blood pressure and lipids, independent of insulin levels.

The hormonal interactions between insulin and catecholamines described above may play influential roles in determining metabolic and cardiovascular responses in societies experiencing modernization. This suggests that modernizing societies will be at high risk of elevated CVD levels due to increased adiposity, hyperinsulinemia and the associated NIDDM, hypertension and hyperlipidemia.

8.5 Samoan research

In this section, I describe attempts to operationalize some of the concepts discussed above in studies of Samoans experiencing rapid social and economic change. Starting in 1975 and lasting about 8 years, The Samoan Studies Project included a number of cross-sectional measures of human biology and health among Samoans in several different locales, including California, Hawaii, American Samoa and Western Samoa (Baker *et al.*, 1986). The summary here describes some of the early studies, and presents findings from fieldwork recently completed (1990–95) in American and Western Samoa. Most of the results described pertain to a cohort of Samoan men and women studied longitudinally over 4 years; 1990–94 in American Samoa and 1991–95 in Western Samoa.

8.5.1 Adiposity and insulin

Modernizing Samoans are characterized by high prevalence and incidence of high body mass index (BMI) and massive adiposity (Bindon & Baker, 1985; McGarvey, 1991; McGarvey, 1994). The level of BMI and prevalence of obesity has increased from the late 1970s to the early 1990s in American Samoa and Western Samoa (McGarvey *et al.*, 1993; Hodge *et al.*, 1994). For example mean BMI among American Samoan men, 35–44 years of age, was 30.4 kg/m^2 in 1976 and 34.0 kg/m^2 in 1990. Similarly, mean BMI among American Samoan women, 35–44 years of age, was 34.8 kg/m^2 in 1976 and 36.6 kg/m^2 in 1990 (McGarvey *et al.*, 1993). In one urban and two rural areas of Western Samoa studied in 1978 and 1991, mean BMI for ages 25–74 years increased for men from 27.5 to 30.4 kg/m^2 and for women from 29.8 to 32.9 kg/m^2 (Hodge *et al.*, 1994).

Longitudinal studies also indicate substantial weight gain continuing among already overweight adults (McGarvey, 1991; Gershater & McGarvey, 1995). For example American Samoan men and women, 18–34 years old, gained approximately 7 kg of weight over the 5-year period of 1976–81 (McGarvey, 1991). Information on BMI from the recently completed 4-year longitudinal cohort from Western and American Samoa provides more evidence on Samoans' massive body size and continual adult fat accumulation (Figures 8.2 and 3). Men from American Samoa have greater BMI at all ages than their Western Samoan counterparts, and had greater 4-year BMI increases, especially in those 25–34 years (Figure 8.2). American Samoan women of each age also have greater BMI than Western Samoan women at each study time, but had similar 4-year BMI increases, i.e., approximately 1 kg/m^2, except for the oldest age group (Figure 8.3). BMI is notably higher in all Samoan women compared to men, with age-specific mean BMI ranging from 27 to 36 kg/m^2 in men, and from 29 to 39 kg/m^2 in women (Figures 8.2 and 3).

The author has conducted three different longitudinal studies with American Samoans focusing on weight, subcutaneous fat and BMI from 1976 to 1994, which allows a comparison of the annual changes in BMI through different periods of American Samoan modernization (Figure 8.4). Annual average BMI change over the period of follow-up is presented for young adults, 25–34 years of age, at the start of the period of observation, whose weight and fatness change the most rapidly over time, for three longitudinal cohorts: a 5-year study from 1976–1981 (McGarvey, 1991); a 14-year study from 1976 to 1990 (Gershater & McGarvey, 1995); and a 4-year study from 1990 to 1994. It appears that BMI in American Samoan men increased over time more slowly early during this period of modernization, but has increased dramatically recently (Figure 8.4). American Samoan women on the other hand, increased BMI early in this period of ongoing modernization with lower annual increases later in the period. The slower rate of BMI increase in women in recent times is probably related to their already extant extreme BMI levels, compared to lower values for men. The reasons for these sex differences in the timing of the rates of BMI increase are not clear but may relate to the earlier reduction in physical activity by women compared to men during economic modernization. Men continued to provide the majority of subsistence work and adopt wage labor with moderate physical activity whereas women appeared to almost stop their traditional household gardening activities and reduce plantation work (Greksa *et al.*, 1986).

Adult females have higher levels of obesity, although in the most recent

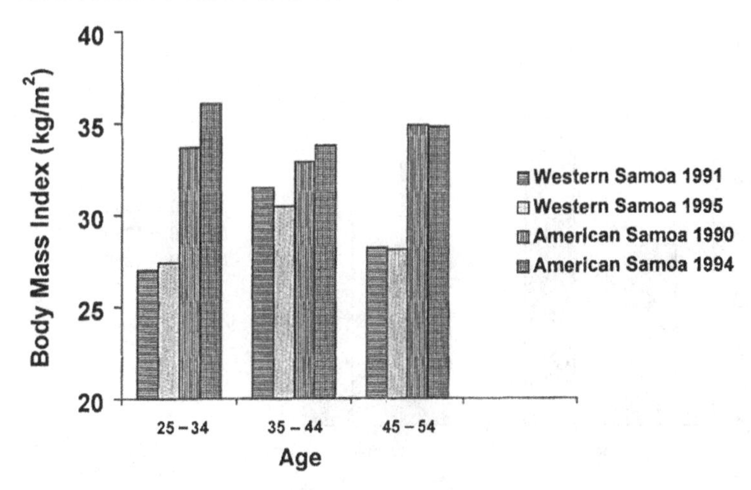

Figure 8.2. BMI of Samoan men, $N = 421$, in the four-year longitudinal cohort.

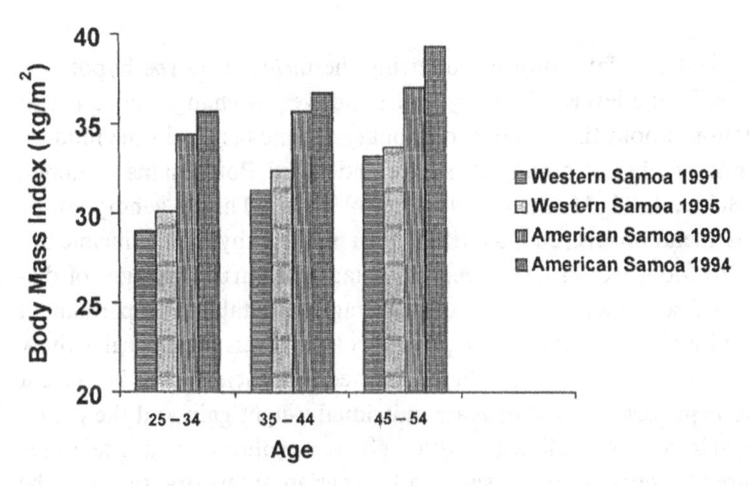

Figure 8.3. BMI of Samoan women, $N = 518$, in the four-year longitudinal cohort.

samples from American Samoa male and female rates of obesity are similar. Between 90 and 92% of men and women of all ages had a BMI > 30 kg/m^2 in American Samoa in 1994, whereas in lesser developed Western Samoa in 1995, approximately 47% of all men, 66% of young women (< 40 years), and 79% of older women had a BMI > 30 kg/m^2.

The temporal trends in diet and physical activity indicate that a chronic positive energy balance among Samoans is largely responsible for levels of

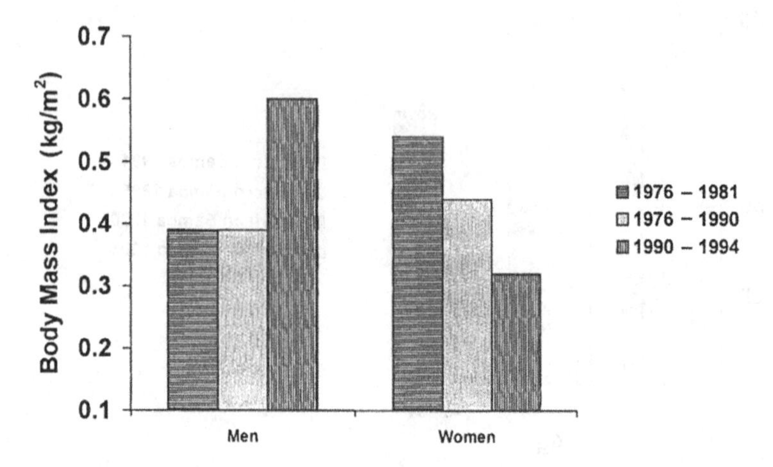

Figure 8.4. Annual change in BMI in men and women in three American Samoan longitudinal cohorts, age 25–34 at baseline: 1976–1981; 1976–1990; and 1990–1994.

body weight and fat; however, following the *thrifty genotype* hypothesis (Neel, 1962), the levels of obesity and rapid weight changes have led to speculations about the role of evolutionary genetic factors in the modern pandemic of obesity among Samoans and other Polynesians (Zimmet. 1979; Baker, 1984; McGarvey *et al.*, 1989, 1994). Thrifty genotypes, or efficient metabolisms, characterized by a relative hyperinsulinemic response to food, would have been advantageous during voyages of discovery and settlement, and in the early stages of establishing permanent communities in Polynesia. In the presence of low levels of physical activity and increased dietary intakes these inherited tendencies towards efficient metabolic processes could increase individual weight gain and the prevalence and incidence of adiposity. Although these evolutionary and adaptive ideas are attractive when considering Polynesian prehistory, they may be seen primarily as stimulating hypotheses for concrete biological and epidemiological studies of diet, activity, and metabolic changes among modernizing Polynesians and other populations (McGarvey, 1994).

We have reported preliminary findings on hyperinsulinemia among Samoans (McGarvey & Levinson, 1992). Fasting insulin levels in adults are directly related to adiposity in all age and sex groups. The insulin levels are highest among the most overweight group, American Samoan females age 25–44 years, and lowest in Western Samoan men of all ages (Figure 8.5). Insulin values for Samoans reflect their adiposity, and those for American Samoans are higher than among population samples of USA residents.

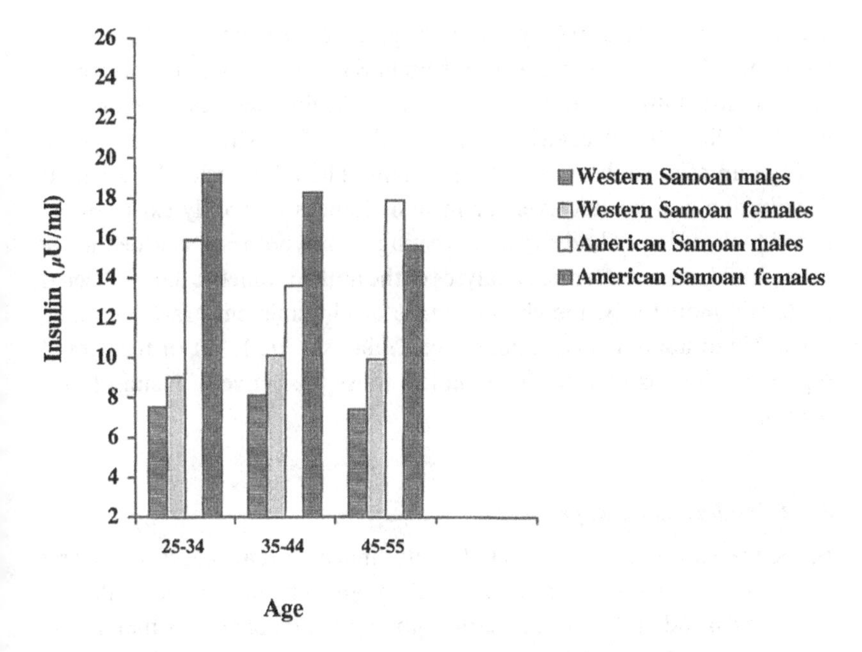

Figure 8.5. Fasting insulin in Samoans, $N = 1155$, in 1990–1991 by age and sex.

When BMI and insulin levels from several populations are examined, Samoan adult insulin levels are comparable to groups with similar BMI levels, such as Mexican–Americans and Pima Indians. Thus, there is no evidence now that Samoans are characterized by a fasting hyperinsulinemia beyond that attributable to adiposity levels.

Analyses of the 4-year follow-up (1990–5) in about 1000 Samoan adults show marked increases over the 4 years in adiposity (Figures 8.2 and 3). Four-year increases in insulin should be most evident among younger adults who have not experienced glucose intolerance, although detailed causal analyses have just started. Several specific hypotheses about insulin and adiposity will be tested in the future analyses. First, what is the relationship between baseline insulin levels and later 2- and 4-year weight and adiposity changes? Prior studies in Pima Indians, Mexican–Americans, and others discussed above would suggest that weight gain will be positively related to insulin levels only among those with lower insulin levels. Second, is there a different shape to the adiposity and insulin relationship in Samoans because of the thrifty genotype processes? If Samoans are susceptible as a group to hyperinsulinemia, we might expect baseline insulin to relate to later weight gain throughout the range of insulin values, excluding those with glucose intolerance and insulin resistance. Third, is there a relationship in Samoans between measures of adipose distribution and insulin levels and insulin changes over time? It may be difficult to detect an independent effect of fat distribution on insulin in a group of already overweight Samoans. The relationship between fat distribution and insulin levels or insulin changes may only exist among individuals of lower body weight. Testing this hypothesis as a deductive process will require intensive analyses of the anthropometric data collected to derive, inductively, the circumference or circumference ratios such as waist–hip or abdominal circumference (Galanis *et al.*, 1995), or the sum of regional skinfolds or ratio of skinfolds, most predictive of insulin levels overall.

8.5.2 Insulin and blood pressure

Blood pressure levels in the recent longitudinal cohort are characterized by sharp four year increases in American Samoan men and women at all ages (Figures 8.6 and 7). The mean systolic blood pressure of approximately 140 mmHg in 1994 among men and women 45–54 years is noteworthy. By comparison Western Samoan blood pressure values are generally lower at each age and time (Figures 8.6 and 7). Samoan adult blood pressure levels are higher than, and not representative of, traditional populations or very recently modernized groups. They indicate chronic exposure to non-traditional diet and activity patterns leading to adiposity (McGarvey & Baker, 1979; McGarvey & Schendel, 1986).

There are many significant correlations between insulin and blood pressure in almost all age and sex groups of Samoans (McGarvey & Levinson, 1993*a*). These associations are confounded by adiposity levels and fat distribution and relationships with insulin and blood pressure. In models controlling for BMI and the waist–hip circumference ratio, there are fewer independent effects of insulin on blood pressure. Insulin is significantly and positively related to age and BMI adjusted systolic blood pressure in American Samoan males of all ages, especially younger males. Diastolic blood pressure is related to insulin in older males, especially those from Western Samoa. In females, insulin and both systolic and diastolic blood pressures are significantly associated. Age-specific analyses show these associations are quite strong in 35–44-year-old females. We have also shown longitudinal insulin and blood pressure associations, primarily among 35–54-year-old females (McGarvey & Levinson, 1994). A one standard deviation difference in baseline insulin among these females

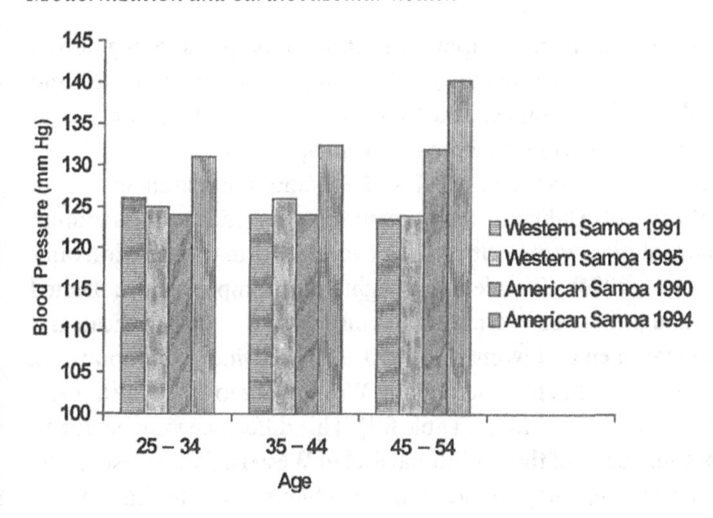

Figure 8.6. Systolic blood pressure of Samoan men, $N = 397$, in the four-year longitudinal cohort.

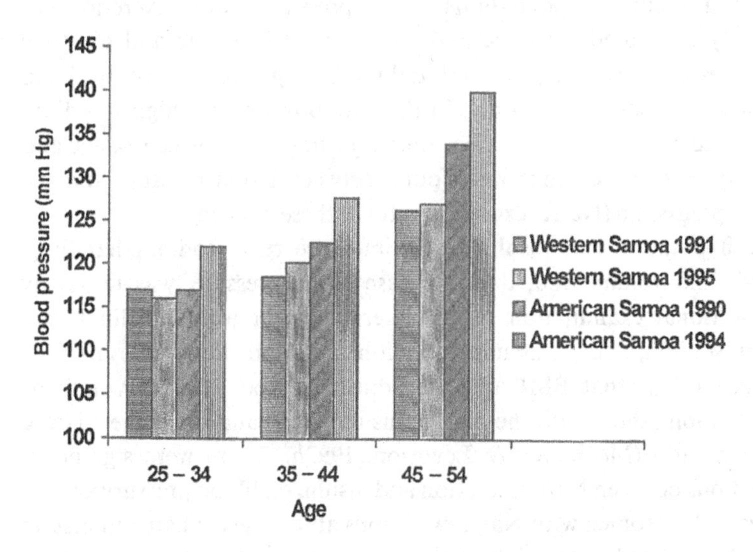

Figure 8.7. Systolic blood pressure of Samoan women, $N = 492$, in the four-year longitudinal cohort.

produces a 2–3 mmHg elevation in blood pressure over 2 years. Ultimately the insulin and blood pressure associations are attributable to both the renal sodium and water retention concomitant with hyperinsulinemia, and insulin's influence on SNS activity, as described above. The cross-sectional and longitudinal insulin and blood pressure associations in middle aged

Samoans are most likely due to their impact on blood pressure regulation mechanisms over 15–20 years of marked adiposity, insulin levels and resistance, glucose intolerance, and hyperlipidemias, and perhaps other metabolic factors such as peri-menopausal changes in women.

As a result of the importance of sodium and potassium intake in evaluating the role of insulin on blood pressure, we first had to assess sodium and potassium intakes by measuring their overnight urinary excretion rates (McGarvey *et al.*, 1994). Complete overnight urine samples were collected and sodium and potassium excretion rates measured for 926 American and Western Samoan men and women in 1990–91. Potassium excretion rates, and thus intake, were much higher among Western Samoans of both sexes, but sodium intakes were similar (Table 8.1). This difference may be due to retention of some parts of the traditional diet in Western Samoa, especially banana consumption, compared to American Samoans. The similarity in sodium intake indicates the widespread use of salt in cooking and at the table throughout the Samoan archipelago. In cross-sectional analysis of the relation of sodium and potassium to blood pressure, sodium excretion was positively associated with age and BMI-adjusted systolic and diastolic blood pressure ($r = 0.24$, $p < 0.001$ and $r = 0.16$, $p < 0.02$, respectively) in American Samoan women only. In this group those with higher sodium intakes had higher blood pressure, after adjusting for the effects of age and adiposity. Similar relations were found between blood pressure and the sodium/potassium (Na/K) excretion ratio in these women.

The hypothesis that insulin levels influence renal sodium handling, expand fluid volume and, thus, increase blood pressure, was tested by cross-sectional examination of the interactions between insulin levels, urinary sodium, and potassium excretion on blood pressure levels. The expectation was that BMI and age adjusted blood pressure would be highest among those with the highest insulin levels and the highest Na/K excretion ratio (McGarvey & Levinson, 1993*b*). There were significant interactions between Na/K excretion and insulin on blood pressure among women only. Women with Na/K excretions above the median and insulin below the median had the highest blood pressures, and among women with relatively lower insulin levels, increased Na/K excretion had a powerful effect on blood pressure. Insulin levels above the median in a population prone to hyperinsulinemia may indicate insulin resistance, glucose intolerance and an altered dose–response curve between insulinemia and renal sodium retention.

Longitudinal studies will test whether there is an interaction between Na/K excretion and insulin on blood pressure in Samoans. Given cross-

Table 8.1. *Mean ($\pm$ s.e.) overnight sodium (Na) and potassium (K) excretion rates (mEq/l per h) in American and Western Samoans by sex*

	Men		Women	
	Na	K	Na	K
N =	133		229	
American Samoans	13.1 $\pm$ 0.8	6.4 $\pm$ 0.4	13.9 $\pm$ 0.5	5.6 $\pm$ 0.3
N =	263		301	
Western Samoans	11.7 $\pm$ 0.8	8.8 $\pm$ 0.5*	13.4 $\pm$ 0.4	8.9 $\pm$ 0.4*

Note: * Significant difference between American and Western Samoans in K excretion.

sectional findings and the evidence of a complex association between insulin and weight change, the hypothesis regarding longitudinal variation is that blood pressure changes will be more pronounced among those with both low to moderate insulin and Na/K excretion levels. As modernizing Samoans have a constellation of elevated cardiovascular risk factors, linear effects among risk factors over time are not to be expected. There may well be curvilinear relationships, e.g., plateau effects, at high levels of certain factors, especially those characterized by development of resistance such as insulin.

8.5.3 *Insulin and lipids*

Total cholesterol levels were lower and HDL-cholesterol levels higher than expected in the early studies of Samoans, considering their adiposity (Pelletier & Hornick, 1986). In our 1990–91 cross-sectional studies with adult Samoans, who average greater body weight than earlier samples, cholesterol levels and HDL-cholesterol levels had worsened, suggesting an increased risk of CVD (McGarvey *et al.*, 1993; Galanis *et al.*, 1995). Preliminary descriptive analyses of the 1994–95 lipid and lipoprotein profiles clearly indicate a potential for CVD morbidity and mortality in modern Samoans (Table 8.2). Although mean total cholesterol levels are slightly less than conventional clinical cut-offs, e.g., 200 mg/dl, approximately 50% of Samoans 40 years and older have elevated total cholesterol. Mean HDL-cholesterol levels are strikingly low, most especially for American Samoans, and in combination with moderately high population mean LDL-cholesterol levels suggest a substantial proportion of the Samoan population may be at increased risk. The lack of differences

Table 8.2. *Mean (± s.e.) serum lipid and lipoprotein levels (mg/dl) in American and Western Samoans by sex*

	American Samoans		Western Samoans	
	Men	Women	Men	Women
N =	145	216	265	295
Age (years)	45.5 ± 0.8	44.0 ± 0.6	43.2 ± 0.5	42.9 ± 0.5
Cholesterol	195.1 ± 2.9	196.3 ± 2.2	197.4 ± 2.3	196.0 ± 2.3
HDL-cholesterol	32.1 ± 0.7*	35.3 ± 0.5**	42.1 ± 0.7	44.1 ± 0.7
LDL-cholesterol	131.7 ± 3.6	138.2 ± 2.2	135.7 ± 2.2	135.8 ± 2.1
Triglycerides	169.3 ± 9.7*	117.2 ± 4.1**	102.8 ± 6.2	84.2 ± 4.3

Note: * Significant difference between American and Western Samoan men.
** Significant difference between American and Western Samoan women.

between American and Western Samoan samples in cholesterol and LDL-cholesterol levels appears, perhaps, counter-intuitive, but is due to specific dietary differences. Dietary cholesterol is higher in American Samoans stemming from consumption of meat products, whereas saturated fat intake is much higher in Western Samoa due to traditionally high coconut consumption (Galanis *et al.*, 1998). The two different sources of fat consumption produce similar lipid and lipoprotein levels. In the context of massive adiposity and high risk of NIDDM, the lipid and lipoprotein profiles of contemporary Samoans require attention and interventions.

In lipid data collected in the late 1970s, insulin was inversely related to high density lipoprotein cholesterol, after adjustment for adiposity, among middle-aged American Samoan men and women studied in 1978 (Hornick & Fellmeth, 1981). In the 1990–91 data, insulin did not appear to be an important mediator between abdominal adiposity and lipid levels (Galanis *et al.*, 1995). Among men, insulin appeared to influence only the association between abdominal fat and total cholesterol, but no other lipid/lipoprotein levels. This may be attributable to the high levels of adiposity and hyperinsulinemia among Samoans, and may reflect a threshold or plateau effect of insulin on lipids in the context of insulin resistance.

8.5.4 Urinary catecholamines

Researchers in Samoa have focused on catecholamine excretion in relation to several modernization processes, including differences between rural and urban residents, individual coping styles, week day versus weekend, and other factors such as physical activity and diet (Hanna *et al.*, 1986; James *et al.*, 1985; James *et al.*, 1987; Pearson *et al.*, 1990; for review see Pearson *et al.*, 1993). Among young adult American Samoans, increased epinephrine:creatinine ratios were found in those who were more self-reliant, made their own decisions and did as they wished, compared to those with more of a group orientation (Hanna *et al.*, 1986). The authors interpreted these findings to suggest that there was greater stress among those with a greater non-traditional individual orientation and less stress among those with the traditional group orientation.

Among Western Samoan young adult men from a rural village, and workers and students from the capital city Apia, James and colleagues found that both epinephrine and norepinephrine excretion rates were lower in rural villagers than in urban residents (James *et al.*, 1985). There was an overall bivariate association between systolic blood pressure and norepinephrine excretion, but it was attenuated by adjustment for residence and occupation. In addition, urinary catecholamine excretion was higher at the end of the work week than at the beginning (James *et al.*, 1987). In a study of three groups of young Samoan men and women, norepinephrine levels were lower in rural Western Samoa, intermediate in American Samoa and highest in Hawaii, although epinephrine levels were generally similar (Pearson *et al.*, 1990).

In the recent longitudinal study of adiposity, insulin, lipids and blood pressure, overnight urinary catecholamine excretion was assessed in the final year, 1995, from Western Samoa (Steele *et al.*, 1998) Norepinephrine excretion rates were higher in young men, ≤ 40 years of age, and women of all ages from peri-urban villages compared to those in rural villages (Figure 8.8). Epinephrine excretion rates were higher in young men from peri-urban villages compared to rural villages. In future work the following hypotheses will also be tested: (1) catecholamine excretion rates will be elevated among those with elevated baseline and 4-year insulin levels, (2) elevated catecholamine levels will be independently related to blood pressure and lipid levels, (3) and individual-level measures of psychosocial stress, such as expression of anger or social support (see below), will be related to catecholamine levels.

8.5.5 Psychosocial factors

Status inconsistency and lifestyle incongruity

Earlier cross-sectional studies during the 1970s and 1980s showed that several psychosocial factors were related to blood pressure. Status incon-

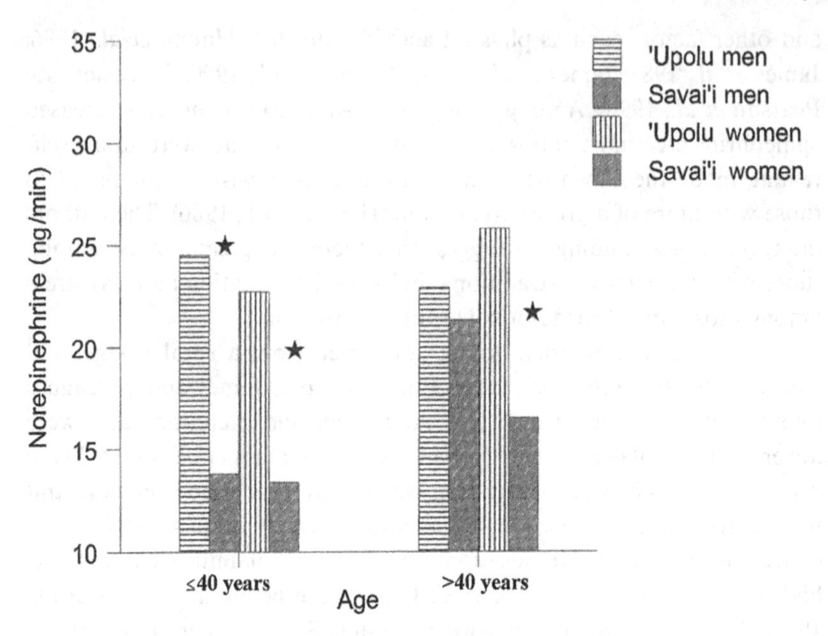

Figure 8.8. Overnight urinary norepinephrine excretion rates in peri-urban, 'Upolu, villages and rural, Savai'i, villages, Western Samoa 1995, $N = 417$. (* = significant difference between 'Upolu and Savai'i within age and sex).

sistency between level of education and occupation was positively related to blood pressure in adult men in the most modernized Pago Pago area of American Samoa. Men who were in managerial occupations or had local community political and religious responsibilities, but had little education, exhibited markedly elevated adiposity and age-adjusted blood pressure (McGarvey & Schendel, 1986). Men with managerial jobs and/or community responsibilities, but without the resources afforded by more education, appeared to experience greater stress related to their jobs and social roles, and had elevated blood pressure.

In modernizing American Samoan communities, such leaders are expected to obtain a greater share of USA federal and territorial resources available for infrastructure improvements and social services. Competition among adult Samoans for traditional social recognition, political titles, and power, have been well described and may be an important context within which modern activities occur (Howard, 1986). Individual achievement orientation and competitive acculturation among contemporary Samoans, thus, may not be purely a novel response to aspects of the modern global economic system, but culturally patterned responses which have been elaborated with exposure to the modern environment. Certainly, greater conceptual and operational work remains to study such factors.

Lifestyle incongruity (LI) as described above has been used in the current Samoan studies. We studied cross-sectionally over 700 adults from Western Samoa and operationalized LI as the mismatch between occupation and style of life, measured by an index of household possessions and media-related consumption (Chin-Hong & McGarvey, 1996). In men, age and BMI adjusted blood pressure was higher in those with higher occupation than lifestyle indices. This was even stronger among young men (< 40 years) and those from the more economically developed island. For example, among young men, age and BMI adjusted systolic blood pressure was highest in the lowest quintile of lifestyle incongruity, i.e., those whose occupational rank was much higher than their lifestyle rank, and lowest in the highest quintile of lifestyle incongruity (Figure 8.9). Among older women (≥ 40 years), however, we found higher adjusted blood pressure among those whose lifestyle most exceeded their occupational level (Figure 8.10).

This seeming contradiction is resolved with a close consideration of the differences in social roles and activities among young modernized men and older women. Among the young men, financial demands from the extended family, which still exerts a strong influence upon those with better jobs, may force a reduction in their individual material lifestyle consumption patterns. These demands may cause stress through increased inability of the young men to control their own wages. Among the men upward socioeconomic mobility, characterized by good jobs but a lag in material lifestyle, may also lead to work stress, regardless of demands from the extended family. Among older women, excess material consumption for enhanced prestige, i.e., living beyond their means, leads to stress elevations in blood pressure. We speculated that these two effects are not unrelated, as in the traditional Samoan system the goods, largely food, produced by young men are commonly taken by older women for domestic use. In modernizing Samoa, this is elaborated because of the increasing importance of prestige enhancement by projection of a nontraditional material display of lifestyle.

As part of the recent study of urinary catecholamine excretion in Western Samoa in 1995 described above, we found an inverse association of norepinephrine and epinephrine levels and LI among young men in peri-urban villages of the island of 'Upolu (Steele *et al.*, 1998). Young men who had higher occupational rank than material lifestyle rank showed evidence of psychophysiological stress. We interpret this in a similar

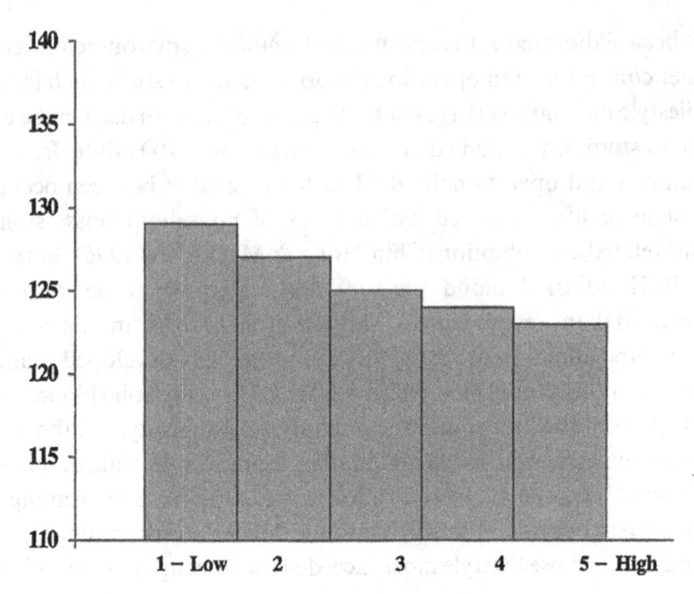

Figure 8.9. Lifestyle incongruity and systolic blood pressure in Western Samoan men < 40 years of age in 1991, $N = 202$.

behavioral and sociocultural manner to the LI and blood pressure association described above. The finding of a catecholamine and LI relationship indicates the validity of using LI as a measure of psychophysiological stress.

Further work will explore LI, including its longitudinal effects on blood pressure, its association with measures of acculturation, the interaction between LI and emotion coping (see below), and a focus on job stress. Based on the cross-sectional findings described above, we hypothesize that those with high LI will develop elevated levels of urinary catecholamines and blood pressure. Although the relationship between acculturation and LI is unknown, we hypothesize that stress of LI may be greatest among those still living in Samoa and with the highest acculturation levels. We will need to perform more qualitative studies to uncover the salient individual and socio-cultural factors which may moderate the stress of modernization in Samoa.

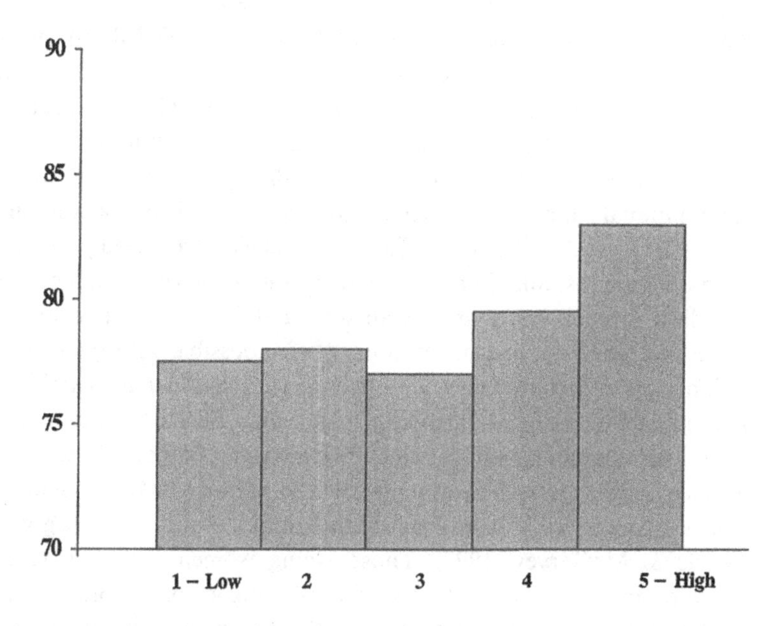

Figure 8.10. Lifestyle incongruity and diastolic blood pressure in Western Samoan women ≥ 40 years of age in 1991, $n = 153$.

Anger expression

The expression of emotion is one of several psychosocial coping domains which allow individuals to buffer the burden of both perceived and unperceived stressors relating to CVD risk factors. Anger expression in Samoans is a key emotional pattern and has been investigated culturally and historically (Howard, 1979; Gerber, 1985; McGarvey & Schendel, 1986; Steele & McGarvey, 1996, 1997). Social control of anger expression is a salient feature of childhood enculturation and anger suppression appears to be a feature of adult social interactions, especially when there are differences in status (Gerber, 1985). In the absence of external controls of anger, such as among migrants, explosive episodes of anger and violence have been reported (Howard, 1979).

In an earlier cross-sectional study we showed that among American Samoan women residing in the most modernized areas, low emotional expression was associated with relatively higher blood pressures, and that the association was strongest among women with less than 7 years of education (McGarvey & Schendel, 1986). We concluded that those with low education may be more traditional and unable to express negative emotions, and that their low education may make it difficult to participate in the new economic, social and material lifestyle opportunities they see in the more modernized areas. Combined with a general Samoan tendency for emotional suppression, these exposures may lead to relatively unbuffered stress, psychophysiological arousal, and elevated blood pressure.

In our recent studies in both American and Western Samoa, we used a standard inventory of anger expression, and found that assessments of anger-out, anger-in, and anger control were possible with good internal reliability (Steele & McGarvey, 1996). American Samoan men and women had greater difficulty controlling anger than Western Samoans, and American Samoan men had greater anger-out scores than Western Samoan men. Anger-out was inversely related to age and BMI adjusted blood pressure among all women from both Samoas, especially young women (Steele & McGarvey, 1997). Those young women with low levels of outward anger expression had significantly higher blood pressure levels (Figure 8.11). We speculated that both role expansion among young women in modernizing Samoa and the learned patterns of anger suppression increase stress levels and lead to blood pressure elevations.

It is noteworthy that both recent and earlier studies found emotional expression and blood pressure associations only among women, those residing in modernizing areas during the late 1970s, and all young women throughout the Samoan islands today. This suggests the need for more detailed studies of women's lives and adjustments to modernization in Samoa.

Future work will focus on the interaction between anger expression and lifestyle incongruity, and other socio-ecological measures of stress, on blood pressure, catecholamine, lipid and lipoprotein levels. I list some of hypotheses to be explored: (1) low anger expression will be associated with higher catecholamine and blood pressure levels, especially among women, and particularly among employed young women, (2) men with high negative lifestyle incongruity, i.e., higher occupational than material possessions status, will have higher catecholamine levels, blood pressure, and lipid profiles, (3) there will be an interaction between stress assessed by lifestyle incongruity and anger expression on blood pressure for men and women. Throughout all the planned analyses, the context of age, sex, social status, and acculturation will be assessed to understand the associations of psychosocial stress and health associations.

Finally the emphasis will be to understand with causal modeling both

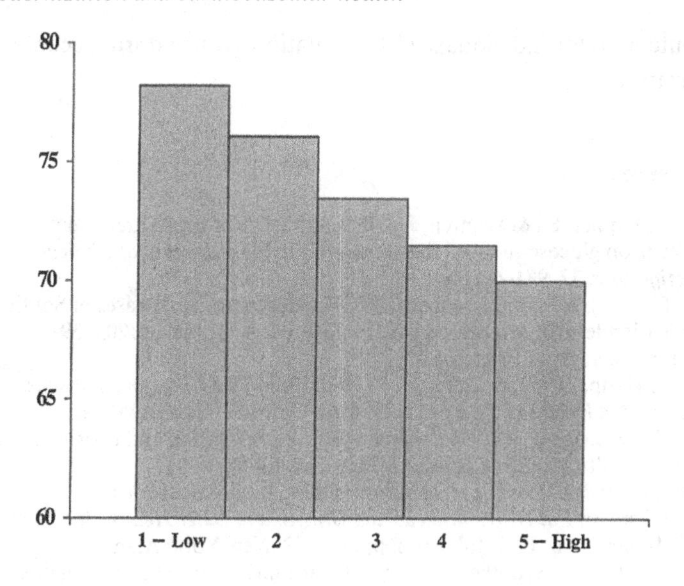

Figure 8.11. Anger expression and diastolic blood pressure in Samoan women < 40 years of age in 1990–1991, $N = 210$.

the independent temporal relationships among biological factors (such as insulin, adiposity, catecholamines) and psychosocial factors (such as lifestyle incongruity and anger expression), and their interactive relationships with CVD risk factors.

8.6 Conclusion

This chapter emphasized the role of insulin and its multiple independent and interactive effects on CVD risk, especially in modernizing populations. This review of the basic laboratory, clinical, and population-based scientific research has convinced the author that a collaborative interdisciplinary perspective is necessary for advances by biological anthropologists studying CVD in modernizing populations. Many of these questions require several investigators sharing ideas, measures, and perspectives. Furthermore, longitudinal research is formally the only design which allows causal process to be understood among multiple variables and their interrelationships. Finally, a mixture of quantitative biological and social science techniques with qualitative and ethnographically informed studies are needed to interpret the contextual nature of adaptive processes and health outcomes for individuals and population groups during economic modernization.

8.7 References

Acheson, K., Jequier, E., & Wahren, J. (1983). Influence of beta-adrenergic blockade on glucose-induced thermogenesis in man. *Journal of Clinical Investigations*, **72**, 981–6.

Baker, P. T. (1984). Migrations, genetics and the degenerative diseases of South Pacific islanders. In *Migration and Mobility*, ed. A. Boyce, pp. 209–39. London: Taylor and Francis Ltd.

Baker, P. T., Hanna, J. M., & Baker, T. S., eds. (1986). *The Changing Samoans: Behavior and Health in Transition*. Oxford: Oxford University Press.

Bindon, J. R., & Baker, P. T. (1985). Modernization, migration and obesity among Samoan adults. *Annals of Human Biology*, **12**, 67–76.

Björnthorp, P. (1988). Possible mechanism relating fat distribution and metabolism. In *Fat Distribution during Growth and Later Health Outcomes*, ed. C. Bouchard & F. E. Johnston, pp. 175–92. New York: Alan R. Liss, Inc.

Björnthorp, P. (1991). Hypothesis. Visceral fat accumulation: The missing link between psychosocial factors and cardiovascular disease? *Journal of International Medicine*, **230**, 195–201.

Bogardus, C., Lillioja, S., Ravussin, E., Abbott, W., Zawadzki, J. K., Young, A., Knowler, W. C., Jacobowitz, R., & Moll, P. P. (1986). Familial dependency of resting metabolic rate. *New England Journal of Medicine*, **315**, 96–100.

Bouchard, C. (1994). Genetics of obesity: Overview and research directions. In *The Genetics of Obesity*, ed. C. Bouchard, pp. 223–34. Boca Raton FL: CRC Press.

Brown, D. E. (1982). Physiological stress and culture change in a group of Filipino-Americans, a preliminary investigation. *Annals of Human Biology*, **9**, 553–63.

Cassel, J. C. (1976). The contribution of the social environment to host resistance. *American Journal of Epidemiology*, **104**, 107–23.

Chin-Hong, P. V., & McGarvey, S. T. (1996). Lifestyle incongruity and adult blood pressure in Western Samoan. *Psychosomatic Medicine*, **58**, 130–7.

Collins, V. R., Dowse, G. K., Finch, C. F., & Zimmet, P. Z. (1990). An inconsistent relationship between insulin and blood pressure in three Pacific island populations. *Journal of Clinical Epidemiology*, **43**, 1369–78.

Crooks, D. L. (1995). American children at risk: Poverty and its consequences for children's health, growth and school achievement. *Yearbook of Physical Anthropology*, **38**, 57–86.

DeFronzo, R. A. (1981). Insulin and renal sodium handling: Clinical implications. *International Journey of Obesity*, **5**, 93–6.

DeFronzo, R., & Ferrannini, E. (1991). Insulin resistance: A multifaceted syndrome responsible for NIDDM, obesity, hypertension, dyslipidemia, and atherosclerotic cardiovascular disease. *Diabetes Care*, **14**, 173–94.

Dewey, K. (1981). Nutritional consequences of the transformation from subsistence to commercial agriculture in Tabasco, Mexico. *Human Ecology*, **9**, 151–67.

Dressler, W. W. (1982). *Hypertension and Culture Change: Acculturation and Disease in the West Indies*. South Salem, NY: Redgrave.

Dressler, W. W. (1985). Psychosomatic symptoms, stress, and modernization: A model. *Cultural Medicine and Psychiatry*, **9**, 257–86.

Dressler, W. W. (1988). Social consistency and psychological distress. *Journal of Health and Social Behavior*. **29**, 79–91.

Dressler, W. W. (1990). Lifestyle, stress, and blood pressure in a Southern black community. *Psychosomatic Medicine*, **52**, 182–98.

Dressler, W. W. (1991). Social support, lifestyle incongruity, and arterial blood pressure in a Southern black community. *Psychosomatic Medicine*, **53**, 608–20.

Dressler, W. W. (1995). Modeling biocultural interactions: Examples from studies of stress and cardiovascular disease. *Yearbook of Physical Anthropology*, **38**, 27–56.

Dressler, W. W., Dos Santos, J. E., Viteri, F., & Gallagher, P. N. (1991). Social and dietary predictors of serum lipids: A Brazilian example. *Social Science and Medicine*, **32**, 1229–35.

Ducimetiere, P., Eschwege, E., Papoz, L., Richard, J., Claude, J., & Rosselin, G. (1980). Relationship of plasma insulin levels to the incidence of myocardial infarction and coronary heart disease mortality in a middle-aged population. *Diabetologia*, **19**, 205–10.

Ernst, N. D., & Harlan, W. R. (1991). Obesity and cardiovascular disease in minority populations. Proceedings of a Conference. *American Journal of Clinical Nutrition*, **53** (Suppl.), 1507S-651S.

Fournier, A. M., Gadia, M. T., Kubrusly, D. B., Skyler, J. S. & Sosenko, J. M. (1986). Blood pressure, insulin and glycemia in non-diabetic subjects. *American Journal of Medicine*, **80**, 861–4.

Franke, R. W. (1987). The effects of colonialism and neocolonialism on the gastronomic patterns of the Third World. In *Food and Evolution: Toward A Theory of Human Food Habits*, ed. M. Harris, & E. B. Ross, pp. 455–80. PA: Temple University Press.

Frenk, J., Bobadilla, J. L., Stern, C., Frejka, T., & Lozano, R. (1991). Elements for a theory of the health transition. *Health Transition Review*, **1**, 21–38.

Friedlaender, J. S. (1987). *The Solomons Island Project. A Long-term Study of Health, Human Biology and Culture Change*. NY: Oxford Press.

Galanis, D. J., McGarvey, S. T., Sobal, J., Bausserman, L., & Levinson, P. D. (1995). Relations of body fat and fat distribution to the serum lipid, apolipoprotein and insulin concentrations of Samoan men and women. *International Journal of Obesity*, **19**, 731–8.

Galanis, D. J., McGarvey, S. T., Quested, C., Sio, B., Agale-Fáamuli, S. (1998). Dietary intake among modernizing Samoans: Implications for risk of cardiovascular disease. *Journal of American Dietetics Association* (in press).

Gates, J., Parpia, B., Campbell, T. C., & Junshi, C. (1996). Association of dietary factors and selected plasma variables with sex hormone-binding globulin in rural Chinese women. *American Journal of Clinical Nutrition*, **63**, 22–31.

Gentry, W. D., Chesney, A., Gary, H., Hall, R., & Harburg, E. (1982). Habitual anger-coping styles. 1. Effect on mean blood pressure and risk for essential hypertension. *Psychosomatic Medicine*, **44**, 195–202.

Gerber, E. R. (1985). Rage and obligation: Samoan emotion in conflict. In *Person Self and Experience: Exploring Pacific Ethnopsychologies*, ed. G. White, & J. Kirkpatrick, pp. 121–67. Berkeley: Univ. of California Press.

Gershater, E., & McGarvey, S. T. (1995). Fourteen-year changes in adiposity and blood pressure in American Samoan adults. *American Journal of Human Biology*, **7**, 597–606.

Greksa, L. P., Pelletier, D. L., & Gage, T. B. (1986). Work in contemporary and traditional Samoa. In *The Changing Samoans. Behavior and Health in Transition*, ed. P. T. Baker, J. M. Hanna, & T. S. Baker, pp. 297–326.NY: Oxford Press.

Gross, D. R., & Underwood, B. (1971). Technological change and calorie costs: Sisal agriculture in Northeastern Brazil. *American Anthropologist*, **73**, 725–40.

Hanna, J. M., James, G. D., & Martz, J. M. (1986). Hormonal measures of stress. In *The Changing Samoans. Behavior and Health in Transition*. ed. P. T. Baker, J. M. Hanna, & T. S. Baker, pp. 203–21. NY: Oxford.

Harburg, E., Blackelock, E., & Roeper, P. (1979). Resentful and reflective coping with arbitrary authority and blood pressure: Detroit. *Psychosomatic Medicine*, **41**, 189–96.

Harrison, G. A. (1995). *The Human Biology of the English Village*. Oxford: Oxford Press.

Harrison, G. A., Palmer, C. D., Jenner, D., & Reynolds, V. (1981). Associations between rates of urinary catecholamine excretion and aspects of lifestyle among adult women in some Oxfordshire villages. *Human Biology*, **53**, 617–33.

Hazuda, H. P., Haffner, S. M., Stern, M. P., & Eifler, C. (1988). Effects of acculturation and socioeconomic status on obesity and diabetes in Mexican Americans. *American Journal of Epidemiology*, **128**, 1289–301.

Henry, J. P., & Cassell, J. C. (1969). Psychosocial factors in essential hypertension: Recent epidemiological and animal experimental evidence. *American Journal of Epidemiology*, **90**, 171–200.

Henry, J. P., & Stephens, P. M. (1977). *Stress, Health, and the Social Environment*. New York: Springer Verlag.

Hoag, S., Marshall, J. A., Jones, R. H., & Hamman, R. F. (1995). High fasting insulin levels associated with lower rates of weight gain in persons with normal glucose tolerance: The San Luis Valley Diabetes Study. *International Journal of Obesity*, **19**, 175–80.

Hodge, A. M., Dowse, G. K., Toelupe, P., Collins, V. R., Imo, T., & Zimmet, P. Z. (1994). Dramatic increase in the prevalence of obesity in Western Samoa over the 13 year period 1978–1991. *International Journal of Obesity*, **18**, 419–28.

Hodge, A. M., Dowse, G. K., & Zimmet, P. Z. (1993). Association of body mass index and waist–hip circumference ratio with cardiovascular disease risk factors in Micronesian Nauruans. *International Journal of Obesity*, **17**, 399–407.

Hornick, C. A., & Fellmeth, B. D. (1981). High density lipoprotein cholesterol, insulin and obesity in Samoans. *Atherosclerosis*, **39**, 321–8.

Howard, A. (1979). Polynesia and micronesia in psychiatric perspective. *Transcultural Psychological Research Review*, **16**, 123–45.

Howard, A. (1986). Samoan coping behavior. In *The Changing Samoans. Behavior and Health in Transition*, ed. P. T. Baker, J. M. Hanna, & T. S. Baker, pp. 394–418. NY: Oxford Press.

James, G. D., Crews, D. E., & Pearson, J. (1989). Catecholamines and stress. In *Human Population Biology*. ed. M. A. Little, & J. D. Haas, pp. 280–95. New York: Oxford Press.

James, G. D., Baker, P. T., Jenner, D. A., & Harrison, G. A. (1987). Variation in lifestyle characteristics and catecholamine excretion rates among young Western Samoan men. *Social Science and Medicine*, **25**, 981–6.

James, G. D., Jenner, D. A., Harrison, G. A. & Baker, P. T. (1985). Differences in catecholamine excretion rates, blood pressure and lifestyle among young Western Samoan men. *Human Biology*, **57**, 635–47.

Jenner, D. A., Harrison, G. A., Day, J. A., Huizinga, J., & Salzano, F. M. (1982). Interpopulation comparisons of urinary catecholamines. A pilot study. *Annals of Human Biology*, **9**, 579–82.

Katzeff, H. L., O'Connell, M., Horton, E. S., Danforth, E. Jr, Young, J. B., & Landsberg, L. (1986). Metabolic studies in human obesity during overnutrition and undernutrition: Thermogenic and hormonal responses to norepinephrine. *Metabolism*, **35**, 166–75.

Kaufman, J. S., Owoaje, E. E., James, S. A., Rotimi, C. N., & Cooper, R. S. (1996). Determinants of hypertension in West Africa: Contribution of anthropometric and dietary factors to urban–rural and socioeconomic gradients. *American Journal of Epidemiology*, **143**, 1203–18.

Krotkiewski, M., Björntorp, P., Sjostrom, L., & Smith, U. (1983). Impact of obesity on metabolism in men and women. Importance or regional adipose tissue distribution. *Journal of Clinical Investigations*, **72**, 1150–62.

Landsberg, L., (1986). Diet, obesity and hypertension: An hypothesis involving insulin, the sympathetic nervous system, and adaptive thermogenesis. *Quarterly Journal of Medicine*, **61**, 1081–90.

Landsberg, L. (1990). Insulin resistance, energy balance and sympathetic nervous system activity. *Clinical and Experimental Hypertension, A*, **12**, 817–30.

Landsberg, L., & Young, J. B. (1985*a*). Insulin-mediated glucose metabolism in the relationship between dietary intake and sympathetic nervous system activity. *International Journal of Obesity*, **5**, (suppl 2), 63–8.

Landsberg, L., & Young, J. B. (1985*b*). The influence of diet on the sympathetic nervous system. In *Neuroendocrine Perspectives*, ed. E. E. Muller, R. M. MacLeod, & L. A. Frohman, pp. 191–218. Amsterdam: Elsevier Science Publishers.

Laville, M., Cornu, C., Normand, S., Mithieux, G., Beylot, M., & Riou, J-P. (1993). Decreased glucose-induced thermogenesis at the onset of obesity. *American Journal of Clinical Nutrition*, **57**, 851–6.

Leung, S. S. F., Lau, J. T.F., Xu, Y. Y., Tse, L. Y., Huen, K. F., Wong, G. W.K., Law, W. Y., Yeung, V. T. F., Yeung, W. K. Y., & Leung, N. K. (1996). Secular changes in standing height, sitting height ansd sexual maturation of Chinese – The Hong Kong Growth Study, 1993. *Annals of Human Biology*, **23**, 297–306.

Marchione. T., & Prior, F. W. (1980). The dynamics of malnutrition in Jamaica. In *Social and Biological Predictors of Nutritional Status, Physical Growth and Neurological Development*, ed. L. Greene, & F. Johnston, pp. 201–22. NY: Academic Press.

Marmot, M. G., & Syme, S. L. (1976). Acculturation and coronary heart disease in Japanese–Americans. *American Journal of Epidemiology*, **104**, 225–47.

McGarvey, S. T. (1991). Obesity in Samoans and a perspective on its etiology in Polynesians. *American Journal of Clinical Nutrition*, **53**, 1586S–94S.

McGarvey, S. T. (1992). Biocultural predictors of age increase in Samoan blood pressure. *American Journal of Human Biology*, **4**, 27–36.

McGarvey, S. T. (1994). The thrifty gene concept and adiposity studies in biological anthropology. *Journal of Polynesian Society*, **103**, 29–42.

McGarvey, S. T., & Baker, P. T. (1979). The effects of modernization and migration on Samoan blood pressure. *Human Biology*, **51**, 461–75.

McGarvey, S. T., & Schendel, D. E. (1986). Blood pressure patterns. In *The Changing Samoans. Behavior and Health in Transition*, ed. P. T. Baker, J. M. Hanna, & T. S. Baker, pp. 351–93. NY: Oxford Press.

McGarvey, S. T., & Levinson, P. D. (1992). Insulin levels, distribution and correlates in adult Samoans. *American Journal of Physical Anthropology*, **86**, Suppl., 120.

McGarvey, S. T., & Levinson, P. D. (1993*a*). Insulin and blood pressure in Samoans. *American Jounal of Physical Anthropology*, Suppl 16, 120.

McGarvey, S. T., & Levinson, P. D. (1993*b*). Insulin and sodium and potassium excretion correlate with blood pressure in Samoan women. *Clinical Research*, **41**, 168A.

McGarvey, S. T., & Levinson, P. D. (1994). Baseline insulin, body mass and change in blood pressure in Samoan adults. *American Journal of Physical Anthropology*, Suppl 18, 143.

McGarvey, S. T., Bindon, J. R., Crews, D. E., & Schendel, D. E. (1989). Modernization and adiposity: causes and consequences. In *Human Population Biology*, ed. M. A. Little, & J. D. Haas, pp. 263–79. New York: Oxford Press.

McGarvey, S. T., Levinson, P. D., Bausserman, L., Galanis, D. J., & Hornick, C. A. (1993). Population change in adult obesity and blood lipids in American Samoa from 1976–1978 to 1990. *American Journal of Human Biology*, **5**, 17–30.

McGarvey, S. T., Levinson, P. D., & Zhai, S. (1994). Sodium and potassium intake and blood pressure in modernizing Samoan adults. *American Journal of Human Biology*, **6**, 128.

Modan, M., Halkin, H., Almog, S., Lusky, A., Eshkol, A., Shefi, M. Shitrit, M. & Fuchs, Z. (1985). Hyperinsulinemia: A link between hypertension, obesity and glucose intolerance. *Journal of Clinical Investigations*, **75**, 809–17.

Nahum, R., Thong, K., & Hillier, S. (1995). Metabolic regulation of androgen production by human thecal cells *in vitro*. *Human Reproduction*, **10**, 75–81.

Neel, J. V. (1962). Diabetes mellitus: a "thrifty" genotype rendered detrimental by "progress"? *American Journal of Human Genetics*, **14**, 353–62.

Olefsky, J. M., Farquhar, J. W., & Reaven, G. M. (1974). Appraisal of the role of insulin in hypertriglyceridemia. *American Journal of Medicine*, **57**, 551–60.

O'Rahilly, S., Turner, R. C., & Matthews, D. R. (1988). Impaired pulsatile secretion of insulin in relatives of patients with non-insulin-dependent diabetes. *New England Journal of Medicine*, **318**, 1225–30.

O'Dea, K. (1995). Overview of the thrifty genotype hypothesis. *Asia Pacific Journal of Clinical Nutrition*, **4**, 339–40.

Ostfeld, A. M., & D'Atri, D. (1977). Rapid sociocultural change and high blood pressure. *Advances in Psychosomatic Medicine*, **9**, 20–32.

Pearson, J. D., Hanna, J. M., Fitzgerald, M. H., & Baker, P. T. (1990). Modernization and catecholamine excretion of young Samoan adults. *Social Science and Medicine*, **31**, 729–36.

Pearson, J. D., James, G. D., & Brown, D. E. (1993). Stress and changing lifestyles in the Pacific: Physiological stress responses of Samoans in rural and urban settings. *American Journal of Human Biology*, **5**, 49–60.

Pelletier, D. L., & Hornick, C. A. (1986). Blood lipid studies. In *The Changing Samoans. Behavior and Health in Transition*, ed. P. T. Baker, J. M. Hanna, & T. S. Baker, pp. 327–49. NY: Oxford Press.

Polonsky, K. S., Given, B. D., Hirsh, L. J., Tillil, H., Shapiro, E. T., Beebe, C., Frank, B. H., Galloway, J. A., & Van Cauter, E. (1988). Abnormal patterns of insulin secretion in non-insulin-dependent diabetes mellitus. *New England Journal of Medicine*, **318**, 1231–9.

Prentice, A. M., Coward, W. A., Davies, H. L., Murgatroyd, P. R., Black, A. E., Goldberg, G. R., Ashford, J., & Sawyer, M. (1985). Unexpectedly low levels of energy expenditure in health women. *Lancet*, **i**, 1419–22.

Pyörälä, K. (1979). Relationship of glucose tolerance and plasma insulin to the incidence of coronary heart disease: results from two population studies in Finland. *Diabetes Care*, **2**, 131–41.

Ravussin, E., Lillioja, S., Anderson, T. E., Christian, L., & Bogardus, C. (1986). Determinants of 24-hour energy expenditure in man. *Journal of Clinical Investigations*, **78**, 1568–78.

Reaven, G. R. (1988). Role in insulin resistance in human disease. *Diabetes*, **37**, 1595–607.

Rebuffe-Scrive, M., Enk, L., Crona, N., Lonnroth, P., Abrahamsson, L., & Bjornthorp, P. (1985). Fat cell metabolism in different regions in women. *Journal of Clinical Investigations*, **75**, 1973–6.

Rowe, J. E., Young, J. B., Minaker, K. L., Stevens, A. L., Pallotta, J., & Landsberg, L. (1981). Effect of insulin and glucose infusions on sympathetic nervous system activity in normal man. *Diabetes*, **30**, 219.

Rutz, H. J., & Orlove, B. S. (1989). *The Social Economy of Consumption. Monographs in Economic Anthropology*, No. 6. Lanham MD: University Press of America.

Saad, M., Lilioja, S., Nyomba, B., Castillo, C., Ferraro, R., De Gregorio, M., Ravussin, E., Knowler, W. C., Bennett, P. H., & Howard, B. V. (1991). Racial differences in the relation between blood pressure and insulin resistance. *New England Journal of Medicine*, **324**, 733–9.

Scotch, N. A. (1963). Sociocultural factors in the epidemiology of Zulu hypertension. *American Journal of Public Health*, **52**, 1205–13.

Shen, T., Habicht, J-P., & Chang, Y. (1996). Effect of economic reforms on child growth in urban and rural areas of China. *New England Journal of Medicine*, **335**, 400–6.

Sims, E. A. H. (1976). Experimental obesity, dietary induced thermogenesis and their clinical implications. In *Clinics in Endocrinology and Metabolism*, ed. M. J. Albring, pp. 377–95. PA: W. B. Saunders.

Smith, T. (1967). Sociocultural incongruity and change: A review of empirical findings. 1967. In *Social Stress and Cardiovascular Disease, ed. S. Syme & L. Reeder. Milbank Memorial Fund Quarterly*, **45**, 23–39.

Stalder, M., Pometta, D., & Suenram, A. (1981). Relationship between plasma insulin levels and high density lipoprotein cholesterol levels in healthy men. *Diabetologia*, **21**, 544–8.

Steele, M. S., & McGarvey, S. T. (1996). Anger expression patterns in modernizing Samoan adults. *Psychological Reports*, **79**, 1339–48.

Steele, M. S., & McGarvey, S. T. (1997). Anger expression and blood pressure in Samoan adults. *Psychosomatic Medicine*, **59**, 632–7.

Steele, M. S., McGarvey, S. T., & Bereiter, D. (1998). Urinary catecholamine excretion and lifestyle incongruity in Western Samoan adults. *American Journal of Human Biology*, in press.

Swinburn, B. A., Nyomba, B. L., Saad, M. F., Zurlo, F., Raz, I., Knowler, W. C., Lillioja, S., Bogardus, C., & Ravussin, E. (1991). Insulin resistance associated with lower rates of weight gain in Pima Indians. *Journal of Clinical Investigations*, **88**, 168–73.

Syme, L. S., Borhani, N., & Buechley, R. (1965). Cultural mobility and coronary heart disease in an urban area. *American Journal of Epidemiology*, **82**, 334–46.
Valdez, R., Mitchell, B. D., Haffner, S. M., Hazuda, H. P., Morales, P. A., Monterrosa, A., & Stem, M. P. (1994). Predictors of weight change in a bi-ethnic population. The San Antonio Heart Study. *International Journal of Obesity*, **18**, 85–91.
Veblen, T. (1899). *Theory of the Leisure Class*. NY: Macmillan.
Weber, M. (1946). Class, status, party. In *From Max Weber*, ed. H. H. Gerth & C. W. Mills, pp. 180–95. New York: Oxford University Press.
Welborn, T. A., & Wearne, K. (1979). Coronary heart disease incidence and cardiovascular mortality in Busselton with reference to glucose and insulin concentration. *Diabetes Care*, **2**, 154–60.
Wessen, A. F. (ed.), Hooper, A., Huntsman, J., Prior, I. A. M., & Salmond, C. E. (1992). *Migration and Health in a Small Society: The Case of Tokelau*. NY: Oxford University Press.
Willis, D., Mason, H., Gilling-Smith, P., & Franks, S. (1996). Modulation by insulin of follicle-stimulating hormone and luteinizing hormone actions in human granulosa cells of normal and polycystic ovaries. *Journal of Clinical Endocrinology and Metabolism*, **81**, 302–9.
World Health Organization. (1990). Diet, nutrition and the prevention of chronic diseases: Report of a WHO study group. *Technical Report Series No. 797*. pp. 1–203. Geneva: WHO Press.
Young, J. B., & Landsberg, L. (1977). Suppression of sympathetic nervous system during fasting. *Science*, **196**, 1473–5.
Zimmet, P. (1979). Epidemiology of diabetes and its macrovascular manifestations in Pacific populations: The medical effects of social progress. *Diabetes Care*, **2**, 144–53.

Index